Lecture Notes in Physics

Volume 877

Founding Editors

W. Beiglböck
J. Ehlers
K. Hepp
H. Weidenmüller

Editorial Board

B.-G. Englert, Singapore, Singapore
U. Frisch, Nice, France
P. Hänggi, Augsburg, Germany
W. Hillebrandt, Garching, Germany
M. Hjort-Jensen, Oslo, Norway
R. A. L. Jones, Sheffield, UK
H. von Löhneysen, Karlsruhe, Germany
M. S. Longair, Cambridge, UK
J.-F. Pinton, Lyon, France
J.-M. Raimond, Paris, France
A. Rubio, Donostia, San Sebastian, Spain
M. Salmhofer, Heidelberg, Germany
D. Sornette, Zurich, Switzerland
S. Theisen, Potsdam, Germany
D. Vollhardt, Augsburg, Germany
W. Weise, Garching, Germany and Trento, Italy

For further volumes:
http://www.springer.com/series/5304

The Lecture Notes in Physics

The series Lecture Notes in Physics (LNP), founded in 1969, reports new developments in physics research and teaching – quickly and informally, but with a high quality and the explicit aim to summarize and communicate current knowledge in an accessible way. Books published in this series are conceived as bridging material between advanced graduate textbooks and the forefront of research and to serve three purposes:

- to be a compact and modern up-to-date source of reference on a well-defined topic
- to serve as an accessible introduction to the field to postgraduate students and nonspecialist researchers from related areas
- to be a source of advanced teaching material for specialized seminars, courses and schools

Both monographs and multi-author volumes will be considered for publication. Edited volumes should, however, consist of a very limited number of contributions only. Proceedings will not be considered for LNP.

Volumes published in LNP are disseminated both in print and in electronic formats, the electronic archive being available at springerlink.com. The series content is indexed, abstracted and referenced by many abstracting and information services, bibliographic networks, subscription agencies, library networks, and consortia.

Proposals should be sent to a member of the Editorial Board, or directly to the managing editor at Springer:

Christian Caron
Springer Heidelberg
Physics Editorial Department I
Tiergartenstrasse 17
69121 Heidelberg / Germany
christian.caron@springer.com

Gary P. Zank

Transport Processes in Space Physics and Astrophysics

Gary P. Zank
CSPAR
University of Alabama in Huntsville
Huntsville, AL, USA

ISSN 0075-8450 ISSN 1616-6361 (electronic)
ISBN 978-1-4614-8479-0 ISBN 978-1-4614-8480-6 (eBook)
DOI 10.1007/978-1-4614-8480-6
Springer New York Heidelberg Dordrecht London

Library of Congress Control Number: 2013946652

© Springer Science+Business Media New York 2014
This work is subject to copyright. All rights are reserved by the Publisher, whether the whole or part of the material is concerned, specifically the rights of translation, reprinting, reuse of illustrations, recitation, broadcasting, reproduction on microfilms or in any other physical way, and transmission or information storage and retrieval, electronic adaptation, computer software, or by similar or dissimilar methodology now known or hereafter developed. Exempted from this legal reservation are brief excerpts in connection with reviews or scholarly analysis or material supplied specifically for the purpose of being entered and executed on a computer system, for exclusive use by the purchaser of the work. Duplication of this publication or parts thereof is permitted only under the provisions of the Copyright Law of the Publisher's location, in its current version, and permission for use must always be obtained from Springer. Permissions for use may be obtained through RightsLink at the Copyright Clearance Center. Violations are liable to prosecution under the respective Copyright Law.
The use of general descriptive names, registered names, trademarks, service marks, etc. in this publication does not imply, even in the absence of a specific statement, that such names are exempt from the relevant protective laws and regulations and therefore free for general use.
While the advice and information in this book are believed to be true and accurate at the date of publication, neither the authors nor the editors nor the publisher can accept any legal responsibility for any errors or omissions that may be made. The publisher makes no warranty, express or implied, with respect to the material contained herein.

Printed on acid-free paper

Springer is part of Springer Science+Business Media (www.springer.com)

*For my parents, Pat and Lyn Zank, and
my daughters, Taryn and Janine*

Acknowledgment

I should like to thank Adele Corona for her constant encouragement and help with the writing of this book. Without her, it would never have been completed. The help provided by Dr Alex Dosch was invaluable. He made numerous excellent suggestions for the organization, exposition, and content of the book, and I benefited enormously from his help and insight. Alex developed an excellent Solutions Manual that is available from the publisher. I should like to thank my many collaborators and teachers over the years, from whom I learned everything I know. Finally, the contribution of the students who grappled with this class over the past few years has to be acknowledged. They caught typos, made suggestions for clearer presentation of ideas and results, and worked the many problems in the book.

Contents

1	**Introduction**	1
	Reference	5
2	**Statistical Background**	7
	2.1 Probability Set Function	7
	2.2 Random or Stochastic Variables	11
	2.3 The Probability Density Function	14
	2.4 The Distribution Function	19
	2.5 Expectations and Moments	23
	2.6 Conditional Probability and Marginal and Conditional Distributions	33
	2.7 Stochastic Independence	42
	2.8 Particular Distributions	47
	2.8.1 The Binomial Distribution	47
	2.8.2 The Poisson Distribution	50
	2.8.3 The Normal or Gaussian Distribution	53
	2.9 The Central Limit Theorem	57
	2.10 Relation Between Microscopic and Macroscopic Descriptions: Particles, the Gibbs Ensemble, and Liouville's Theorem	61
	2.11 The Language of Fluid Turbulence	64
	References	70
3	**The Boltzmann Transport Equation**	71
	3.1 Derivation of the Boltzmann Transport Equation	71
	3.2 The Boltzmann Collision Operator	74
	3.2.1 Collision Dynamics	76
	3.3 Conservation Laws, the H-Theorem, and the Maxwell-Boltzmann Distribution Function	82
	3.4 The Boltzmann Equation and the Fluid Equations	87
	3.5 The Relaxation Time Approximation	91
	3.6 The Chapman-Enskog Expansion	91

	3.7	Application 1: Structure of Weak Shock Waves	98
		3.7.1 Weak Solutions and the Rankine-Hugoniot Relations	102
	3.8	Application 2: The Diffusion and Telegrapher Equations	110
	References	119
4	**Charged Particle Transport in a Collisional Magnetized Plasma**	121
	4.1	The Kinetic Equation and Moments for a Magnetized Plasma	121
	4.2	Markov Processes, the Chapman-Kolmogorov Equation, and the Fokker-Planck Equation	127
		4.2.1 A More Formal Derivation of the Chapman-Kolmogorov Equation and the Fokker-Planck Equation	131
	4.3	Collision Dynamics, the Rosenbluth Potentials, and the Landau Collision Operator	135
	4.4	Electron-Proton Collisions	142
	4.5	Collisions with a Maxwellian Background	145
	4.6	Collision Operator for Fast Ions	150
	4.7	Proton-Electron Collisions	152
	4.8	Transport Equations for a Collisional Electron-Proton Plasma	155
	4.9	Application 1: Transport Perpendicular to a Mean Magnetic Field	161
	4.10	Application 2: The Equations of Magnetohydrodynamics	166
	4.11	Application 3: MHD Shock Waves	175
	References	184
5	**Charged Particle Transport in a Collisionless Magnetized Plasma**	185
	5.1	Transport Equations for Non-relativistic Particles Scattered by Plasma Fluctuations	186
		5.1.1 The Focussed Transport Equation	186
		5.1.2 The Diffusive Transport Equation	192
	5.2	Transport Equation for Relativistic Charged Particles	204
		5.2.1 Derivation of the Focussed Transport Equation	204
	5.3	The Magnetic Correlation Tensor	215
	5.4	Quasi-linear Transport Theory of Charged Particle Transport: Derivation of the Scattering Tensor	226
	5.5	Diffusion Perpendicular to the Mean Magnetic Field: The Nonlinear Guiding Center Theory	235
	5.6	Hydrodynamic Description of Energetic Particles	240
	5.7	Application 1: Diffusive Shock Acceleration	250
	5.8	Application 2: The Modulation of Cosmic Rays by the Solar Wind	257
	References	259

6 The Transport of Low Frequency Turbulence 261
6.1 Basic Description of Low-Frequency Turbulence 263
6.2 Mean Field Description of MHD Fluctuations 267
6.3 The Transport Equation for the Magnetic Energy Density 272
6.4 Modeling the Dissipation Terms 275
References 279

Index 281

Chapter 1
Introduction

Understanding the self-consistent coupling and flow of energy, momentum, matter, and electromagnetic fields is necessary for every aspect of solar, space, and astrophysics. Transport of energy from the center of Sun or a star to the photosphere determines its stellar evolution, for example. The transport of convective energy and its translation to magnetic energy and the subsequent transport of magnetic energy into the solar corona is thought to be responsible for the unexpected heating of the solar corona to its very high temperatures. The transport of mass, momentum, and energy is necessary to understand the physics of expanding flows, such as the solar wind and stellar winds, the interstellar medium, and almost every conceivable flow found in a solar, space, and astrophysics context.

To illustrate the fascinating complexity of transport processes within a space physics setting, consider the interaction of the solar wind with the local interstellar medium. The transport of matter from the solar atmosphere is manifest in the supersonic solar wind that expands non-adiabatically past the Earth and planets to eventually interact with the local interstellar medium at some 120 Astronomical Units (AU). Over this vast distance, the solar wind begins in a highly collisional state deep in the atmosphere of the Sun, becoming essentially collisionless as it expands away from the corona, and is then mediated weakly beyond the planets by charge-exchange processes with neutral interstellar hydrogen that flows into the heliosphere.[1] In the case of the large-scale heliosphere, despite the coupling of plasma and neutral H through a collisional process, the associated creation of so-called pickup ions introduces an important new collisionless transport element into the system as well. Specifically, in the supersonic solar wind, neutral H drifts through the supersonic solar wind, which has a radially expanding velocity of $\sim$350 to $\sim$700 km/s in an opposite direction with a speed of $\sim$20 km/s. Charge exchange

---

[1]The large scale structure of the bubble of solar material carved out by the solar wind expanding into the partially ionized local interstellar medium is called the heliosphere. See the review by Zank (1999) for a discussion of the large-scale heliosphere and its coupling through collisional charge-exchange processes to the local interstellar medium.

between a solar wind proton and an interstellar H atom creates a new ion (the "pickup ion) which responds immediately to the motional electric field of the solar wind, accelerating the pickup ion to co-move with the solar wind. The newly created pickup ions have an energy corresponding to that of the solar wind speed in which they were "picked up, i.e., of the order of about 1 keV, and form an unstable ring-beam distribution. The ring-beam instability excites Alfvén waves that scatter the initial unstable pickup ion distribution to a more stable bispherical distribution, which is well approximated by a simple shell distribution that co-moves with the background solar wind. The pickup ions come to dominate the solar wind thermally in the distant heliosphere, although the ram pressure carried by the relatively cool solar wind remains the energetically dominant process. Furthermore, the excited Alfvén waves act as a new source of turbulence in the outer heliosphere, and therefore modify the transport of magnetic turbulence in the outer heliosphere. It transpires that the driving and transport and dissipation of low frequency magnetohydrodynamic turbulence in the heliosphere is necessary to explain the heating of the solar wind and the observed temperature profile. This in turn affects the modulation, i.e., the transport, of galactic and anomalous cosmic rays. In summary, the apparently simple coupling through collisional charge exchange of neutral interstellar hydrogen and the solar wind plasma yields an enormously complicated set of transport processes, both collisional and collisionless, including thermal, suprathermal (pickup ions), and superthermal (cosmic rays) particles and low-frequency magnetic fields. The purpose behind this book is to develop some of the theoretical techniques necessary to understand the underlying transport modeling of various processes in systems such as the outer heliosphere-local interstellar medium system.

The study of transport processes in gases by Maxwell and Boltzmann laid the foundations for using kinetic processes to understand and develop a macroscopic description of non-equilibrium phenomena. Within the context of a collisional gas, a fairly systematic and complete theory of transport in dilute or moderately dense neutral gases was completed, predicting transport coefficients for diffusion, heat conduction, and viscosity that appear to be in reasonable accord with observations. Nonetheless, this remains an active field of research with numerous applications. Transport phenomena in low density gases are based on the idea that frequent collisions, essentially binary, between individual elements of the gas (molecules or atoms) drive irreversibility and dissipation because of the subsequent randomization of individual particle trajectories–the essence of "molecular chaos. The random collisions of the particles are balanced by the free streaming of the individual particles in a macroscopic inhomogeneous medium subject to the possible presence of an external force field such as gravity. The balancing of collisions, streaming, and the response to an external force, results in the system of particles reaching a quasi-steady state for which a macroscopic description of fluxes of mass, momentum, and energy can be deduced. The macroscopic fluxes at a first approximation are linearly related to the gradients of density, temperature/pressure, and velocity. Such relations allow one to derive transport coefficients whose detailed form depends on assumptions about the nature of the inter-particle forces and particle structure.

1 Introduction

Despite the success in modeling neutral dilute gases in the collisional regime, charged gases (plasmas) have proved to be immensely challenging. A *classical collisional theory for plasmas* was developed over the past 50 plus years by many authors, based effectively on extensions of the Chapman-Enskog-Grad formalism, but predications of transport phenomena did not compare very well to observations, sometimes different by several orders of magnitude. The difficulties stem in part from two effects. On the one hand, collisions in a fully ionized plasma cannot be treated as binary interactions since the Coulomb force is long range. The second is that charged particles respond to the magnetic field of the medium. Large-scale magnetic field structure induces particle drifts of various forms, particle trapping, and so forth. The combination of classical scattering models of plasma and magnetic field geometry improves the predictions of transport phenomena in a plasma, and this is sometimes referred to as the *neoclassical transport theory for plasmas*. However, despite some successes, the neoclassical transport theory remains largely unsuccessful when applied to realistic plasma configurations. The reason for this is that a plasma is almost never in a quiescent state. The long-range Coulomb force ensures that the collective response of a plasma to fluctuations and variation governs its behavior. Under many circumstances in space and astrophysical plasmas, particle collisions play a relatively minor (although, as illustrated in the solar wind interstellar medium interaction example above, not an unimportant) role in the dynamics of a plasma. Collective processes in both collisional and collisionless plasmas results in the particles organizing themselves as waves, vortices, advected fluctuations, streamers, coherent structures, etc. that can be mutually coupled through processes such as wave-wave coupling or turbulence. Such coupling of fluctuations and coherent structures is often much more effective in transporting mass, momentum, and energy than individual particle action. The transport theory for collisionless plasmas is often described as *anomalous transport theory*. This theory has been particularly intensively studied in an astrophysical context to describe the transport of energetic particles in a low-frequency magnetically turbulent inhomogeneous flow.

In this book, we present a systematic treatment of classical transport theory as applied to gases. The application of the Chapman-Enskog approach is simplified in that we assume a relaxation time scattering operator rather than using a full collisional integral. This allows us to address the detail of the derivation of the transport coefficients without becoming too mired in tedious algebra. We also used a simplified form of the gyrotropic-averaged transport equation in Chap. 2 without derivation (although this is derived in Chap. 4) as an example of the use of polynomial expansion techniques to derive simpler transport equations. In this case, we use the gyrotropic-averaged problem to derive both a telegrapher transport equation and a diffusion transport equation for an isotropic distribution of particles. The polynomial approach allows us to discuss the connection between the telegrapher and diffusion equations. The derivation of the viscous and heat conduction transport coefficients allows us to apply a multiple-scales perturbation analysis to the Navier-Stokes equations to derive evolution equations for linear and non-linear waves, the latter providing a model for weak shock waves. This leads to a more general discussion of weak solutions and shock waves.

Chapter 3 extends the classical transport theory of gases to a collisional plasma. The analysis is based about the Chapman-Kolmogorov equation and the derivation of the Fokker-Planck transport equation. The formalism of the Rosenbluth potentials is particularly useful in deriving the transport coefficients for a variety of collisional processes, and the Landau form of the collision operator is obtained. The collisional processes that are considered are electron-proton, and proton-electron collisions, collisions with a Maxwellian background, and fast particle collisions. This allows for the derivation of the transport equations for a proton-electron plasma, which are used to describe the diamagnetic heat flux. The two-fluid proton-electron transport equations can be simplified further to obtain the equations of magnetohydrodynamics (MHD). Some discussion about conservation laws associated with MHD is presented in Chap. 3, followed by a brief presentation of MHD shock waves and their classification based on the shock polar relation.

Chapter 4 considers anomalous transport of energetic particles in a space and astrophysics context. Specifically, we address the transport of energetic particles that experience pitch-angle scattering in a turbulent fluctuating magnetic field. Both the gyrophase-averaged transport equation for non-relativistic and relativistic energy particles is derived. The gyrophase-averaged transport equation is also known as the focused transport equation and is appropriate to non-isotropic particle distributions. The derivation of the focused transport equations is presented in more detail than is found typically in the literature. The focused transport equation is still very complicated and we introduce a Legendre polynomial technique to consider distributions that are nearly isotropic. This approach, which is very systematic, yields the classical cosmic ray transport equation derived originally by E.N. Parker, W.I. Axford, and L.J. Gleeson. Because the interaction between energetic particles and turbulence is so critical to the derivation of the transport coefficients, we address in some detail the magnetic correlation tensor in Chap. 4. The use of the magnetic correlation tensor allows the derivation of the momentum and spatial diffusion tensors using a quasi-linear methodology. To evaluate the perpendicular component of the spatial diffusion tensor, we utilize the non-linear guiding center approach that has become popular. Chapter 4 concludes with some applications of the cosmic ray transport theory. The cosmic ray transport equation is further reduced by invoking a fluid moment closure to yield the cosmic ray two-fluid equations. Some discussion about linear and nonlinear waves and weak shock structure is presented. A very important application of the energetic particle transport equation is to the energization of particles at a shock wave the mechanism of diffusive shock acceleration. This is discussed from both steady and time-dependent perspectives. We conclude with a brief analytic treatment of cosmic ray modulation in the heliosphere.

The final chapter, Chap. 5, considers the transport of fields. In this case, we address the transport of low frequency magnetohydrodynamic turbulence in an expanding flow. The MHD turbulence transport models extend classical turbulence transport models used successfully for hydrodynamics, including in engineering applications. A basic description of MHD turbulence is presented, based on either a Kolmogorov or Irishnikov-Kraichnan phenomenology. Adopting one of these phenomenologies allows us to develop an energy-containing transport theory for MHD turbulence based on a mean-field decomposition of the incompressible inhomogeneous MHD equations. A comparison with solar wind observations of the energy density in magnetic field fluctuations with the theoretical transport turbulence model illustrates the considerable success that these models have enjoyed.

The book is aimed at advanced undergraduates, graduate students, and new postdoctoral students. The general background that is needed for students taking a class based on this book is basic mathematical statistics, although Chap. 1 provides much of the needed background. It is useful for students to have introductory classes in plasma physics, statistical mechanics, and mathematical methods.

A course can be tailored in different ways, depending on the interests and needs of the person teaching the class. In classes given at the University of Alabama in Huntsville, I have tended to cover most of the book in a single semester, omitting some sections in each chapter depending on the interests of the students. Graduate students have found the pace challenging, particularly since I have required the students to do most of the problems listed in the book. Problems are distributed quite liberally throughout the book. A detailed solutions manual is available and has been published by A. Dosch and G.P. Zank and is available from the publisher. If a class has the requisite background in mathematical statistics, most of Chap. 1 can be omitted or relegated to background reading. Chapter 2 is important even for students familiar with the H-theorem because of the systematic development from the Chapman-Enskog approach of the hydrodynamic equations (Euler and Navier-Stokes). The subsections on weak shocks, perturbative techniques, and shock waves are important for later sections in Chaps. 3 and 4. The section on the telegrapher equation in Chap. 2 can be neglected. The first sections of Chap. 3 should be included, but the alternative, more formal derivation of the Chapman-Kolmogorov equation can be omitted. The development of the Rosenbluth potentials should be addressed, which includes the Landau scattering operator. Some of the particular scattering processes can be included in a course but not all sections need to be included. Derivation of the MHD equations from the two-fluid proton-electron transport equations is probably necessary given their wide application, and the section on MHD shock waves is also particularly useful. Chapter 4 provides a comprehensive treatment of energetic particle transport in a collisionless plasma. Depending on interests, much of the derivation of the transport coefficients from the magnetostatic correlation tensor may be neglected but the hydrodynamic description of a coupled thermal and energetic particle plasma system is useful. The wide applicability of diffusive shock acceleration theory makes this section important. Finally, Chap. 5 can be neglected if the purpose of the class is to address transport equations for particles. However, it provides the needed background to understand simple models of turbulence and an introduction to the transport of turbulence that is not found elsewhere.

Reference

G.P. Zank, Modeling the interaction of the solar wind with the local interstellar medium: A theoretical perspective. Space Sci. Rev. **89**(3–4), 413–687 (1999)

Chapter 2
Statistical Background

2.1 Probability Set Function

For many interesting physical problems, we need to describe the "long-term" behavior of systems governed by macroscopic laws and microscopic randomness. A random event has an outcome that is uncertain and unpredictable. Sometimes small changes in initial conditions can result in a substantially different outcome – this is the essence of chaos. Quantities that change randomly in time and space are called stochastic processes. Physical systems that are subject to stochastic driving will have a random component and the variables that describe the system are also stochastic processes. Examples of physical problems include the behavior of gases in the presence of microscopic collisions of the constituent particles, the collective propagation of energetic charged particles in a magnetically turbulent medium, the collective behavior of dust particles in an accretion disk subject to coagulation and destruction, the evolution of a gas of charged protons and electrons (a plasma), etc.

Before considering specific physical problems, some basic statistical concepts need to be reviewed. There are numerous excellent texts that introduce the basic elements of probability theory and stochastic theory and this introductory chapter draws heavily from these. The classic treatise is that of Feller (1968), from which almost all introductory texts draw. Much of this chapter is based on the books by Hogg and Craig (1978) and Gibra (1973).

Suppose we perform n independent experiments under identical conditions. If an outcome A results n_A times, then the probability that A occurs is

$$P(A) = \lim_{n \to \infty} \frac{n_A}{n}.$$

More formally, let $\mathcal{C}$ be the set of all possible outcomes of a random experiment. $\mathcal{C}$ is the *sample space*. An outcome is a point or an element in the sample space. Thus, a sample space $\mathcal{C}$ is a set of elements or points, each of which corresponds to an outcome of an experiment or observation. A sample space can be finite or infinite.

Example. Toss a coin twice and denote the outcomes as H (head) and T (tail). The possible outcomes are

$$\{(H,H),(H,T),(T,H),(T,T)\},$$

and so the sample space is given by

$$\mathcal{C} = \{(H,H),(H,T),(T,H),(T,T)\}.$$

The subset E corresponding to the *event* of heads occurring on the first toss contains two elements

$$E = \{(H,H),(H,T)\}.$$

Another subset is the event F where a head occurs on the first and second toss, given by

$$F = \{(H,H)\}.$$

An event E is defined as a set of outcomes, and an event has occurred if the outcome of the experiment corresponds to an element of subset E. A *null event* corresponds to the empty set $\emptyset$, i.e., the set of no outcomes. If the subset E consists of all possible outcomes of the experiment, then E is the sample space (and obviously an event).

Define a *probability set function* $P(C)$ such that if $C \subset \mathcal{C}$, then $P(C)$ is the probability that the outcome of the random experiment is an element of C. We take $P(C)$ to be the number about which the relative frequency n_A/n converges after many experiments. The properties that we want of the probability set function may be defined as follows.

Definition. If $P(C)$ is defined for a subset C of the space $\mathcal{C}$, and $C_1, C_2, C_3, \ldots$ are disjoint subsets of $\mathcal{C}$, then $P(C)$ is called the *probability set function* of the outcome of the random experiment if

(i) $P(C) \geq 0$,
(ii) $P(C_1 \cup C_2 \cup C_3 \cdots) = P(C_1) + P(C_2) + P(C_3) + \cdots$,
(iii) $P(\mathcal{C}) = 1$.

Theorem 1. *For each $C \subset \mathcal{C}$, $P(C) = 1 - P(C^*)$, where C^* denotes the complement of C.*

Proof. Since $\mathcal{C} = C \cup C^*$ and $C \cap C^* = \emptyset$, $1 = P(C) + P(C^*)$.

Theorem 2. $P(\emptyset) = 0$.

2.1 Probability Set Function

Proof. In Theorem 1, take $C = \emptyset$ so that $C^* = \mathcal{C}$. Hence

$$P(\emptyset) = 1 - P(\mathcal{C}) = 1 - 1 = 0.$$

Theorem 3. *If C_1 and C_2 are subsets of $\mathcal{C}$ such that $C_1 \subset C_2$, then $P(C_1) \leq P(C_2)$.*

Proof. $C_2 = C_1 \cup (C_1^* \cap C_2)$ and $C_1 \cap (C_1^* \cap C_2) = \emptyset$. Hence

$$P(C_2) = P(C_1) + P(C_1^* \cap C_2).$$

But $P(C_1^* \cap C_2) \geq 0$, hence $P(C_2) \geq P(C_1)$.

Theorem 4. *For each $C \subset \mathcal{C}$, $0 \leq P(C) \leq 1$.*

Proof. Since $\emptyset \subset C \subset \mathcal{C}$, $P(\emptyset) \leq P(C) \leq P(\mathcal{C})$ or $0 \leq P(C) \leq 1$.

Theorem 5. *If $C_1 \subset \mathcal{C}$ and $C_2 \subset \mathcal{C}$, then*

$$P(C_1 \cup C_2) = P(C_1) + P(C_2) - P(C_1 \cap C_2).$$

Proof. Since

$$C_1 \cup C_2 = C_1 \cup (C_1^* \cap C_2) \quad \text{and} \quad C_2 = (C_1 \cap C_2) \cup (C_1^* \cap C_2),$$

we have

$$P(C_1 \cup C_2) = P(C_1) + P(C_1^* \cap C_2)$$

and

$$P(C_2) = P(C_1 \cap C_2) + P(C_1^* \cap C_2).$$

Hence

$$P(C_1 \cup C_2) = P(C_1) + P(C_2) - P(C_1 \cap C_2).$$

Example. Two coins are tossed and the ordered pairs form the sample space

$$\mathcal{C} = \{c : c = (H, H), (H, T), (T, H), (T, T)\},$$

and let $P(c \in \mathcal{C}) = \frac{1}{4}$. Suppose C_1 is the event that a head is tossed with the first coin and C_2 the event that a head is tossed with the second coin. The events C_1 and C_2 therefore correspond to the subsets

$$C_1 = \{c : c = (H, H), (H, T)\} \quad \text{and} \quad C_2 = \{c : c = (H, H), (T, H)\}.$$

To find the probability that the first coin toss corresponds to a head or the second to head, we compute

$$P(C_1) = P(C_2) = \frac{1}{2},$$

and the probability that tossing the two coins each results in a head is given by

$$P(C_1 \cap C_2) = \frac{1}{4}.$$

The probability that a head is tossed by one or the other coin is given by

$$P(C_1 \cup C_2) = \frac{1}{2} + \frac{1}{2} - \frac{1}{4} = \frac{3}{4}.$$

Exercises

1. A positive integer from 1 to 6 is chosen by casting a die. Thus $C = \{c : c = 1, 2, 3, 4, 5, 6\}$. Let $C_1 = \{c : c = 1, 2, 3, 4\}$, $C_2 = \{c : c = 3, 4, 5, 6\}$. If $P(c \in C) = \frac{1}{6}$, find $P(C_1)$, $P(C_2)$, $P(C_1 \cap C_2)$, and $P(C_1 \cup C_2)$.
2. Draw a number without replacement from the set $\{1, 2, 3, 4, 5\}$, i.e., choose a number, and then a second from the remaining numbers, etc. Assume that all 20 possible results have the same probability. Find the probability that an odd digit will be selected (a) the first time, (b) the second time, and (c) both times.
3. Draw cards from an ordinary deck of 52 cards and suppose that the probability set function assigns a probability of $\frac{1}{52}$ to each of the possible outcomes. Let C_1 denote the collection of 13 hearts and C_2 the collection of 4 kings. Compute $P(C_1)$, $P(C_2)$, $P(C_1 \cap C_2)$ and $P(C \cup C_2)$.
4. A coin is tossed until a head results. The elements of the sample space C are therefore H, TH, TTH, TTTH, TTTTH, etc. The probability set function assigns probabilities $\frac{1}{2}, \frac{1}{4}, \frac{1}{8}, \frac{1}{16}$, etc. Show that $P(C) = 1$. Suppose $C_1 = \{c : c \text{ is } H, TH, TTH, TTTH, \text{ or } TTTTH\}$ and $C_2 = \{c : c \text{ is } TTTTH \text{ or } TTTTTH\}$. Find $P(C_1)$, $P(C_2)$, $P(C_1 \cap C_2)$, and $P(C_1 \cup C_2)$.
5. A coin is tossed until for the first time the same result appears twice in succession. Let the probability for each outcome requiring n tosses be $1/2^{n-1}$. Describe the sample space, and find the probability of the events (a) the tosses end before the sixth toss, (b) an even number of tosses is required.
6. Find $P(C_1 \cap C_2)$ if the sample space is $C = C_1 \cup C_2$, $P(C_1) = 0.8$ and $P(C_2) = 0.5$.
7. Suppose $C \subset \mathcal{C} = \{c : 0 < c < \infty\}$ with $C = \{c : 4 < c < \infty\}$ and $P(C) = \int_C e^{-x} dx$. Determine $P(C)$, $P(C^*)$, and $P(C \cup C^*)$.
8. If $C \subset \mathcal{C}$ is a set for which $\int_C e^{-|x|} dx$ exists, $\mathcal{C} = \{c : -\infty < x < \infty\}$, then show that this set function is *not* a probability set function. What constant should the integral be multiplied by to make it a probability set function?
9. If $C_1 \subseteq \mathcal{C}$ and $C_2 \subseteq \mathcal{C}$ of the sample space $\mathcal{C}$, show that

$$P(C_1 \cap C_2) \leq P(C_1) \leq P(C_1 \cup C_2) \leq P(C_1) + P(C_2).$$

2.2 Random or Stochastic Variables

As discussed above, elements of a sample space C may not be numbers, being outcomes such as "heads" or "tails". Since we are typically interested in quantifying the outcome of an experiment, we formulate a rule by which elements $c \in C$ may be represented by numbers x, pairs or n-tuples of numbers $(x_1, x_2, \ldots, x_n)$.

Definition. Consider a random experiment with a sample space C. A function X that assigns to each $c \in C$ one and only one real number $x = X(c)$ is a *random variable*, and the *space* of X is the set of real numbers $\mathcal{A} = \{x : x = X(c), c \in C\}$.

Example. Coin toss: $C = \{c : \text{where } c \text{ is } T \text{ or } H\}$, and $T \equiv$ Tail, $H \equiv$ Head. Define a function X such that

$$X = \begin{cases} 0 \text{ if } & c = T \\ 1 \text{ if } & c = H \end{cases}$$

Therefore X is a real-valued function defined on C which maps $c \in C$ to a set of real numbers $\mathcal{A} = \{x : x = 0, 1\}$. X is a random variable and the space associated with X is $\mathcal{A}$.

Sometimes the set C has elements that are real numbers, so that if we write $X(c) = c$, then $\mathcal{A} = C$.

Two forms of random variable can be defined. (1) *Discrete random variables* are those that take on a finite or denumerably infinite number of distinct values. (2) *Continuous random variables* are those that take on a continuum of values within the given range. Random variables are generally denoted by capital Latin letters such as X, Y, Z. Some examples of discrete and continuous random variables are the daily demand for coffee at a Starbucks (discrete), the number of customers at a checkout per hour (discrete), the daily number of absences from a company (discrete), the waiting time of a passenger for a train at a particular train station (continuous), the daily consumption of gas by your car (continuous), and the annual snowfall in Alabama (continuous).

Example. A vendor at a rugby game buys "koeksisters" (a form of unfilled donut from South Africa) for $1.00 each and sells them for $2.50 each. Unsold koeksisters cannot be returned. Suppose the demand during a game is a random variable Y. Suppose the vendor orders a quantity X of koeksisters. Let F denote the profit after the game, which can then be computed as

$$F(Y) = \begin{cases} \$1.50X & Y \geq X \\ \$1.50Y - \$1.00(X - Y) & Y < X \end{cases}$$

where $Y = 1, 2, \ldots$.

Example. If a die is rolled twice, the random variable X that describes the sum of the values is $X(c) = c$ where $c = 2, 3, \ldots, 12$.

Example. A sample of five items is drawn randomly from a lot. The random variable X that describes the number of defective items in the sample is $X(c) = c$, where $c = 0, 1, 2, 3, 4, 5$. If the random variable Y is the number of non-defective items in the sample, define the random variable $Z = |X - Y|$. Thus, the random variable $Z(c) = c$ where $c = 1, 3, 5$.

Just as we refer to an "event C" with $C \subset \mathcal{C}$, we can introduce an event A. Like the definition for the probability $P(C)$, we define the probability of the event A, $P(X \in A)$. With $A \subset \mathcal{A}$, let $C \subset \mathcal{C}$ such that $C = \{c : c \in \mathcal{C} \text{ and } X(c) \in A\}$. Thus C has as its elements all outcomes in $\mathcal{C}$ for which the random variable X has a value that is in A. This means that we can define $P(A)$ to be equal to be $P(C)$ where $C = \{c : c \in \mathcal{C} \text{ and } X(c) \in A\}$. This allows us to use the same notation without confusion.

That $P(A)$ is a probability set function can be seen as follows. For condition (i) above, $P(A) = P(C) \geq 0$.

Consider two mutually exclusive events A_1 and A_2. Here

$$P(A_1 \cup A_2) = P(C),$$

where $C = \{c : c \in \mathcal{C} \text{ and } X(c) \in A_1 \cup A_2\}$. However,

$$C = \{c : c \in \mathcal{C} \text{ and } X(c) \in A_1\} \cup \{c : c \in \mathcal{C} \text{ and } X(c) \in A_2\} = C_1 \cup C_2,$$

say. Since C_1 and C_2 disjoint, we have

$$P(C) = P(C_1) + P(C_2) = P(A_1) + P(A_2),$$

which is condition (ii).

Finally, since $C = \{c : c \in \mathcal{C} \text{ and } X(c) \in \mathcal{A}\}$, it implies that $P(\mathcal{A}) = P(\mathcal{C}) = 1$.

Example. Let a coin be tossed twice, and consider the number of heads observed. The sample space is

$$\mathcal{C} = \{c : \text{where } c = TT, TH, HT, HH\}$$

$$X(c) = \begin{cases} 0 \text{ if } c = TT \\ 1 \text{ if } c = TH \text{ or } HT \\ 2 \text{ if } c = HH \end{cases}$$

Hence, $\mathcal{A} = \{x : X = 0, 1, 2\}$.

Let $A \subset \mathcal{A}$ such that $A = \{x : x = 1\}$. What is $P(A)$?

Since $X(c) = 1$ if c is TH or HT,

$$\Rightarrow C \subset \mathcal{C} \quad \text{such that} \quad C = \{c : c = TH \text{ or } HT\}$$

$$\Rightarrow P(A) = P(C) \quad \text{i.e.,} \quad P(X = 1) = P(C).$$

2.2 Random or Stochastic Variables

Define

$$C_1 = \{c : c = TT\}$$
$$C_2 = \{c : c = TH\}$$
$$C_3 = \{c : c = HT\}$$
$$C_4 = \{c : c = HH\}$$

Suppose $P(C)$ assigns a probability of $\frac{1}{4}$ to each C_i. Then $P(C_1) = \frac{1}{4}$, $P(C_2 \cup C_3) = \frac{1}{2}$, and $P(C_4) = \frac{1}{4}$, so that $P(X = 0) = \frac{1}{4}$, $P(X = 1) = \frac{1}{2}$, and $P(X = 2) = \frac{1}{4}$.

Example. An experiment yields a random value in the interval $(0, 1)$, so the sample space is $\mathcal{C} = \{c : 0 < c < 1\}$. Let the probability set function be given by the "length"

$$P(C) = \int_C dc,$$

so, for example, if $C = \{c : \frac{1}{3} < c < \frac{2}{3}\}$ then

$$P(C) = \int_{1/3}^{2/3} dc = \frac{1}{3}.$$

Define a random variable $X = X(c) = 2c + 1$, so that the space $\mathcal{A} = \{x : 1 < x < 3\}$. For $A \subset \mathcal{A}$, such that e.g., $A = \{x : a < x < b, a > 1, b < 3\}$, we have $C = \{c : (a-1)/2 < c < (b-1)/2, a > 1, b < 3\}$. Hence

$$P(A) = P(C) = \int_{(a-1)/2}^{(b-1)/2} dc = \int_a^b \frac{1}{2} dx.$$

Typically, however, we assume a probability distribution for the random variable X rather than introducing the sample space $\mathcal{C}$ and the probability set function $P(C)$.

Example. Suppose the probability set function $P(A)$ of a random variable X is

$$P(A) = \int_A f(x)dx \quad \text{where} \quad f(x) = 2x, \quad x \in \mathcal{A} = \{x : 0 < x < 1\}.$$

$A_1 = \{x : 0 < x < \frac{1}{4}\}$ and $A_2 = \{x : \frac{1}{2} < x < \frac{3}{4}\}$ are subsets of $\mathcal{A}$. Then

$$P(A_1) = P(X \in A_1) = \int_0^{1/4} 2x \, dx = \frac{1}{16}$$

and

$$P(A_2) = P(X \in A_2) = \int_{1/2}^{3/4} 2x\,dx = \frac{5}{16}.$$

Hence, it follows that since $A_1 \cap A_2 = \emptyset$, $P(A_1 \cup A_2) = P(A_1) + P(A_2) = \frac{3}{8}$.

Example. Consider two random variables X and Y, and let $\mathcal{A} = \{(x, y) : 0 < x < y < 1\}$ be the 2-space. Suppose the probability set function is

$$P(A) = \int\int_A 2\,dx\,dy.$$

If $A_1 = \{(x, y) : \frac{1}{2} < x < y < 1\}$, then

$$P(A_1) = P[(X, Y) \in A_1] = \int_{1/2}^{1}\int_{1/2}^{y} 2\,dx\,dy = \int_{1/2}^{1}(2y-1)\,dy = y^2 - y\big|_{1/2}^{1} = \frac{1}{4}.$$

Suppose $A_2 = \{(x, y) : x < y < 1, 0 < x \le \frac{1}{2}\}$, then $A_2 = A_1^*$, and

$$P(A_2) = P[(X, Y) \in A_2] = P(A_1^*) = 1 - P(A_1) = \frac{3}{4}.$$

Exercises

1. Select a card from a standard deck of 52 playing cards with outcome c. Let $X(c) = 4$ if c is an ace, $X(c) = 3$ for a king, $X(c) = 2$ for a queen, $X(c) = 1$ for a jack, and $X(c) = 0$ otherwise. Suppose $P(C)$ assigns a probability $\frac{1}{52}$ to each outcome c. Calculate the probability $P(A)$ on the space $\mathcal{A} = \{x : x = 0, 1, 2, 3, 4\}$ of the random variable X.
2. Suppose the probability set function $P(A)$ of the random variable X is $P(A) = \int_A f(x)\,dx$ where $f(x) = 2x/9$, $x \in \mathcal{A} = \{x : 0 < x < 3\}$. For $A_1 = \{x : 0 < x < 1\}$ and $A_2 = \{x : 2 < x < 3\}$, compute $P(A_1)$, $P(A_2)$, and $P(A_1 \cup A_2)$.
3. Suppose that the random variable X has space $\mathcal{A} = \{x : 0 < x < 1\}$. If $A_1 = \{x : 0 < x < \frac{1}{2}\}$ and $A_2 = \{x : \frac{1}{2} \le x < 1\}$, find $P(A_2)$ if $P(A_1) = \frac{1}{4}$.

2.3 The Probability Density Function

The distribution of the random variable X refers to the distribution of probability, and this applies even when more than one random variable is involved. We discuss some random variables whose distributions can be described by a *probability density function* of both the *discrete* and *continuous* type. Consider first probability

2.3 The Probability Density Function

distribution functions (pdfs) of one random variable. Suppose X denotes a random variable with one-dimensional space $\mathcal{A}$ such that $\mathcal{A}$ is a set of discrete points. Let $f(x)$ be a one-to-one function $f(x) > 0$, $x \in \mathcal{A}$ with

$$\sum_{\mathcal{A}} f(x) = 1.$$

Whenever a probability set function $P(A)$, $A \subset \mathcal{A}$, can be expressed as

$$P(A) = \sum_{A} f(x),$$

then X is a random variable of the *discrete type*, and X has a *discrete distribution*.

Example. Consider a discrete random variable X with space $\mathcal{A} = \{x : x = 0, 1, 2, 3\}$, and let

$$P(A) = \sum_{A} f(x),$$

where

$$f(x) = \frac{3!}{x!(3-x)!} \left(\frac{1}{2}\right)^3, \qquad x \in \mathcal{A},$$

(recall $0! = 1$). If $A = \{x : x = 0, 1, 2\}$, then

$$\Rightarrow P(X \in A) = \frac{3!}{0!3!}\left(\frac{1}{2}\right)^3 + \frac{3!}{1!2!}\left(\frac{1}{2}\right)^3 + \frac{3!}{2!1!}\left(\frac{1}{2}\right)^3 = \frac{7}{8}.$$

Note that $P(\mathcal{A}) = 1$.

Example. Consider a discrete random variable X with space $\mathcal{A} = \{x : x = 0, 1, 2, 3, \ldots\}$, and let

$$f(x) = \left(\frac{1}{2}\right)^x, \qquad x \in \mathcal{A}.$$

Thus, $P(X \in A) = \sum_{A} f(x)$. For $A = \{x : x = 1, 3, 5, 7, \ldots\}$,

$$P(X \in A) = \frac{1}{2} + \left(\frac{1}{2}\right)^3 + \left(\frac{1}{2}\right)^5 + \cdots = \frac{2}{3}.$$

Suppose that the one-dimensional Riemann integral over the space $\mathcal{A}$ satisfies

$$\int_{\mathcal{A}} f(x)dx = 1,$$

where $f(x)$ is a one-to-one function $f(x) > 0$, $x \in \mathcal{A}$ with at most a finite number of discontinuities in every finite subset (interval) of $\mathcal{A}$. Whenever a probability set function $P(A)$, $A \subset \mathcal{A}$, can be expressed as

$$P(A) = P(X \in A) = \int_{A} f(x)dx,$$

then X is a random variable of the *continuous type*, and X has a *continuous distribution*.

Example. Let $\mathcal{A} = \{x : 0 < x < \infty\}$ and $f(x) = ae^{-3x}$, $x \in \mathcal{A}$. The probability set function is

$$P(X \in A) = \int_{A} f(x)dx = \int_{A} ae^{-3x}dx.$$

Since $P(\mathcal{A}) = 1$,

$$P(\mathcal{A}) = \int_{0}^{\infty} ae^{-3x}dx = 1 \implies a = 3.$$

If $A = \{x : 0 < x < 1\}$, then

$$P(X \in A) = \int_{0}^{1} 3e^{-3x}dx = 1 - e^{-3}.$$

The probability $P(A)$ is determined completely by the *probability density function (pdf)* $f(x)$, whether or not X is a discrete or continuous random variable.

The concept of the pdf of one random variable is readily extended to the pdf of multiple random variables. For example, suppose the two random variables X and Y are discrete or continuous and have a distribution such that the probability set function $P(A)$, $A \subset \mathcal{A}$ can be expressed as

$$P(A) = P[(X, Y) \in A] = \sum\sum_{A} f(x, y),$$

or

$$P(A) = P[(X, Y) \in A] = \int\int_{A} f(x, y)dxdy.$$

2.3 The Probability Density Function

In either case, $f(x, y)$ is the pdf of the two random variables X and Y. Of course, $P(\mathcal{A}) = 1$.

Suppose that the space of a continuous random variable X is $\mathcal{A} = \{x : 0 < x < \infty\}$ and that the pdf is $3e^{-3x}$, $x \in \mathcal{A}$. We can write

$$f(x) = \begin{cases} 3e^{-3x} & 0 < x < \infty \\ 0 & \text{elsewhere} \end{cases}$$

and $f(x)$ is the pdf of X. Hence

$$\int_{-\infty}^{\infty} f(x)dx = \int_{-\infty}^{0} 0 dx + \int_{0}^{\infty} 3e^{-3x} dx = 1.$$

If $A \subset \mathcal{A}$ such that $A = \{x : a < x < b\}$, then

$$P(A) = P(a < x < b) = \int_{a}^{b} f(x)dx.$$

If $A = \{x : x = a\}$, then $P(A) = 0$, which implies that $P(a < x < b) = P(a \leq x \leq b)$.

Example. Suppose the random variable X has pdf

$$f(x) = \begin{cases} 3x^2, & 0 < x < 1 \\ 0, & \text{elsewhere} \end{cases}$$

To find $P(\frac{1}{4} < X < \frac{1}{2})$, we evaluate

$$P(\frac{1}{4} < X < \frac{1}{2}) = \int_{1/4}^{1/2} f(x)dx = \int_{1/4}^{1/2} 3x^2 dx = \frac{3}{32}.$$

Similarly,

$$P(-\frac{1}{2} < X < \frac{1}{2}) = \int_{-1/2}^{1/2} f(x)dx = \int_{-1/2}^{0} 0 dx + \int_{0}^{1/2} 3x^2 dx = \frac{1}{8}.$$

Example. Let

$$f(x, y) = \begin{cases} 6x^2 y & 0 < x < 1, \ 0 < y < 1 \\ 0 & \text{elsewhere} \end{cases}$$

be the pdf of two random variables X, Y. For example,

$$P(0 < X < \frac{3}{4}, \frac{1}{3} < Y < 2) = \int_{1/3}^{2} \int_{0}^{3/4} f(x,y) dx dy$$

$$= \int_{1/3}^{1} \int_{0}^{3/4} 6x^2 y \, dx dy + \int_{1}^{2} \int_{0}^{3/4} 0 \, dx dy$$

$$= \frac{3}{8} + 0 = \frac{3}{8}.$$

Exercises

1. Find the constant a that ensures that $f(x)$ is a pdf of the random variable X: (a) $f(x) = a \left(\frac{2}{3}\right)^x$, $x = 1, 2, 3, \ldots$, 0 elsewhere. (b) $f(x) = axe^{-x}$, $0 < x < \infty$, 0 elsewhere.
2. Consider a function of the random variable X such that

$$f(x) = \begin{cases} ax & 0 \le x < 10 \\ a(20-x) & 10 \le x < 20 \\ 0 & \text{elsewhere} \end{cases}$$

   Find a so that $f(x)$ is a pdf and sketch the graph of the pdf. Compute $P(X \ge 10)$ and $P(15 \le X \le 20)$.
3. Let $f(x) = x/15$, $x = 1, 2, 3, 4, 5$, 0 elsewhere, be the pdf of X. Find $P(X = 1 \text{ or } 2)$, $P(\frac{1}{2} < X < \frac{5}{2})$, and $P(1 \le X \le 2)$.
4. Compute $P(|X| < 1)$ and $P(X^2 < 9)$ for the following pdfs of X, (a) $f(x) = x^2/18$, $-3 < x < 3$, 0 elsewhere. (b) $f(x) = (x+2)/18$, $-2 < x < 4$, 0 elsewhere.
5. Given $P(X > a) = e^{-\lambda a}(\lambda a + 1)$, $\lambda > 0$, $a \ge 0$, find the pdf of X and $P(X > \lambda^{-1})$.
6. Let $f(x) = x^{-2}$ $1 < x < \infty$, 0 elsewhere, be the pdf of X. If $A_1 = \{x : 1 < x < 2\}$ and $A_2 = \{x : 4 < x < 5\}$, find $P(A_1 \cup A_2)$ and $P(A_1 \cap A_2)$.
7. Let $f(x,y) = 4xy$, $0 < x < 1$, $0 < y < 1$, 0 elsewhere, be the pdf of X and Y. Find $P(0 < X < \frac{1}{2}, \frac{1}{4} < Y < 1)$, $P(X = Y)$, $P(X < Y)$, and $P(X \le Y)$.
8. Given that the random variable X has the pdf

$$f(x) = \begin{cases} \frac{5}{a} & -0.1a < x < 0.1a \\ 0 & \text{elsewhere} \end{cases}$$

and $P(|X| < 2) = 2P(|X| > 2)$, find the value of a.

2.4 The Distribution Function

Suppose a random variable X has the probability set function $P(A)$, and A is a 1D set. For a real number x, let $A = \{y : -\infty < y \leq x\}$, so that $P(A) = P(X \in A) = P(X \leq x)$. The probability is thus a function of x, say $F(x) = P(X \leq x)$. The function $F(x)$ is the *distribution function* of the random variable X. Hence, if $f(x)$ is the pdf of X, we have for a discrete random variable X

$$F(x) = \sum_{y \leq x} f(y),$$

and for a continuous random variable X

$$F(x) = \int_{y \leq x} f(y) dy.$$

The distribution function is therefore of a discrete or continuous type.

Example. A die is rolled once. The sample space is $\mathcal{C} = \{1, 2, 3, 4, 5, 6\}$. To find the probability that the value on the upturned face is less than or equal to three, let X be the random variable whose value is less than or equal to three. Thus, the event $X \leq 3$ is the subset $C = \{1, 2, 3\} \subset \mathcal{C}$. The distribution function

$$F(3) = P(X \leq 3) = P(C) = P(1) + P(2) + P(3) = \frac{1}{6} + \frac{1}{6} + \frac{1}{6} = \frac{1}{2}.$$

Sketch the graph of the distribution function.

A distribution function $F(x)$ possesses the following properties.

(i) $0 \leq F(x) \leq 1$ because $0 \leq P(X \leq x) \leq 1$.
(ii) $F(x)$ is a non-decreasing function of x. This can be seen as follows. Suppose $x_1 < x_2$. Then

$$\{x : x \leq x_2\} = \{x : x \leq x_1\} \cup \{x : x_1 < x \leq x_2\}$$

and

$$P(X \leq x_2) = P(X \leq x_1) + P(x_1 < X \leq x_2),$$

from which it follows that

$$F(x_2) - F(x_1) = P(x_1 < X \leq x_2) \geq 0.$$

(iii) $F(\infty) = 1$ and $F(-\infty) = 0$ ($\{x : x \leq -\infty\} = \emptyset$).
(iv) If $a < b$, then, from (ii),

$$P(a < X < b) = F(b) - F(a).$$

Without proof,
$$P(X = b) = F(b) - F(b-),$$
where $F(b-)$ denotes the left-hand limit of $F(x)$ at $x = b$. If the distribution is continuous at $x = b$, then $P(X = b) = 0$.

(v) Also without proof, $F(x)$ is continuous to the right at each point x.

Example. The distribution function of a random variable X is given by
$$F(x) = \begin{cases} 0 & x < 0 \\ \frac{1}{4} & 0 \leq x < 1 \\ \frac{2}{3} & 1 \leq x < 2 \\ 1 & x \geq 2 \end{cases}$$

Thus,
$$P(X \leq 1) = F(1) = \frac{2}{3};$$
$$P(0 < X \leq 2) = F(2) - F(0) = 1 - \frac{1}{4} = \frac{3}{4};$$
$$P(1 < X \leq 2) = F(2) - F(1) = 1 - \frac{2}{3} = \frac{1}{3};$$
$$P(1 \leq X \leq 2) = F(2) - F(0) = 1 - \frac{1}{4} = \frac{3}{4}.$$

Example. Suppose the random variable X is continuous with pdf $f(x) = e^{-x}$, $0 < x < \infty$, 0 elsewhere. The distribution function of X is therefore
$$F(x) = \int_{-\infty}^{x} 0 \, dy, \quad x < 0$$
$$= \int_{0}^{x} e^{-y} dy = 1 - e^{-x}, \quad 0 \leq x$$

Sketch this function. $F(x)$ is a continuous function for all real x, and the derivative exists at all points except $x = 0$.

Example. Let a distribution function be given by (sketch this function)
$$F(x) = \begin{cases} 0, & x < 0 \\ \frac{x+1}{2}, & 0 \leq x < 1 \\ 1, & 1 \leq x \end{cases}$$

so that
$$P(-3 < X \leq \frac{1}{2}) = F\left(\frac{1}{2}\right) - F(-3) = \frac{3}{4} - 0 = \frac{3}{4},$$

2.4 The Distribution Function

for example, and

$$P(X = 0) = F(0) - F(0-) = \frac{1}{2} - 0 = \frac{1}{2}.$$

Note that $F(x)$ is a mixture of both discrete and continuous type distributions.

Suppose X is a random variable with space $\mathcal{A}$, and $Y = g(X)$ is a function of X. $Y = g(X)$ is a random variable with space $\mathcal{B} = \{y : y = g(x), x \in \mathcal{A}\}$ and a probability set function. If $y \in \mathcal{B}$, the event $Y = g(X) \leq y$ occurs when and only when the event $X \in A \subset \mathcal{A}$ occurs, where $A = \{x : g(x) \leq y\}$. The distribution function of Y is therefore

$$G(y) = P(Y \leq y) = P(g(X) \leq y) = P(A).$$

Example. Suppose a random variable X has the pdf $f(x) = x + \frac{1}{2}, 0 < x < 1, 0$ elsewhere. The distribution function of X is given by

$$F_X(x) = \int_0^x \left(y + \frac{1}{2}\right) dy = \frac{1}{2}(x^2 + x),$$

so that

$$F_X(x) = \begin{cases} 0, & x < 0 \\ \frac{x}{2}(1+x), & 0 \leq x \leq 1 \\ 1, & x \geq 1 \end{cases}$$

Consider now the composite random variable $Y = X^2$. The distribution function of Y is

$$F_Y(a) = P(Y \leq a) = P(X^2 \leq a) = P(-\sqrt{a} \leq X \leq \sqrt{a}).$$

Y is always non-negative,

$$F_Y(a) = P(X \leq \sqrt{a}) = F_X(\sqrt{a}),$$

therefore

$$F_Y(a) = \begin{cases} 0, & a < 0 \\ \frac{\sqrt{a}}{2}(1 + \sqrt{a}), & 0 \leq a \leq 1 \\ 1, & a \geq 1 \end{cases}$$

Differentiation of F_Y yields the pdf as

$$f_Y(y) = \begin{cases} \frac{1}{2}\left(1 + \frac{1}{2\sqrt{y}}\right), & 0 \leq y \leq 1 \\ 0. & \text{elsewhere} \end{cases}$$

Finally, for example,

$$P(Y > 0.49) = 1 - P(Y \leq 0.49) = 1 - F_Y(0.49) = 1 - \frac{1}{2}(0.49 - \sqrt{0.49}) = 0.405.$$

Extending these ideas to two or more random variables e.g., X and Y, is readily accomplished. Let $P(A)$ be the probability set function of X and Y, where A is a 2D set. If $A = \{(u, v) : u \leq x, v \leq y\}$, x, y real numbers, then

$$P(A) = P[(X, Y) \in A] = P(X \leq x, Y \leq y).$$

The *distribution function* of X and Y is

$$F(x, y) = P(X \leq x, Y \leq y).$$

If X and Y are continuous type random variables with pdf $f(x, y)$, then

$$F(x, y) = \int_{-\infty}^{y} \int_{-\infty}^{x} f(u, v) du dv,$$

so that

$$\frac{\partial^2 F(x, y)}{\partial x \partial y} = f(x, y),$$

at continuous points.

Example. The random variables X, Y, and Z have the pdf $f(x, y, z) = e^{-(x+y+z)}$, $0 < x, y, z < \infty$, 0 elsewhere. The distribution function of X, Y, and Z, is

$$F(x, y, z) = P(X \leq x, Y \leq y, Z \leq z)$$
$$= \int_0^z \int_0^y \int_0^x e^{-(u+v+w)} du dv dw$$
$$= (1 - e^{-x})(1 - e^{-y})(1 - e^{-z}), \qquad 0 \leq x, y, z < \infty,$$

and 0 elsewhere.

Exercises

1. Let $f(x)$ be the pdf of a random variable X. Find the distribution function $F(x)$ of X and sketch the graph

    $f(x) = 1$, $x = 0$, 0 elsewhere.
    $f(x) = 1/3$, $x = -1, 0, 1$, 0 elsewhere.
    $f(x) = x/15$, $x = 1, 2, 3, 4, 5$, 0 elsewhere.
    $f(x) = 3(1 - x)^2$, $0 < x < 1$, 0 elsewhere.
    $f(x) = x^{-2}$, $1 < x < \infty$, 0 elsewhere.
    $f(x) = 1/3$, $0 < x < 1$ and $2 < x < 4$, 0 elsewhere.

2. Given the distribution function

$$F(x) = \begin{cases} 0, & x < -1 \\ \frac{x+2}{4}, & -1 \le x < 1 \\ 1, & 1 \le x \end{cases}$$

Sketch $F(x)$ and compute $P(-\frac{1}{2} < X \le \frac{1}{2})$, $P(X = 0)$, $P(X = 1)$, and $P(2 < X \le 3)$.

3. Suppose the random variable X has a distribution function $F(x) = 1 - e^{-0.1x}$, $x \ge 0$. Find (a) the pdf of X, (b) $P(X > 100)$, and (c) let $Y = 2X + 5$ and find the corresponding distribution function F_Y.

4. Let $f(x) = 1$, $0 < x < 1$, 0 elsewhere be the pdf of X. Find the distribution function and pdf of $Y = \sqrt{X}$.

5. Let $f(x) = x/6$, $x = 1, 2, 3$, 0 elsewhere, be the pdf of X. Find the distribution function and pdf of $Y = X^2$.

6. Suppose that a random variable X has the pdf

$$f(x) = \begin{cases} \frac{1}{2}, & -1 \le x \le 1 \\ 0, & \text{elsewhere} \end{cases}$$

Determine (a) the distribution function $F(x)$; (b) the probability density function of the random variable $Y = X^2$, and (c) compute $P(Y > 0.36)$.

2.5 Expectations and Moments

One of the most important concepts needed for the transport theory of almost any physical system is that of moments. This section provides the foundation for the remaining chapters. Suppose X is a continuous or discrete random variable with pdf $f(x)$ and let $u(X)$ be a function of X such that

$$E[u(x)] = \int_{-\infty}^{\infty} u(x) f(x) dx, \quad \text{or} \quad E[u(x)] = \sum_{x} u(x) f(x)$$

exists, then $E[u(x)]$ is called the *mathematical expectation* or *expected value* of $u(x)$.

More generally, if $X_1, X_2, \ldots, X_n$ are continuous random variables with *joint pdf* $f(x_1, x_2, \ldots x_n)$, then for the function $u(X_1, X_2, \ldots X_n)$, the expectation is defined by

$$E[u(x_1, x_2, \ldots x_n)] = \int_{-\infty}^{\infty} \cdots \int_{-\infty}^{\infty} u(x_1, x_2, \ldots x_n) f(x_1, x_2, \ldots x_n) dx_1 dx_2 \ldots dx_n.$$

The case for discrete random variables is similarly defined.

The expectation possesses the following properties:

1. For a constant k, $E(k) = \int_{-\infty}^{\infty} kf(x)dx = k\int_{-\infty}^{\infty} f(x)dx = k$.
2. For a constant k, $E(ku) = kE(u)$.
3. For constants k_1, k_2, $E(k_1 u_1 + k_2 u_2) = k_1 E(u_1) + k_2 E(u_2)$.
4. For constants c and k, $E[(ku)^c] = k^c E(u^c)$ or $E(kX)^c = k^c E(X^c)$.

Properties (1)–(3) demonstrate that E is a linear operator.

Example. Suppose X has the pdf

$$f(x) = \begin{cases} 2x, & 0 < x < 1 \\ 0, & \text{elsewhere} \end{cases}$$

Corresponding expectations are

$$E(X) = \int_{-\infty}^{\infty} xf(x)dx = \int_0^1 x2xdx = \frac{2}{3}$$

$$E(X^2) = \int_{-\infty}^{\infty} x^2 f(x)dx = \int_0^1 x^2 2xdx = \frac{1}{2}$$

$$E(X + 2X^2) = \frac{2}{3} + 2\frac{1}{2} = \frac{5}{3}.$$

Example. Suppose X and Y have the joint pdf

$$f(x, y) = \begin{cases} x + y, & 0 < x < 1, \quad 0 < y < 1 \\ 0, & \text{elsewhere} \end{cases}$$

Then

$$E(XY) = \int_0^1 \int_0^1 xy(x + y)dxdy$$

$$= \int_0^1 \left(\frac{1}{3}y + \frac{1}{2}y^2\right) dy = \frac{1}{3}.$$

Similarly

$$E(XY^2) = \int_{-\infty}^{\infty} \int_{-\infty}^{\infty} xy^2 f(x, y)dxdy$$

$$= \int_0^1 \int_0^1 xy^2(x + y)dxdy = \frac{17}{12}.$$

Example. Suppose a rod of length three units is randomly divided into two parts. If X is the length of the left-hand piece, we might assume that X has the pdf

2.5 Expectations and Moments

$$f(x) = \begin{cases} \frac{1}{3}, & 0 < x < 3 \\ 0, & \text{elsewhere} \end{cases}$$

Note that

$$\int_0^3 f(x)dx = 1.$$

Hence the expected value of the length X is

$$E(X) = \int_0^3 x\frac{1}{3}dx = \frac{3}{2} \quad \text{and} \quad E(3 - X) = \frac{3}{2}.$$

However, note that the expected product of the two lengths is

$$E[X(3 - X)] = \int_0^3 x(3-x)\frac{1}{3}dx = \frac{9}{6} = \frac{3}{2} \neq \left(\frac{3}{2}\right)^2,$$

illustrating that the expected value of a product is not the product of expected values.

For a random variable X that is either continuous or discrete with pdf $f(x)$, let $u(X) = X$. This defines the *mean value* μ of X with

$$\mu = E(X) = \begin{cases} \int_{-\infty}^{\infty} xf(x)dx & \text{continuous} \\ \sum_x xf(x) & \text{discrete} \end{cases}$$

The *variance* σ^2 of X is obtained by taking $u(X) = (X - \mu)^2$ i.e.,

$$\sigma^2 = E[(X - \mu)^2] = \int_{-\infty}^{\infty} (x - \mu)^2 f(x)dx$$

for a continuous random variable (an analogous definition for a discrete random variable holds). The *standard deviation* of X is simply σ. Observe that since E is a linear operator,

$$\sigma^2 = E[(X - \mu)^2] = E(X^2 - 2\mu X + \mu^2)$$
$$= E(X^2) - 2\mu E(X) + \mu^2$$
$$= E(X^2) - \mu^2.$$

The variance σ^2 of a random variable represents a measure of the variability of observations or fluctuations about the mean μ. If a random variable has a small variance or standard deviation, then most of the values are grouped around the mean. We may therefore expect the probability that a random variable assumes a value within an interval about the mean is greater than for a similar random variable with a larger variance. A useful but rather weak result for estimating the probability of a random variable falling within n standard deviations of its mean is the following inequality.

Theorem (Chebyshev's inequality). *The probability that any random variable X falls within n standard deviations of the mean is at least $(1 - n^{-2})$, or equivalently*

$$P(\mu - n\sigma < X < \mu + n\sigma) \geq 1 - n^{-2}.$$

Proof. Since

$$\sigma^2 = E[(X - \mu)^2] = \int_{-\infty}^{\infty} (x - \mu)^2 f(x) dx$$

$$= \int_{-\infty}^{\mu-n\sigma} (x - \mu)^2 f(x) dx + \int_{\mu-n\sigma}^{\mu+n\sigma} (x - \mu)^2 f(x) dx + \int_{\mu+n\sigma}^{\infty} (x - \mu)^2 f(x) dx$$

$$\geq \int_{-\infty}^{\mu-n\sigma} (x - \mu)^2 f(x) dx + \int_{\mu+n\sigma}^{\infty} (x - \mu)^2 f(x) dx$$

since the middle integral is non-negative. Because $|x - \mu| \geq n\sigma$ whenever $x \geq \mu + n\sigma$ or $x \leq \mu - n\sigma$, we have $(x - \mu)^2 \geq n^2\sigma^2$ in both remaining integrals. Thus,

$$\sigma^2 \geq \int_{-\infty}^{\mu-n\sigma} n^2\sigma^2 f(x) dx + \int_{\mu+n\sigma}^{\infty} n^2\sigma^2 f(x) dx$$

from which we obtain

$$\int_{-\infty}^{\mu-n\sigma} f(x) dx + \int_{\mu+n\sigma}^{\infty} f(x) dx \leq n^{-2}.$$

Hence we have established

$$P(\mu - n\sigma < X < \mu + n\sigma) = \int_{\mu-n\sigma}^{\mu+n\sigma} f(x) dx \geq 1 - n^{-2}.$$

By way of example, for $n = 2$, the random variable X has a probability of at least $1 - (2)^{-2} = 3/4$ of being within two standard deviations of the mean, or equivalently, that 3/4 or more of the observations of any distribution fall in the interval $\mu \pm 2\sigma$.

Example. Suppose a random variable X with an unknown probability distribution has a mean $\mu = 8$, a variance $\sigma^2 = 9$. Then, for example,

$$P(-4 < X < 20) = P[8 - (4)(3) < X < 8 + (4)(3)] \geq \frac{15}{16}.$$

2.5 Expectations and Moments

Consider instead

$$P(|X - 8| \geq 6) = 1 - P(|X - 8| < 6) = 1 - P(-6 < X - 8 < 6)$$
$$= 1 - P[8 - (2)(3) < X < 8 + (2)(3)] \leq \frac{1}{4}.$$

These values are lower bounds only.

Another important concept is the covariance and the related correlation function since it enables the introduction of the idea of statistical independence and stationarity of a random variable. These concepts are particularly important for the statistical description of turbulence, for example. Let X, Y, and Z be random variables with joint pdf $f(x, y, z)$. The mathematical expectation

$$E[(X - \mu_1)(Y - \mu_2)] = E(XY - \mu_2 X - \mu_1 Y + \mu_1 \mu_2)$$
$$= E(XY) - \mu_2 E(X) - \mu_1 E(Y) + \mu_1 \mu_2$$
$$= E(XY) - \mu_1 \mu_2,$$

is the *covariance* of X and Y. Here μ_1, μ_2, μ_3, σ_1^2, σ_2^2, and σ_3^2 denote the means and variances of X, Y, and Z respectively. Similarly, the covariance of X and Z is $E[(X - \mu_1)(Z - \mu_3)]$ and the covariance of Y and Z is $E[(Y - \mu_2)(Z - \mu_3)]$. If the standard deviations $\sigma_1 > 0$ and $\sigma_2 > 0$, we define the *correlation coefficient* of X and Y as

$$\rho_{12} = \frac{E[(X - \mu_1)(Y - \mu_2)]}{\sigma_1 \sigma_2}.$$

In general, if the standard deviations are positive, the correlation coefficient of any two random variables is the covariance of the two random variables divided by the product of the their standard deviations.

Example. The random variables X and Y have the joint pdf

$$f(x, y) = \begin{cases} x + y, & 0 < x < 1, 0 < y < 1, \\ 0, & \text{elsewhere} \end{cases}$$

To compute the correlation coefficient, we need

$$\mu_1 = E(X) = \int_0^1 \int_0^1 x(x + y) dx dy = \frac{7}{12},$$
$$\mu_2 = E(Y) = \int_0^1 \int_0^1 y(x + y) dx dy = \frac{7}{12},$$

and

$$\sigma_1^2 = E(X^2) - \mu_1^2 = \int_0^1 \int_0^1 x^2(x+y)dxdy - \left(\frac{7}{12}\right)^2 = \frac{11}{144},$$

$$\sigma_2^2 = E(Y^2) - \mu_2^2 = \int_0^1 \int_0^1 y^2(x+y)dxdy - \left(\frac{7}{12}\right)^2 = \frac{11}{144}.$$

The covariance of X and Y is

$$E(XY) - \mu_1\mu_2 = \int_0^1 \int_0^1 xy(x+y)dxdy - \left(\frac{7}{12}\right)^2 = -\frac{1}{144},$$

yielding the correlation coefficient as

$$\rho = \frac{-\frac{1}{144}}{\sqrt{\left(\frac{11}{144}\right)\left(\frac{11}{144}\right)}} = -\frac{1}{11}.$$

An important expectation is the *moment generating function* of a random variable X, either continuous or discrete. Suppose there exists a finite real number t for which the expectation

$$E(e^{tX}) = \int_{-\infty}^{\infty} e^{tx} f(x)dx \quad \text{or} \quad E(e^{tX}) = \sum_x e^{tx} f(x)$$

(continuous or discrete respectively) exists. Then, $M(t) = E(e^{tX})$ is the moment generating function. Setting $t = 0 \Rightarrow M(0) = 1$. Not every distribution has a (real-valued) moment generating function, but if it does, then the moment-generating function is unique and completely determines the distribution of the random variable.

For example, let X be a continuous random variable with

$$M(t) = (1-t)^{-2} = \int_{-\infty}^{\infty} e^{tx} f(x)dx, \quad t < 1.$$

It is not obvious how to determine $f(x)$. Consider a distribution with pdf

$$f(x) = \begin{cases} xe^{-x}, & 0 < x < \infty \\ 0 & \text{elsewhere} \end{cases}$$

Then

$$M(t) = \int_0^\infty e^{tx} xe^{-x} dx = \int_0^\infty xe^{-(1-t)x} dx$$

2.5 Expectations and Moments

$$= \int_0^\infty \frac{e^{-(1-t)x}}{1-t} dx = (1-t)^{-2}.$$

Thus, the pdf has the moment generating function $M(t) = (1-t)^{-2}, t < 1$.

An important property of moment generating functions is derivatives of all orders exist at $t = 0$,

$$\frac{dM(t)}{dt} = M'(t) = \int_{-\infty}^\infty x e^{tx} f(x) dx$$

and analogously for a discrete random variable X. Hence, for both cases, $t = 0$ implies

$$M'(0) = E(X) = \mu.$$

Similarly, the second derivative

$$M''(t) = \int_{-\infty}^\infty x^2 e^{tx} f(x) dx,$$

yields $M''(0) = E(X^2)$. Hence,

$$\sigma^2 = E(X^2) - \mu^2 = M''(0) - [M'(0)]^2.$$

Thus, using the previous example $M(t) = (1-t)^{-2}, t < 1$, yields

$$M'(t) = 2(1-t)^{-3}, \quad \text{and} \quad M''(t) = 6(1-t)^{-4},$$

so that $\mu = M'(0) = 2$ and $\sigma^2 = M''(0) - \mu^2 = 2$.

In general, for $m > 0$ an integer, the mth derivative of the moment generating function generates the mth moment of the distribution,

$$M^m(0) = E(X^m) = \int_{-\infty}^\infty x^m f(x) dx \quad \text{or} \quad \sum_x x^m f(x),$$

and this is used to define macroscopic quantities for e.g., gases, plasmas, collections of stars, etc.

Note that a mean, or any other higher-order moments or expectations, need not exist even for a well-defined pdf, as the following example illustrates.

Example. Suppose the random variable X has pdf

$$f(x) = \begin{cases} x^{-2}, & 1 < x < \infty \\ 0, & \text{elsewhere} \end{cases}$$

Then

$$\int_1^\infty xx^{-2}dx = \lim_{c\to\infty}\int_1^c \frac{dx}{x} = \lim_{c\to\infty}(\ln c - ln1),$$

does not exist, and hence neither does the mean value of X, nor other higher-order expectations.

Example. Consider the moment generating function $M(t) = e^{t^2/2}$, $-\infty < x < \infty$. We can use an alternative approach to computing moments rather than simply differentiating. We may write

$$e^{t^2/2} = 1 + \frac{1}{1!}\left(\frac{t^2}{2}\right) + \frac{1}{2!}\left(\frac{t^2}{2}\right)^2 + \cdots + \frac{1}{k!}\left(\frac{t^2}{2}\right)^k + \cdots$$

$$= 1 + \frac{1}{2!}t^2 + \frac{(3)(1)}{4!}t^4 + \cdots + \frac{(2k-1)(2k-3)\cdots(3)(1)}{(2k)!}t^{2k} + \cdots$$

The MacLaurin's series for $M(t)$ is

$$M(t) = M(0) + \frac{M'(0)}{1!}t + \frac{M''(0)}{2!}t^2 + \cdots + \frac{M^m(0)}{m!}t^m + \cdots$$

$$= 1 + \frac{E(X)}{1!}t + \frac{E(X^2)}{2!}t^2 + \cdots + \frac{E(X^m)}{m!}t^m + \cdots$$

Thus the coefficient of $(t^m/m!)$ in the MacLaurin expansion of $M(t)$ is $E(X^m)$. For the example above, we therefore have

$$E(X^{2k}) = (2k-1)(2k-3)\cdots(3)(1) = \frac{(2k)!}{2^k k!},$$

$k = 1, 2, 3, \ldots$, and $E(X^{2k-1}) = 0, k = 1, 2, 3, \ldots$.

Before completing this section, we should point out that many functions do not have real-valued moment generating functions. However, if we define $\phi(t) = E(e^{itX})$, t an arbitrary real number, then this expectation exists for every distribution and is called the *characteristic function* of the distribution. In fact, the characteristic function may be defined from the pdf $f(x)$ using the Fourier integral

$$\phi(t) = \langle e^{itx}\rangle = \int_{-\infty}^\infty e^{itx} f(x)dx,$$

and of course the inverse Fourier transform yields the pdf

2.5 Expectations and Moments

$$f(x) = \frac{1}{2\pi} \int_{-\infty}^{\infty} e^{-itx} \phi(t) dt.$$

By expanding the exponential in the Fourier integral, we can compute moments as before,

$$\phi(t) = \int_{-\infty}^{\infty} \left(1 + itx - \frac{1}{2}t^2 x^2 + \cdots \right) f(x) dx$$

$$= 1 + iE(X)t - \frac{1}{2} E(X^2) t^2 + \cdots$$

Exercises

1. Suppose X has the pdf

$$f(x) = \begin{cases} \frac{x+2}{18}, & -2 < x < 4 \\ 0, & \text{elsewhere} \end{cases}$$

   Find $E(X)$, $E[(X+2)^3]$, and $E[6X - 2(X+2)^3]$.

2. The *median* of a random variable X is the value x such that the distribution function $F(x) = \frac{1}{2}$. Compute the median of the random variable X for the pdf

$$f(x) = \begin{cases} 2x, & 0 < x < 1 \\ 0, & \text{elsewhere} \end{cases}$$

3. The *mode* of a random variable X is the value that occurs most frequently - sometimes called the *most probable value*. The value a is the mode of the random variable X if

$$f(a) = max f(x),$$

   (for a continuous pdf). The mode is not necessarily unique. Compute the mode and median of a random variable X with pdf

$$f(x) = \begin{cases} \frac{2}{3} x, & 0 \le x \le 1 \\ \frac{1}{3}, & 1 < x \le 3. \end{cases}$$

4. Suppose X and Y have the joint pdf

$$f(x,y) = \begin{cases} e^{-x-y}, & 0 < x < \infty, \quad 0 < y < \infty \\ 0, & \text{elsewhere} \end{cases}$$

   and that $u(X,Y) = X$, $v(X,Y) = Y$, and $w(X,Y) = XY$. Show that $E[u(X,Y)] \cdot E[v(X,Y)] = E[w(X,Y)]$.

5. If X and Y are two exponentially distributed random variable with pdfs
$$f_X(x) = 2e^{-2x}, \quad x \geq 0; \quad f_Y(y) = 4e^{-4y}, \quad y \geq 0,$$
calculate $E(X+Y)$.
6. Suppose X and Y have the joint pdf
$$f(x,y) = \begin{cases} 2, & 0 < x < y, \quad 0 < y < 1 \\ 0 & \text{elsewhere} \end{cases}$$
and that $u(X,Y) = X$, $v(X,Y) = Y$, and $w(X,Y) = XY$. Show that $E[u(X,Y)] \cdot E[v(X,Y)] \neq E[w(X,Y)]$.
7. Let X have a pdf $f(x)$ that is positive at $x = -1, 0, 1$ and zero elsewhere. (a) If $f(0) = \frac{1}{2}$, find $E(X^2)$. (b) If $f(0) = \frac{1}{2}$, and if $E(X) = \frac{1}{6}$, determine $f(-1)$ and $f(1)$.
8. A random variable X with an unknown probability distribution has a mean $\mu = 12$ and a variance $\sigma^2 = 9$. Use Chebyshev's inequality to bound $P(6 < X < 18)$ and $P(3 < X < 21)$.
9. Two distinct integers are chosen randomly without replacement from the first six positive integers. What is the expected value of the absolute value of the difference of these two numbers?
10. Assume that the random variable X has mean μ, standard deviation σ, and moment generating function $M(t)$. Show that
$$E\left(\frac{X-\mu}{\sigma}\right) = 0; \quad E\left[\left(\frac{X-\mu}{\sigma}\right)^2\right] = 1,$$
and
$$E\left\{\exp\left[t\left(\frac{X-\mu}{\sigma}\right)\right]\right\} = e^{-\mu t/\sigma} M\left(\frac{t}{\sigma}\right).$$

11. Suppose that $E[(x-b)^2]$ exists for a random variable X for all real b. Show that $E[(x-b)^2]$ is a minimum when $b = E(X)$.
12. Suppose that $R(t) = E(e^{t(X-b)})$ exists for a random variable X. Show that $R^m(0)$ is the mth moment of the distribution about the point b, where m is a positive integer.
13. Let $\Psi(t) = \ln M(t)$, where $M(t)$ is the moment generating function of a distribution. Show that $\Psi'(0) = \mu$ and $\Psi''(0) = \sigma^2$.
14. Suppose X is a random variable with mean μ and variance σ^2, and assume that the third moment $E[(X-\mu)^3]$ exists. The ratio $E[(X-\mu)^3]/\sigma^3$ is a measure of the *skewness* of the distribution. Graph the following pdfs and show that the skewness is negative, zero, and positive respectively:

   (a) $f(x) = (x+1)/2, -1 < x < 1, 0$ elsewhere.
   (b) $f(x) = 1/2, -1 < x < 1, 0$ elsewhere.
   (c) $f(x) = (1-x)/2, -1 < x < 1, 0$ elsewhere.

15. Suppose X is a random variable with mean μ and variance σ^2, and assume that the fourth moment $E[(X-\mu)^4]$ exists. The ratio $E[(X-\mu)^4]/\sigma^4$ is a measure of the *kurtosis* of the distribution. Graph the following pdfs and show that the kurtosis is smaller for the first distribution

    (a) $f(x) = 1/2, -1 < x < 1, 0$ elsewhere.
    (b) $f(x) = 3(1-x^2)/4, -1 < x < 1, 0$ elsewhere.

2.6 Conditional Probability and Marginal and Conditional Distributions

One is sometimes interested only in outcomes that are elements of a subset C_1 of the sample space C. Thus, the subset becomes effectively the new sample space. Let $P(C)$ be the probability set function defined on C and let $C_1 \subset C$ such that $P(C_1) > 0$. Suppose $C_2 \subset C$. We want to define the probability of the event C_2 relative to the hypothesis of the event C_1. This is called the *conditional probability* of the event C_2 relative to the event C_1, or simply the conditional probability of C_2 given C_1, denoted by $P(C_2|C_1)$. Specifically, we define $P(C_2|C_1)$ such that

$$P(C_2|C_1) = \frac{P(C_1 \cap C_2)}{P(C_1)}$$

and $P(C_1) > 0$. We then have

1. $P(C_2|C_1) \geq 0$.
2. $P(C_2 \cup C_3 \cup \cdots | C_1) = P(C_2|C_1) + P(C_3|C_1) + \cdots$, provided $C_2, C_3, \ldots$ are mutually disjoint sets.
3. $P(C_1|C_1) = 1$.

Properties (1) and (3) are obvious and (2) is an exercise. These properties of course ensure that $P(C_2|C_1)$ is a probability set function defined for subsets of C_1 – called the conditional probability set function given C_1.

Consider now a subset A of the event space $\mathcal{A}$ of one or more random variables defined on the sample space C. If P is the probability set function of the induced probability on $\mathcal{A}$, and $A_1 \subset \mathcal{A}$ and $A_2 \subset \mathcal{A}$, then the conditional probability of the event A_2 given the event A_1 is

$$P(A_2|A_1) = \frac{P(A_2 \cap A_1)}{P(A_1)}$$

provided $P(A_1) > 0$.

Note that the above definition yields the multiplication rule for probabilities

$$P(C_1 \cap C_2) = P(C_1)P(C_2|C_1).$$

Example. Suppose we draw cards successively and randomly without replacement from an ordinary deck of cards. Given that three spades were drawn in the first six drawings, what is the probability that the seventh draw will yield a spade? Let C_1 denote the event of three spades in the first six draws and C_2 the event of a spade on the seventh drawing. We want to compute $P(C_1 \cap C_2)$. We therefore have

$$P(C_1) = \frac{\binom{13}{3}\binom{39}{3}}{\binom{52}{6}},$$

and

$$P(C_2|C_1) = \frac{10}{46}.$$

Hence, using

$$P(C_1 \cap C_2) = P(C_1)P(C_2|C_1) \simeq 0.028.$$

Note that the multiplication rule can be extended to multiple events quite straightforwardly. For three events, we have

$$\begin{aligned}P(C_1 \cap C_2 \cap C_3) &= P\left[(C_1 \cap C_2) \cap C_3\right] \\ &= P(C_1 \cap C_2)P(C_3|C_1 \cap C_2) \\ &= P(C_1)P(C_2|C_1)P(C_3|C_1 \cap C_2).\end{aligned}$$

Let $f(x_1, x_2)$ be the joint pdf of two random variables X_1 and X_2. Consider the event $a < X_1 < b$, $a < b$. This event can occur when and only when $a < X_1 < b, -\infty < X_2 < \infty$.

$$P(a < X_1 < b, -\infty < X_2 < \infty) = \int_a^b \int_{-\infty}^{\infty} f(x_1, x_2) dx_2 dx_1$$

for the continuous case (the extension to the discrete case is obvious). Now $\int_{-\infty}^{\infty} f(x_1, x_2) dx_2$ is a function of x_1 only, say $f_1(x_1)$. Hence, for every $a < b$,

$$P(a < X_1 < b) = \int_a^b f_1(x_1) dx_1$$

2.6 Conditional Probability and Marginal and Conditional Distributions

so that $f_1(x_1)$ is a function of X_1 only. $f_1(x_1)$ results from integrating (or summing) the joint pdf $f(x_1, x_2)$ over all x_2 for a fixed x_1. The function $f_1(x_1)$ is called the *marginal pdf* for X_1. A marginal pdf for X_2 is defined by

$$f_2(x_2) = \int_{-\infty}^{\infty} f(x_1, x_2) dx_1.$$

Example. Suppose the joint pdf of X_1 and X_2 is

$$f(x_1, x_2) = \begin{cases} \frac{x_1 + x_2}{8}, & 0 \le x_1 \le 2, \quad 0 \le x_2 \le 2 \\ 0, & \text{elsewhere} \end{cases}$$

The marginal pdf of X_1 is

$$f_1(x_1) = \int f(x_1, x_2) dx_2 = \frac{1}{8} \int_0^2 (x_1 + x_2) dx_2 = \frac{x_1 + 1}{4},$$

and the marginal pdf of X_2 is

$$f_2(x_2) = \int f(x_1, x_2) dx_1 = \frac{1}{8} \int_0^2 (x_1 + x_2) dx_1 = \frac{x_2 + 1}{4}.$$

Note that

$$\int_0^2 f_1(x_1) dx_1 = 1 = \int_0^2 f_2(x_2) dx_2.$$

Example. Suppose the joint pdf of X_1 and X_2 is given by

$$f(x_1, x_2) = \begin{cases} \frac{x_1 + x_2}{21}, & x_1 = 1, 2, 3, x_2 = 1, 2 \\ 0, & \text{elsewhere} \end{cases}$$

We have, for example,

$$P(X_1 = 3) = f(3, 1) + f(3, 2) = \frac{3}{7} \text{ and } P(X_2 = 2)$$

$$= f(1, 2) + f(2, 2) + f(3, 2) = \frac{4}{7}.$$

The marginal pdf of X_1 is

$$f_1(x_1) = \sum_{x_2=1}^{2} \frac{x_1 + x_2}{21} = \frac{2x_1 + 3}{21}, \qquad x_1 = 1, 2, 3$$

and 0 elsewhere. The marginal pdf of X_2 is

$$f_2(x_2) = \sum_{x_1=1}^{3} \frac{x_1 + x_2}{21} = \frac{6 + 3x_2}{21}, \quad x_2 = 1, 2$$

and 0 elsewhere. The preceding probabilities can be computed directly from the marginals as $P(X_1 = 3) = f_1(3) = \frac{3}{7}$ and $P(X_2 = 2) = f_2(2) = \frac{4}{7}$.

Consider now the moment generating function (if it exists) $M(t_1, t_2) = E(e^{t_1 X + t_2 Y})$, t_1, t_2 finite, of the pdf $f(x, y)$ of the random variables X and Y. The moment-generating function, like the single random variable case, completely determines the joint distribution of X and Y, and hence the marginal distributions of X and Y. This follows from

$$M(t_1, 0) = E(e^{t_1 X}) = M(t_1), \quad \text{and} \quad M(0, t_2) = E(e^{t_2 Y}) = M(t_2).$$

For continuous random variables,

$$\frac{\partial^{k+m} M(t_1, t_2)}{\partial t_1^k t_2^m} = \int_{-\infty}^{\infty} \int_{-\infty}^{\infty} x^k y^m e^{t_1 x + t_2 y} f(x, y) dx dy,$$

implying that

$$\left. \frac{\partial^{k+m} M(t_1, t_2)}{\partial t_1^k t_2^m} \right|_{t_1 = t_2 = 0} = \int_{-\infty}^{\infty} \int_{-\infty}^{\infty} x^k y^m f(x, y) dx dy = E(X^k Y^m).$$

This yields the following set of useful relations,

$$\mu_1 = E(X) = \frac{\partial M(0, 0)}{\partial t_1}, \quad \mu_2 = E(Y) = \frac{\partial M(0, 0)}{\partial t_2};$$

$$\sigma_1^2 = E(X^2) - \mu_1^2 = \frac{\partial^2 M(0, 0)}{\partial t_1^2} - \mu_1^2;$$

$$\sigma_2^2 = E(Y^2) - \mu_2^2 = \frac{\partial^2 M(0, 0)}{\partial t_2^2} - \mu_2^2;$$

$$E[(X - \mu_1)(Y - \mu_2)] = \frac{\partial^2 M(0, 0)}{\partial t_1 \partial t_2} - \mu_1 \mu_2.$$

Thus, the covariance and correlation functions can be computed using the moment generating function of the joint pdf.

Example. Continuous random variables X and Y have the joint pdf

2.6 Conditional Probability and Marginal and Conditional Distributions

$$f(x, y) = \begin{cases} e^{-y}, & 0 < x < y < \infty, \\ 0 & \text{elsewhere} \end{cases}$$

The moment generating function is given by

$$M(t_1, t_2) = \int_0^\infty \int_x^\infty \exp(t_1 x + t_2 y - y) \, dy \, dx$$

$$= \frac{1}{(1 - t_1 - t_2)(1 - t_2)},$$

provided $t_1 + t_2 < 1$ and $t_2 < 1$. From the moment-generating formulae above, we can derive (Exercise: Check)

$$\mu_1 = 1, \qquad \mu_2 = 2,$$
$$\sigma_1^2 = 1, \qquad \sigma_2^2 = 2,$$
$$E[(X - \mu_1)(Y - \mu_2)] = 1.$$

Hence, the correlation coefficient of X and Y is $\rho = 1/\sqrt{2}$. The moment generating functions of the marginal distributions of X and Y are

$$M(t_1, 0) = \frac{1}{1 - t_1}, \quad t_1 < 1; \quad M(0, t_2) = \frac{1}{(1 - t_2)^2}, \quad t_2 < 1.$$

The corresponding marginal pdfs are

$$f_1(x) = \begin{cases} \int_x^\infty e^{-y} dy = e^{-x}, & 0 < x < \infty \\ 0 & \text{elsewhere} \end{cases},$$

and

$$f_2(y) = \begin{cases} \int_0^y e^{-y} dx = y e^{-y}, & 0 < y < \infty \\ 0 & \text{elsewhere} \end{cases}.$$

Let X_1 and X_2 denote continuous random variables with joint pdf $f(x_1, x_2)$ and marginal pdfs $f_1(x_1)$ and $f_2(x_2)$. Provided $f_1(x_1) > 0$, we define the *conditional pdf* of the continuous random variable X_2 as

$$f(x_2|x_1) = \frac{f(x_1, x_2)}{f_1(x_1)}.$$

It is easily seen that $f(x_2|x_1)$ has the properties of a pdf (Exercise: Check!), which means that it can be used to compute probabilities and expectations. Thus, the conditional probability that $a < X_2 < b$, given that $X_1 = x_1$ is given by

$$P(a < X_2 < b | X_1 = x_1) = \int_a^b f(x_2|x_1) dx_2.$$

Similarly, $P(c < X_1 < d | X_2 = x_2) = P(c < X_1 < d | x_2) = \int_c^d f(x_1|x_2) dx_1$.

The expectation of the function $u(X_2)$ of X_2

$$E[u(X_2)|x_1] = \int_{-\infty}^{\infty} u(x_2) f(x_2|x_1) dx_2$$

is the *conditional expectation* of $u(X_2)$ given $X_1 = x_1$. Specifically, $E(X_2|x_1)$ is the mean and $E([X_2 - E(X_2|x_1)]^2|x_1)$ the variance of the conditional distribution of X_2 given $X_1 = x_1$. These are sometimes referred to as the conditional mean and variance. Obviously,

$$E[(X_2 - E(X_2|x_1))^2|x_1] = E(X_2^2|x_1) - [E(X_2|x_1)]^2.$$

Example. Suppose the random variables X_1 and X_2 have the joint pdf

$$f(x_1, x_2) = \begin{cases} 6x_1, & 0 < x_1 < x_2 < 1 \\ 0, & \text{elsewhere} \end{cases}$$

The marginal pdfs are

$$f_1(x_1) = \int_{x_1}^{1} 6x_1 dx_2 = 6x_1(1 - x_1), \qquad 0 \le x_1 \le 1$$

$$f_2(x_2) = \int_0^{x_2} 6x_1 dx_1 = 3x_2^2, \qquad 0 \le x_2 \le 1$$

Note that

$$\int_0^1 f_1(x_1) dx_1 = \int_0^1 6x_1(1 - x_1) dx_1 = 1$$

$$\int_0^1 f_2(x_2) dx_2 = \int_0^1 3x_2^2 dx_2 = 1.$$

The conditional pdf of X_1 given $X_2 = x_2$ is

$$f(x_1|x_2) = \frac{f(x_1, x_2)}{f_2(x_2)} = \frac{6x_1}{3x_2^2} = \frac{2x_1}{x_2^2},$$

and the conditional mean is

$$E(X_1|x_2) = \int_0^{x_2} x_1 f(x_1|x_2) dx_1 = \int_0^{x_2} \frac{2x_1^2}{x_2^2} dx_1 = \frac{2}{3} x_2.$$

2.6 Conditional Probability and Marginal and Conditional Distributions

Example. The random variables X_1 and X_2 have the joint pdf

$$f(x_1, x_2) = \begin{cases} 2, & 0 < x_1 < x_2 < 1 \\ 0, & \text{elsewhere} \end{cases}$$

The marginal pdfs are thus

$$f_1(x_1) = \begin{cases} \int_{x_1}^1 2 dx_2 = 2(1-x_1), & 0 < x_1 < 1 \\ 0, & \text{elsewhere} \end{cases},$$

and

$$f_2(x_2) = \begin{cases} \int_0^{x_2} 2 dx_1 = 2x_2, & 0 < x_2 < 1 \\ 0, & \text{elsewhere} \end{cases}.$$

The conditional pdf of X_1, given $X_2 = x_2$, is

$$f(x_1|x_2) = \begin{cases} \frac{2}{2x_2} = x_2^{-1}, & 0 < x_2 < 1 \\ 0, & \text{elsewhere} \end{cases}$$

The conditional mean and variance of X_1 given $X_2 = x_2$ are therefore

$$E(X_1|x_2) = \int_{-\infty}^{\infty} x_1 f(x_1|x_2) dx_1$$

$$= \int_0^{x_2} \frac{x_1}{x_2} dx_1 = \frac{1}{2} x_2, \quad 0 < x_2 < 1$$

and

$$E([X_1 - E(X_1|x_2)]^2|x_2) = \int_0^{x_2} \left(x_1 - \frac{1}{2}x_2\right)^2 x_2^{-1} dx_1$$

$$= \frac{x_2^2}{12}, \quad 0 < x_2 < 1.$$

Note that

$$P(0 < X_1 < 1/2 | X_2 = 3/4) = \int_0^{1/2} f(x_1|3/4) dx_1 = \int_0^{1/2} \frac{4}{3} dx_1 = \frac{2}{3},$$

and

$$P(0 < X_1 < 1/2) = \int_0^{1/2} f_1(x_1) dx_1 = \int_0^{1/2} 2(1-x_1) dx_1 = \frac{3}{4}.$$

The definitions introduced above are all extended in an obvious way to multi-variables. Thus, for continuous random variables $X_1, X_2, \ldots, X_n$ with the joint pdf $f(x_1, x_2, \ldots, x_n)$, we have the following definitions:

1. The marginal pdfs $f_1(x_1), f_2(x_2) \ldots, f_n(x_n)$ are defined by $(n-1)$-fold integrals

$$f_i(x_i) = \int_{-\infty}^{\infty} \cdots \int_{-\infty}^{\infty} f(x_1, x_2, \ldots, x_n) dx_1 \ldots dx_{i-1} dx_{i+1} \ldots dx_n,$$

$$1 \le i \le n.$$

2. We can also define a marginal pdf of a set of k variables, $k < n$. For example, suppose $n = 4$, i.e., X_1, X_2, X_3, X_4, and consider the subset $<= X_2$ and X_4, i.e., $k = 2$. The marginal pdf of X_2 and X_4 is the joint pdf of the two variables

$$\int_{-\infty}^{\infty} \int_{-\infty}^{\infty} f(x_1, x_2, x_3, x_4) dx_1 dx_3.$$

This is extended in an obvious way to any subset of n random variables.

3. Provided $f_i(x_i) > 0$, the joint conditional pdf of $X_1, \ldots, X_{i-1}, X_{i+1}, \ldots X_n$ given $X_i = x_i$ is

$$f(x_1, \ldots x_{i-1}, x_{i+1}, \ldots x_n | x_i) = \frac{f(x_1, \ldots, x_n)}{f_i(x_i)}.$$

4. As above, more generally, the joint conditional pdf of $n - k$ of the variables for given values of the remaining k variables is defined as the joint pdf of the n variables divided by the marginal pdf of the group of k variables provided it is positive.

5. Provided $f_i(x_i) > 0$, the conditional expectation of $u(X_1, \ldots, X_{i-1}, X_{i+1}, \ldots X_n)$ given $X_i = x_i$ is defined by

$$E[u(X_1, \ldots, X_{i-1}, X_{i+1}, \ldots X_n) | x_i] = \int_{-\infty}^{\infty} \cdots \int_{-\infty}^{\infty} u(x_1, \ldots, x_{i-1}, x_{i+1}, \ldots, x_n)$$
$$f(x_1, \ldots, x_n | x_i) dx_1, \ldots dx_{i-1} dx_{i+1} \ldots dx_n.$$

Corresponding definitions for discrete random variables $X_1, X_2, \ldots, X_n$ with the joint pdf $f(x_1, x_2, \ldots, x_n)$ hold, now using sums instead of integrals.

Exercises

1. Consider the joint pdf

$$f(x_1, x_2) = \begin{cases} \frac{1}{4} x_1 (1 + 3x_2^2), & 0 < x_1 < 2, \quad 0 < x_2 < 1 \\ 0, & \text{elsewhere} \end{cases}$$

2.6 Conditional Probability and Marginal and Conditional Distributions

Show that $\int f(x_1, x_2) dx_1 dx_2 = 1$. Find $P[(X_1, X_2) \in A]$ where $A = \{f(x_1, x_2) | 0 < x_1 < 1, \frac{1}{4} < x_2 < \frac{1}{2}\}$. Determine $f_1(x_1)$, $f_2(x_2)$, $f(x_1|x_2)$, and $P(1/4 < X_1 < 1/2 | X_2 = 1/3)$.

2. The random variables X_1 and X_2 have the joint pdf

$$f(x_1, x_2) = \begin{cases} x_1 + x_2, & 0 < x_1 < 1, \ 0 < x_2 < 1 \\ 0, & \text{elsewhere} \end{cases}$$

Find the conditional mean and variance of X_2 given $X_1 = x_1$, $0 < x_1 < 1$.

3. Suppose the conditional pdf of X_1 given $X_2 = x_2$ is

$$f(x_1|x_2) = \begin{cases} c_1 \frac{x_1}{x_2^2}, & 0 < x_1 < x_2, \ 0 < x_2 < 1 \\ 0, & \text{elsewhere} \end{cases}$$

and the marginal pdf of X_2 is

$$f_2(x_2) = \begin{cases} c_2 x_2^4, & 0 < x_2 < 1, \\ 0, & \text{elsewhere} \end{cases}$$

Find (i) the constants c_1 and c_2; (ii) the joint pdf of X_1 and X_2; (iii) $P(1/4 < X_1 < 1/2 | X_2 = 5/8)$; and (iv) $P(1/4 < X_1 < 1/2)$.

4. Suppose that the joint pdf of X_1 and X_2 is

$$f(x_1, x_2) = \begin{cases} c x_1^2 x_2, & x_1^2 \leq x_2 < 1 \\ 0, & \text{elsewhere} \end{cases}$$

Determine the value of the constant c and then $P(X_1 \geq X_2)$. Evaluate $f_1(x_1)$ and $f_2(x_2)$. (Hint: sketch the region where $f(x_1, x_2) \geq 0$.)

5. Let $\Psi(t_1, t_2) \equiv \ln M(t_1, t_2)$, where $M(t_1, t_2)$ is the moment generating function of X and Y. Show that

$$\frac{\partial \Psi(0,0)}{\partial t_k}, \quad \frac{\partial^2 \Psi(0,0)}{\partial t_k^2} \quad (k = 1, 2), \quad \frac{\partial^2 \Psi(0,0)}{\partial t_1 \partial t_2},$$

yields the means, the variances, and the covariance of the two random variables.

6. Given the joint pdf of X_1 and X_2,

$$f(x_1, x_2) = \begin{cases} 21 x_1^2 x_2^3, & 0 < x_1 < x_2 < 1, \\ 0 & \text{elsewhere} \end{cases}$$

find the conditional mean and variance of X_1 given $X_2 = x_2$, $0 < x_2 < 1$.

7. Five cards are drawn at random without replacement from a deck of cards. The random variables X_1, X_2, and X_3 denote the number of spades, the

number of hearts, and the number of diamonds that appear among the 5 cards respectively. Determine the joint pdf of X_1, X_2, and X_3. Find the marginal pdfs of X_1, X_2, and X_3. What is the joint conditional pdf of X_2 and X_3 given that $X_1 = 3$?

8. Suppose that the joint pdf of X and Y is given by

$$f(x, y) = \begin{cases} 2, & 0 < x < y, 0 < y < 1, \\ 0, & \text{elsewhere}. \end{cases}$$

Show that the conditional means are $(1 + x)/2$, $0 < x < 1$ and $y/2$, $0 < y < 1$, and the correlation function of X and Y is $\rho = 1/2$. Show also that the variance of the conditional distribution of Y given $X = x$ is $(1 - x)^2/12$, $0 < x < 1$, and that the variance of the conditional distribution of X given $Y = y$ is $y^2/12$, $0 < y < 1$.

9. Let $f(t)$ and $F(t)$ be the pdf and distribution function of the random variable T. The conditional pdf of T given $T > t_0$, t_0 a fixed time, is defined by $f(t|T > t_0) = f(t)/[1 - F(t_0)]$, $t > t_0$, 0 elsewhere. This kind of pdf is used in survival analysis i.e., problems of time until death, given survival until time t_0. Show that $f(t|T > t_0)$ is a pdf. Let $f(t) = e^{-t}$, $0 < t < \infty$, 0 elsewhere, and compute $P(T > 2|T > 1)$.

2.7 Stochastic Independence

Consider two random variables X_1 and X_2 with joint pdf $f(x_1, x_2)$. The joint pdf may be expressed as

$$f(x_1, x_2) = f(x_2|x_1) f_1(x_1).$$

Suppose that $f(x_2|x_1)$ does not depend on x_1. Then the marginal pdf of X_2 (assuming X_2 is continuous) is

$$f_2(x_2) = \int_{-\infty}^{\infty} f(x_2|x_1) f_1(x_1) dx_1$$

$$= f(x_2|x_1) \int_{-\infty}^{\infty} f_1(x_1) dx_1$$

$$= f(x_2|x_1).$$

Hence,

$$f_2(x_2) = f(x_2|x_1) \quad \text{and} \quad f(x_1, x_2) = f_1(x_1) f_2(x_2),$$

2.7 Stochastic Independence

when $f(x_2|x_1)$ is independent of x_1. Thus, if the conditional distribution of X_2 given $X_1 = x_1$ is independent of x_1, then $f(x_1, x_2) = f_1(x_1)f_2(x_2)$.

Definition. Let the random variables X_1 and X_2 have the joint pdf $f(x_1, x_2)$ and marginal pdfs $f_1(x_1)$ and $f_2(x_2)$. Then the random variables X_1 and X_2 are *stochastically independent* if and only if $f(x_1, x_2) = f_1(x_1)f_2(x_2)$. Otherwise, they are stochastically dependent.

Example. Let the joint pdf of the random variables X_1 and X_2 be

$$f(x_1, x_2) = \begin{cases} \frac{1}{2}, & 0 \leq x_1 \leq 2, 0 << x_2 \leq 1, \\ 0, & \text{elsewhere}. \end{cases}$$

The marginal pdfs are

$$f_1(x_1) = \int_0^1 \frac{1}{2} dx_2 = \frac{1}{2}, \qquad 0 \leq x_1 \leq 2$$

$$f_2(x_2) = \int_0^2 \frac{1}{2} dx_1 = 1, \qquad 0 \leq x_2 \leq 1$$

and 0 elsewhere. Thus,

$$f(x_1, x_2) = \frac{1}{2} \times 1 = \frac{1}{2}$$

implies that the random variables X_1 and X_2 are stochastically independent.

Example. Consider the random variables X_1 and X_2 with joint pdf

$$f(x_1, x_2) = \begin{cases} 12 x_1 x_2 (1 - x_2), & 0 < x_1 < 1, 0 < x_2 < 1, \\ 0, & \text{elsewhere}. \end{cases}$$

Since the marginal pdfs are given by

$$f_1(x_1) = \int_0^1 12 x_1 x_2 (1 - x_2) dx_2 = 2x_1;$$

$$f_2(x_2) = \int_0^1 12 x_1 x_2 (1 - x_2) dx_1 = 6x_2(1 - x_2),$$

we have $f(x_1, x_2) = f_1(x_1) f_2(x_2)$ and thus that X_1 and X_2 are stochastically independent.

Some useful theorems follow. These (i) provide a means of determining whether random variables are stochastically independent without computing the marginal pdfs; and (ii) show that the product property of independence carries over to probabilities, expectations, and moment generating functions.

Theorem. *Let the random variables X_1 and X_2 have joint pdf $f(x_1, x_2)$. Then X_1 and X_2 are stochastically independent if and only if*

$$f(x_1, x_2) \equiv g(x_1)h(x_2),$$

where $g(x_1) > 0$, $\forall x_1 \in \mathcal{A}_1$, 0 elsewhere, and $h(x_2) > 0$, $\forall x_2 \in \mathcal{A}_2$, 0 elsewhere.

Proof. If X_1 and X_2 are stochastically independent, then $f(x_1, x_2) = f_1(x_1)f_2(x_2)$, where $f_1(x_1)$ and $f_2(x_2)$ are marginal pdfs. Thus the condition $f(x_1, x_2) = g(x_1)h(x_2)$ holds.

Conversely, if $f(x_1, x_2) = g(x_1)h(x_2)$ holds, then

$$f_1(x_1) = \int_{-\infty}^{\infty} g(x_1)h(x_2)dx_2 = g(x_1)\int_{-\infty}^{\infty} h(x_2)dx_2 = c_1 g(x_1),$$

and

$$f_2(x_2) = \int_{-\infty}^{\infty} g(x_1)h(x_2)dx_1 = h(x_2)\int_{-\infty}^{\infty} g(x_1)dx_1 = c_2 h(x_2).$$

Here c_1 and c_2 are constants. However,

$$\int_{-\infty}^{\infty}\int_{-\infty}^{\infty} f(x_1, x_2)dx_1 dx_2 = 1 = \int_{-\infty}^{\infty} g(x_1)dx_1 \int_{-\infty}^{\infty} h(x_2)dx_2 = c_1 c_2.$$

Hence

$$f(x_1, x_2) \equiv g(x_1)h(x_2) = f_1(x_1)f_2(x_2),$$

and X_1 and X_2 are stochastically independent. The related proof for discrete random variables is similar.

Theorem. *If X_1 and X_2 are stochastically independent random variables with marginal pdfs $f_1(x_1)$ and $f_2(x_2)$, then*

$$P(a < X_1 < b, c < X_2 < d) = P(a < X_1 < b)P(c < X_2 < d),$$

for all constants a, b, c, d satisfying $a < b$ and $c < d$.

Proof.

$$P(a < X_1 < b, c < X_2 < d) = \int_a^b \int_c^d f_1(x_1)f_2(x_2)dx_1 dx_2$$

$$= \int_a^b f_1(x_1)dx_1 \int_c^d f_2(x_2)dx_2$$

$$= P(a < X_1 < b)P(c < X_2 < d).$$

2.7 Stochastic Independence

Theorem. *Suppose X_1 and X_2 are stochastically independent random variables with marginal pdfs $f_1(x_1)$ and $f_2(x_2)$, and $u(X_1)$ and $v(X_2)$ are functions of X_1 and X_2 respectively. Then, the expectation*

$$E[u(X_1)v(X_2)] = E[u(X_1)]E[v(X_2)].$$

Proof. This follows immediately from the definition of stochastic independence.

$$\begin{aligned} E[u(X_1)v(X_2)] &= \int_{-\infty}^{\infty}\int_{-\infty}^{\infty} u(x_1)v(x_2)f_1(x_1)f_2(x_2)dx_1 dx_2 \\ &= \int_{-\infty}^{\infty} u(x_1)f_1(x_1)dx_1 \int_{-\infty}^{\infty} v(x_2)f_2(x_2)dx_2 \\ &= E[u(X_1)]E[v(X_2)]. \end{aligned}$$

Example. Suppose X and Y are stochastically independent random variables with means μ_1 and μ_2 and variances σ_1^2 and σ_2^2 respectively. Then, we have the important result that the correlation coefficient of X and Y is zero because the covariance

$$\sigma_{12} = \frac{E[(X-\mu_1)(Y-\mu_2)]}{\sigma_1 \sigma_2} = \frac{E[(X-\mu_1)]E[(Y-\mu_2)]}{\sigma_1 \sigma_2} = 0.$$

Theorem. *Suppose X_1 and X_2 are stochastically independent random variables with joint pdf $f(x_1, x_2)$ and marginal pdfs $f_1(x_1)$ and $f_2(x_2)$. If the moment generating function $M(t_1, t_2)$ of the distribution exists, then X_1 and X_2 are stochastically independent if and only if $M(t_1, t_2) = M(t_1, 0)M(0, t_2)$.*

Proof. If X_1 and X_2 are stochastically independent, then

$$\begin{aligned} M(t_1, t_2) &= E(e^{t_1 X_1 + t_2 X_2}) \\ &= E(e^{t_1 X_1} e^{t_2 X_2}) \\ &= E(e^{t_1 X_1})E(e^{t_2 X_2}) = M(t_1, 0)M(0, t_2). \end{aligned}$$

Hence, stochastic independence of X_1 and X_2 implies that the moment generating function factors into a product of the two marginal moment generating functions of the marginal distributions.

Suppose now instead that $M(t_1, t_2) = M(t_1, 0)M(0, t_2)$. Since

$$M(t_1, 0) = \int_{-\infty}^{\infty} e^{t_1 x_1} f_1(x_1) dx_1 \quad \text{and} \quad M(0, t_2) = \int_{-\infty}^{\infty} e^{t_2 x_2} f_2(x_2) dx_2,$$

we have

$$M(t_1,0)M(0,t_2) = \int_{-\infty}^{\infty} e^{t_1 x_1} f_1(x_1) dx_1 \int_{-\infty}^{\infty} e^{t_2 x_2} f_2(x_2) dx_2$$

$$= \int_{-\infty}^{\infty} \int_{-\infty}^{\infty} e^{t_1 x_1 + t_2 x_2} f_1(x_1) f_2(x_2) dx_1 dx_2$$

$$= M(t_1, t_2).$$

But

$$M(t_1, t_2) = \int_{-\infty}^{\infty} \int_{-\infty}^{\infty} e^{t_1 x_1 + t_2 x_2} f(x_1, x_2) dx_1 dx_2,$$

which implies that

$$f(x_1, x_2) = f_1(x_1) f_2(x_2),$$

and hence that X_1 and X_2 are stochastically independent random variables. The case for discrete random variables is similar.

The n random variables $X_1, X_2, \ldots, X_n$, with joint pdf $f(x_1, x_2, \ldots, x_n)$ and marginal pdfs $f_1(x_1), f_2(x_2), \ldots, f_n(x_n)$, are *mutually stochastically independent* if and only if $f(x_1, x_2, \ldots, x_n) = f_1(x_1) f_2(x_2) \ldots f_n(x_n)$. The theorems above can be suitably generalized.

Exercises

1. Let the joint pdf of X_1 and X_2 be

$$f(x_1, x_2) = \begin{cases} x_1 + x_2, & 0 < x_1 < 1, 0 < x_2 < 1, \\ 0, & \text{elsewhere}. \end{cases}$$

Show that the random variables X_1 and X_2 are stochastically dependent.

2. Show that the random variables X and Y with joint pdf

$$f(x,y) = \begin{cases} 2e^{-x-y}, & 0 < x < y, 0 < y < \infty, \\ 0, & \text{elsewhere}. \end{cases}$$

are stochastically dependent.

3. Consider the joint pdf of random variables X and Y,

$$f(x,y) = \begin{cases} \frac{x(1+3y^2)}{4}, & 0 < x < 2, 0 < y < 1, \\ 0, & \text{elsewhere}. \end{cases}$$

Are the random variables X and Y stochastically independent? Compute $f(x|y)$ and hence $P(1/4 < X < 1/2 | Y = 3)$.

4. The random variables X and Y have joint pdf

$$f(x, y) = \begin{cases} 4x(1-y), & 0 < x < 1, 0 < y < 1, \\ 0, & \text{elsewhere.} \end{cases}$$

Find $P(0 < X < 1/3, 0 < Y < 1/3)$.

5. Let X_1, X_2, and X_3 be three stochastically independent random variables, each with pdf

$$f(x) = \begin{cases} e^{-x}, & x > 0, \\ 0, & \text{elsewhere} \end{cases}.$$

Find $P(X_1 < 2, 1 < X_2 < 3, X_3 > 2)$.

6. Show that the random variables X and Y with joint pdf

$$f(x, y) = \begin{cases} e^{-x-y}, & 0 < x < \infty, 0 < y < \infty, \\ 0, & \text{elsewhere} \end{cases},$$

are stochastically independent, and that

$$E(e^{t(X+Y)}) = (1-t)^{-2}, \quad t < 1.$$

2.8 Particular Distributions

We consider three of the most important probability distribution functions, the *binomial distribution*, the *Poisson distribution*, and the *normal* or *Gaussian distribution*, and the latter's connection to the *Maxwell-Boltzmann* distribution.

2.8.1 The Binomial Distribution

The binomial theorem is expressed as

$$(a+b)^n = \sum_{x=0}^{n} \binom{n}{x} b^x a^{n-x},$$

for $n > 0$ an integer. Recall, $\binom{n}{x} = \frac{n!}{x!(n-x)!}$. By analogy, introduce the function

$$f(x) = \binom{n}{x} p^x (1-p)^{n-x}, \qquad x = 0, 1, 2, 3, \ldots, n$$
$$= 0 \quad \text{elsewhere}$$

for $n > 0$ an integer and $0 < p < 1$. Clearly, $f(x) \geq 0$ and

$$\sum_x f(x) = \sum_{x=0}^{n} \binom{n}{x} p^x (1-p)^{n-x} = [(1-p) + p]^n = 1.$$

Hence, the function $f(x)$ is the pdf of a discrete random variable X. A random variable with this pdf has a *binomial distribution*, and $f(x)$ is a *binomial pdf*. n and p are the parameters of the binomial distribution, often denoted by $B(n, p)$. For example, $B(4, 1/3)$ has the binomial pdf

$$f(x) = \binom{4}{x} \left(\frac{1}{3}\right)^x \left(\frac{2}{3}\right)^{4-x}, \qquad x = 0, 1, 2, 3, 4$$
$$= 0 \quad \text{elsewhere}$$

The binomial distribution is a very useful model for any experiment or system that admits an outcome drawn from two possibilities only, such as heads or tails in a coin toss, life or death, red or green, etc. If the experiment is repeated n independent times or a system produces n independent outcomes and the probability of "success" is p on each occasion, then the probability for "failure" is $1 - p$. Define the random variable X_i, $i = 1, 2 \ldots, n$ to be 0 if the outcome of the ith performance is a failure and 1 if the outcome is a success. Thus $P(X_i = 0) = 1 - p$ and $P(X_i = 1) = p$, $i = 1, 2 \ldots, n$. The random variables X_i are mutually stochastically independent since the experiment is repeated n independent times. $Y = X_1 + X_2 + \cdots + X_n$ is the number of successes through the n repetitions of the experiment. Let y be an element of the set $\{y : y = 0, 1, 2, \ldots, n\}$. Then $Y = y$ if and only if y of the random variables X_i have the value 1 and $n - y$ have value 0. The number of combinations of y 1's that can be assigned to the X_i is just $\binom{n}{y}$. The probability for each of these possible combinations is simply $p^y (1-p)^{n-y}$ because the X_i are mutually stochastically independent. The $P(Y = y)$ is the sum of the $\binom{n}{y}$ mutually exclusive events, i.e.,

$$\binom{n}{y} p^y (1-p)^{n-y}, \qquad y = 0, 1, 2, 3, \ldots, n$$

and 0 elsewhere. This is the pdf of a binomial distribution.

2.8 Particular Distributions

The moment generating function can be evaluated from

$$M(t) = \sum_x e^{tx} f(x) = \sum_{x=0}^n e^{tx} \binom{n}{x} p^x (1-p)^{n-x}$$

$$= \sum_{x=0}^n \binom{n}{x} (pe^t)^x (1-p)^{n-x}$$

$$= \left[(1-p) + pe^t\right]^n, \qquad \forall t \in \Re.$$

The mean and variance are therefore

$$M'(t) = n\left[(1-p) + pe^t\right]^{n-1} pe^t \implies \mu = M'(0) = np,$$

and

$$M''(t) = n\left[(1-p) + pe^t\right]^{n-1} pe^t + n(n-1)\left[(1-p) + pe^t\right]^{n-2} (pe^t)^2$$
$$\implies \sigma^2 = M''(0) - \mu^2 = np + n(n-1)p^2 - n^2 p^2 = np(1-p).$$

Example. The binomial distribution with pdf

$$f(x) = \binom{4}{x} \left(\frac{1}{2}\right)^x \left(\frac{1}{2}\right)^{4-x}, \qquad x = 0, 1, 2, 3, 4$$

$$= 0 \quad \text{elsewhere}$$

and random variable X has moment generating function

$$M(t) = \left(\frac{1}{2} + \frac{1}{2}e^t\right)^4,$$

and mean $\mu = np = 2$ and variance $\sigma^2 = np(1-p) = 1$. We can compute, for example,

$$P(0 \le X \le 1) = \sum_{x=0}^1 f(x) = \frac{1}{16} + \frac{4}{16} = \frac{5}{16},$$

and

$$P(X = 3) = f(3) = \frac{4!}{3!1!} \left(\frac{1}{2}\right)^3 \left(\frac{1}{2}\right)^1 = \frac{1}{4}.$$

Exercises

1. If the moment generating function of a random variable X is

$$\left(\frac{1}{3} + \frac{2}{3}e^t\right)^5$$

find $P(X = 2 \text{ or } 3)$.

2. The moment generating function of a random variable X is

$$\left(\frac{2}{3} + \frac{1}{3}e^t\right)^9.$$

Show that

$$P(\mu - 2\sigma < X < \mu + 2\sigma) = \sum_{x=1}^{5} \binom{9}{x} \left(\frac{1}{3}\right)^x \left(\frac{2}{3}\right)^{9-x}.$$

3. The probability that a patient recovers from heart surgery is 0.4. If 15 people have had heart surgery, what is the probability that (i) at least 10 survive, (ii) from 3 to 8 survive, (iii) exactly 5 survive? Using Chebyshev's inequality, find and interpret the interval $\mu \pm 2\sigma$.

4. If the random variable X has a binomial distribution with parameters n and p, show that

$$E\left(\frac{X}{n}\right) = p \quad \text{and} \quad E\left[\left(\frac{X}{n} - p\right)^2\right] = \frac{p(1-p)}{n}.$$

2.8.2 The Poisson Distribution

For all values of p, the series

$$1 + p + \frac{p^2}{2!} + \frac{p^3}{3!} + \cdots = \sum_{x=0}^{\infty} \frac{p^x}{x!}, \quad x = 0, 1, 2, \ldots$$

converges to e^p. This motivates the introduction of the function $f(x)$

$$f(x) = \frac{p^x e^{-p}}{x!}, \quad x = 0, 1, 2, \ldots$$
$$= 0, \quad \text{elsewhere} \quad \forall p > 0.$$

2.8 Particular Distributions

Since $p > 0$, $f(x) \geq 0$ and

$$\sum_x f(x) = \sum_{x=0}^{\infty} \frac{p^x e^{-p}}{x!} = e^{-p} \sum_{x=0}^{\infty} \frac{p^x}{x!} = e^{-p} e^p = 1.$$

Hence, $f(x)$ is a pdf of a discrete random variable. A random variable X that has the pdf $f(x)$ is a *Poisson distribution* and $f(x)$ is a *Poisson pdf*.

Examples of Poisson distributions include the random variable X that denotes the number of alpha particles emitted by a radioactive substance that enter some region in a prescribed time interval, or the number of defects in a manufactured article. Even the number of automobile accidents during some unit time is often assumed to be random variable with a Poisson distribution. A process that leads to a Poisson distribution is called a *Poisson process*. The assumptions that underly a Poisson process are essentially that the probability of a change during a sufficiently short interval is independent of changes in other non-overlapping intervals, and is approximately proportional to the length of the interval, and the probability of more than one change during a short interval is essentially zero. One can formalize these assumptions and derive a simple ordinary differential equation that shows that the probability of changes X in an interval of some length has a Poisson distribution.

The moment generating function of a Poisson distribution is

$$M(t) = \sum_x e^{tx} f(x) = \sum_{x=0}^{\infty} e^{tx} \frac{p^x e^{-p}}{x!}$$

$$= e^{-p} \sum_{x=0}^{\infty} \frac{(pe^t)^x}{x!}$$

$$= e^{-p} e^{pe^t} = e^{p(e^t - 1)}, \quad \forall t \in \Re$$

The mean and variance are found to be

$$M'(t) = \exp\left[p\left(e^t - 1\right)\right] pe^t \implies \mu = M'(0) = p$$

$$M''(t) = \exp\left[p\left(e^t - 1\right)\right] pe^t + \exp\left[p\left(e^t - 1\right)\right] \left(pe^t\right)^2$$

$$\implies \sigma^2 = M''(0) - \mu^2 = p + p^2 - p^2 = p,$$

i.e., a Poisson distribution has $\mu = \sigma^2 = p > 0$. This allows us to express the Poisson pdf as

$$f(x) = \frac{\mu^x e^{-\mu}}{x!}, \quad x = 0, 1, 2, \ldots$$

$$= 0 \quad \text{elsewhere} \quad \forall \mu > 0.$$

Example. Consider a random variable X with a Poisson distribution with $\mu = 3$ and $\sigma^2 = 3$, i.e.,

$$f(x) = \frac{3^x e^{-3}}{x!}, \quad x = 0, 1, 2, \ldots$$
$$= 0 \quad \text{elsewhere.}$$

For example,

$$P(1 \leq X) = 1 - P(X = 0) = 1 - f(0) = 1 - e^{-3} = 0.95.$$

Example. If the moment generating function of a random variable X is

$$M(t) = \exp\left[2\left(e^t - 1\right)\right],$$

then X has a Poisson distribution with $\mu = 2$. For example,

$$P(X = 3) = f(3) = \frac{2^3 e^{-2}}{3!} = \frac{4}{3} e^{-2} = 0.18.$$

Example. The average number of radioactive particles passing through a counter during 1 millisecond in an experiment is 4. What is the probability that six particles enter the counter in a given millisecond?

We may assume a Poisson distribution with $x = 6$ and $\mu = 4$, so that

$$f(6) = \frac{4^6 e^{-4}}{6!} = 0.104.$$

Exercises

1. If the random variable X has a Poisson distribution such that $P(X = 1) = P(X = 2)$, find $P(X = 4)$.
2. Given that $M(t) = \exp[4(e^t - 1)]$ is the moment generating function of a random variable X, show that $P(\mu - 2\sigma < X < \mu + 2\sigma) = 0.931$.
3. Suppose that during a given rush hour Wednesday, the number of accidents on a certain stretch of highway has a Poisson distribution with mean 0.7. What is the probability that there will be at least three accidents on that stretch of highway at rush hour on Wednesday?
4. Compute the measures of skewness and kurtosis of the Poisson distribution with mean μ.
5. Suppose the random variables X and Y have the joint pdf

$$f(x, y) = \frac{e^{-2}}{x!(y-x)!}, \quad y = 0, 1, 2, \ldots; x = 0, 1, \ldots, y,$$
$$= 0, \quad \text{elsewhere.}$$

(i) Find the moment generating function $M(t_1, t_2)$ of the joint pdf. (ii) Compute the means, variances, and correlation coefficient of X and Y. (iii) Determine the conditional mean $E(X|y)$.

2.8.3 The Normal or Gaussian Distribution

Known by both terms, depending on the context (mathematical statistics, plasma physics, statistical physics), this is the most familiar and important of the many distribution functions that exist. We can evaluate the integral

$$I = \int_{-\infty}^{\infty} \exp\left(-x^2/2\right) dx,$$

by noting that $I > 0$ and that I^2 may be written as

$$I^2 = \int_{-\infty}^{\infty} \int_{-\infty}^{\infty} \exp\left(-\frac{x^2 + y^2}{2}\right) dx dy.$$

Introducing polar coordinates $x = r \cos \theta$ and $y = r \sin \theta$ yields

$$I^2 = \int_0^{2\pi} \int_0^{\infty} e^{-r^2/2} r \, dr \, d\theta = \int_0^{2\pi} d\theta = 2\pi,$$

which shows that $I = \sqrt{2\pi}$ and so

$$\frac{1}{\sqrt{2\pi}} \int_{-\infty}^{\infty} \exp\left(-x^2/2\right) dx = 1.$$

If we replace x by

$$\frac{x-a}{b}, \quad b > 0,$$

we have

$$\frac{1}{b\sqrt{2\pi}} \int_{-\infty}^{\infty} \exp\left(-\frac{(x-a)^2}{2b^2}\right) dx = 1.$$

Consequently, since $b > 0$, the function

$$f(x) = \frac{1}{b\sqrt{2\pi}} \exp\left(-\frac{(x-a)^2}{2b^2}\right), \quad -\infty < x < \infty,$$

is the pdf of a continuous random variable, and the random variable has a *normal* or *Gaussian* distribution and $f(x)$ is a normal pdf.

The moment generating function for a normal distribution is

$$M(t) = \frac{1}{b\sqrt{2\pi}} \int_{-\infty}^{\infty} e^{tx} \exp\left(-\frac{(x-a)^2}{2b^2}\right) dx$$

$$= \frac{1}{b\sqrt{2\pi}} \int_{-\infty}^{\infty} \exp\left(-\frac{-2b^2tx + x^2 - 2ax + a^2}{2b^2}\right) dx$$

$$= \exp\left(-\frac{a^2 - (a+b^2t)^2}{2b^2}\right) \frac{1}{b\sqrt{2\pi}} \int_{-\infty}^{\infty} \exp\left(-\frac{(x-a-b^2t)^2}{2b^2}\right) dx$$

$$= \exp\left(at + \frac{b^2t^2}{2}\right),$$

after completing the square and since the last integrand is a normal distribution with $a + b^2 t$. The mean and variance can then be computed as

$$M'(t) = M(t)(a + b^2t) \Longrightarrow \mu = M'(0) = a,$$

and

$$M''(t) = M(t)b^2 + M(t)(a + b^2t)^2,$$
$$\Longrightarrow \sigma^2 = M''(0) - \mu^2 = b^2 + a^2 - a^2 = b^2.$$

The normal or Gaussian pdf can therefore be written as

$$f(x) = \frac{1}{\sigma\sqrt{2\pi}} \exp\left(-\frac{(x-\mu)^2}{2\sigma^2}\right), \quad -\infty < x < \infty,$$

and the moment generating function as

$$M(t) = \exp\left(\mu t + \frac{\sigma^2 t^2}{2}\right).$$

Example. If X has the moment generating function

$$M(t) = \exp\left(3t + 16t^2\right),$$

then X is normally distributed with mean $\mu = 3$ and variance $\sigma^2 = 32$.

The normal distribution is often expressed simply as $n(\mu, \sigma^2)$, thus, for example, $n(0, 1)$ implies the pdf of X has mean 0 and variance 1 and is given by

2.8 Particular Distributions

$$f(x) = \frac{1}{\sqrt{2\pi}} e^{-x^2/2}, \quad -\infty < x < \infty, \quad M(t) = e^{t^2/2}.$$

The graph of the normal distribution is the familiar "bell shape," symmetric about $x = \mu$ with a maximum there of $1/\sigma\sqrt{2\pi}$. There are points of inflection at $x = \mu \pm \sigma$ (Exercise: Check!).

A useful "renormalization" of the Gaussian distribution is the following

Theorem. *If the random variable X is $n(\mu, \sigma^2)$, $\sigma^2 > 0$, then the random variable $W = (X - \mu)/\sigma$ is $n(0, 1)$.*

Proof. Since $\sigma^2 > 0$, the distribution function $G(w)$ of W is

$$G(w) = P\left(\frac{X - \mu}{\sigma} \le w\right) = P(X \le w\sigma + \mu),$$

which corresponds to

$$G(w) = \frac{1}{\sigma\sqrt{2\pi}} \int_{-\infty}^{w\sigma+\mu} \exp\left(-\frac{(x-\mu)^2}{2\sigma^2}\right) dx.$$

On setting $y = (x - \mu)/\sigma$, we have

$$G(w) = \frac{1}{\sqrt{2\pi}} \int_{-\infty}^{w} e^{-y^2/2} dy,$$

implying that the pdf $g(w) = G'(w)$ of the continuous random variable W is

$$g(w) = \frac{1}{\sqrt{2\pi}} e^{-w^2/2},$$

meaning that W is $n(0, 1)$.

This theorem is very useful for calculating probabilities of normally distributed variables. Suppose X is $n(\mu, \sigma^2)$. Then if $a < b$, we have

$$P(a < X < b) = P(X < b) - P(X < a)$$
$$= P\left(\frac{X - \mu}{\sigma} < \frac{b - \mu}{\sigma}\right) - P\left(\frac{X - \mu}{\sigma} < \frac{a - \mu}{\sigma}\right)$$
$$= \frac{1}{\sqrt{2\pi}} \int_{-\infty}^{(b-\mu)/\sigma} e^{-w^2/2} dw - \frac{1}{\sqrt{2\pi}} \int_{-\infty}^{(a-\mu)/\sigma} e^{-w^2/2} dw$$

because $W = (X - \mu)/\sigma$ is $n(0, 1)$. Integrals of this form cannot be evaluated so tables are used, based on the $n(0, 1)$ distribution i.e., if

$$N(x) \equiv \frac{1}{\sqrt{2\pi}} \int_{-\infty}^{x} e^{-w^2/2} dw,$$

and X is $n(\mu, \sigma^2)$, then

$$P(a < X < b) = P\left(\frac{X-\mu}{\sigma} < \frac{b-\mu}{\sigma}\right) - P\left(\frac{X-\mu}{\sigma} < \frac{a-\mu}{\sigma}\right)$$

$$= N\left(\frac{b-\mu}{\sigma}\right) - N\left(\frac{a-\mu}{\sigma}\right).$$

Note that it can be shown that $N(-x) = 1 - N(x)$.

Example. For X $n(2, 25)$,

$$P(0 < X < 10) = N\left(\frac{10-2}{5}\right) - N\left(\frac{0-2}{5}\right)$$

$$= N(1.6) - N(-0.4) = 0.945 - (1 - 0.655) = 0.600$$

where the last steps involved looking up a table of normal values. In similar fashion,

$$P(\mu - 2\sigma < X < \mu + 2\sigma) = N\left(\frac{\mu + 2\sigma - \mu}{\sigma}\right) - N\left(\frac{\mu - 2\sigma - \mu}{\sigma}\right)$$

$$= N(2) - N(-2) = 0.977 - (1 - 0.977) = 0.954.$$

Exercises

1. If

$$N(x) = \frac{1}{\sqrt{2\pi}} \int_{-\infty}^{x} e^{-y^2/2} dy,$$

show that $N(-x) = 1 - N(x)$.
2. If X is $n(75, 100)$, find $P(X < 60)$ and $P(70 < X < 100)$.
3. If X is $n(\mu, \sigma^2)$, find a so that $P(-a < (X-\mu)/\sigma < a) = 0.90$.
4. If X is $n(\mu, \sigma^2)$, show that $E(|X-\mu|) = \sigma\sqrt{2/\pi}$.
5. Show that the pdf $n(\mu, \sigma^2)$ has points of inflection at $x = \mu \pm \sigma$.
6. Suppose a random variable X has pdf

$$f(x) = \frac{2}{\sqrt{2\pi}} e^{-x^2/2}, \qquad 0 < x < \infty$$

$$= 0, \qquad \text{elsewhere.}$$

Find the mean and variance of X.
7. Let X_1 and X_2 be two stochastically independent normally distributed random variables with means μ_1 and μ_2 and variances σ_1^2 and σ_2^2. Show that $X_1 + X_2$ is

normally distributed with mean $(\mu_1 + \mu_2)$ and variance $(\sigma_1^2 + \sigma_2^2)$. (Hint: use the uniqueness of the moment generating function.)
8. Compute $P(1 < X^2 < 9)$ if X is $n(1, 4)$.
9. Suppose the random variable X is normally distributed with $n(\mu, \sigma^2)$. What will the distribution be if $\sigma^2 = 0$?

2.9 The Central Limit Theorem

The central limit theorem shows that under certain conditions, probability distributions will converge to the normal or Gaussian distribution. We will show briefly the relationship to the Maxwell-Boltzmann distribution. Many different versions exist with different conditions and convergence properties. We will describe the simplest (and most restrictive) version.

Before establishing the central limit theorem, we will need the following useful result. Consider a limit of the form

$$\lim_{n \to \infty} \left[1 + \frac{a}{n} + \frac{\phi(n)}{n} \right]^{bn},$$

for a and b independent of n and where $\lim_{n \to \infty} \phi(n) = 0$. Then

$$\lim_{n \to \infty} \left[1 + \frac{a}{n} + \frac{\phi(n)}{n} \right]^{bn} = \lim_{n \to \infty} \left[1 + \frac{a}{n} \right]^{bn} = e^{ab}.$$

For example,

$$\lim_{n \to \infty} \left[1 - \frac{w^2}{n} + \frac{w}{n^4} \right]^{2n} = \lim_{n \to \infty} \left[1 + \frac{-w^2}{n} + \frac{w/n^3}{n} \right]^{2n} = e^{-2w^2},$$

for all fixed values of w.

Theorem. *Suppose that X_i, $i = 1, 2, \ldots, n$ is a random sample from a distribution that has mean μ and variance σ^2. Then the random variable $Y_n = \left(\sum_{i=1}^{n} X_i - n\mu \right) / \sqrt{n} \sigma = \sqrt{n} \left(\bar{X}_n - \mu \right) / \sigma$ has a limiting distribution that is normal with mean 0 and variance 1.*

Comment By establishing this theorem, the central limit theorem, it implies that whenever the conditions of the theorem are satisfied, the random variable $\sqrt{n} \left(\bar{X}_n - \mu \right) / \sigma$ has, for a fixed n, an approximate normal distribution with $\mu = 0$ and $\sigma^2 = 1$.

Proof. Let us assume the existence of a moment generating function $M(t) = E(e^{tX})$ for finite values of t for the distribution. (An alternative more general proof would be based on the characteristic function $\phi(t) = E(e^{itX})$ instead.) Introduce the moment generating function for $X - \mu$,

$$m(t) \equiv E\left(e^{t(X-\mu)}\right) = e^{-\mu t} M(t).$$

Hence, $m(0) = 1$, $m'(0) = E(X - \mu) = 0$, $m''(0) = E[(X - \mu)^2] = \sigma^2$. The function $m(t)$ can be expanded using Taylor's formula, for $0 < \xi < t$, such that

$$m(t) = m(0) + m'(0)t + \frac{1}{2}m''(\xi)t^2$$

$$= 1 + \frac{1}{2}m''(\xi)t^2$$

$$= 1 + \frac{1}{2}\sigma^2 t^2 + \frac{1}{2}\left(m''(\xi) - \sigma^2\right)t^2.$$

Now consider $M(t; n)$, where

$$M(t; n) = E\left[\exp\left(t\frac{\sum X_i - n\mu}{\sigma\sqrt{n}}\right)\right]$$

$$= E\left[\exp\left(t\frac{X_1 - \mu}{\sigma\sqrt{n}}\right)\exp\left(t\frac{X_2 - \mu}{\sigma\sqrt{n}}\right)\cdots\exp\left(t\frac{X_n - \mu}{\sigma\sqrt{n}}\right)\right]$$

$$= E\left[\exp\left(t\frac{X_1 - \mu}{\sigma\sqrt{n}}\right)\right]E\left[\exp\left(t\frac{X_2 - \mu}{\sigma\sqrt{n}}\right)\right]\cdots E\left[\exp\left(t\frac{X_n - \mu}{\sigma\sqrt{n}}\right)\right]$$

$$= \left\{E\left[\exp\left(t\frac{X - \mu}{\sigma\sqrt{n}}\right)\right]\right\}^n$$

$$= \left[m\left(\frac{t}{\sigma\sqrt{n}}\right)\right]^n.$$

Hence,

$$m\left(\frac{t}{\sigma\sqrt{n}}\right) = 1 + \frac{t^2}{2n} + \frac{\left(m''(\xi) - \sigma^2\right)t^2}{2n\sigma^2}, \quad 0 < \xi < \frac{t}{\sigma\sqrt{n}}.$$

Thus,

$$M(t; n) = \left[1 + \frac{t^2}{2n} + \frac{\left(m''(\xi) - \sigma^2\right)t^2}{2n\sigma^2}\right]^n.$$

Since $m''(t)$ is continuous at $t = 0$, and since $\xi \to 0$ as $n \to \infty$, we have

2.9 The Central Limit Theorem

$$\lim_{n\to\infty} \left(m''(\xi) - \sigma^2\right) = 0.$$

Hence, using the result above, $a = t^2/2$ and $b = 1$, and so

$$\lim_{n\to\infty} M(t;n) = e^{t^2/2} \quad \forall t \in \Re.$$

It therefore follows that the random variable $Y_n = \sqrt{n}\left(\bar{X}_n - \mu\right)/\sigma$ has a limiting normal distribution with $\mu = 0$ and variance $\sigma^2 = 1$.

Example. Suppose $\bar{X}$ denotes the mean of a random sample of size 75 from a distribution that has the pdf

$$f(x) = 1, \quad 0 < x < 1$$
$$= 0, \quad \text{elsewhere.}$$

Hence, $\mu = 1/2$ and $\sigma^2 = 1/12$. The limiting distribution of $Y_n = \sqrt{n}\left(\bar{X}_n - \mu\right)/\sigma$ is normally distributed, allowing us to compute for example the $P(0.45 < \bar{X} < 0.55)$ by means of

$$P(0.45 < \bar{X} < 0.55) = P\left[\frac{\sqrt{n}(0.45 - \mu)}{\sigma} < \frac{\sqrt{n}(\bar{X}_n - \mu)}{\sigma} < \frac{\sqrt{n}(0.55 - \mu)}{\sigma}\right]$$

$$= P\left[\frac{\sqrt{75}(0.45 - 1/2)}{\sqrt{1/12}} < \frac{\sqrt{75}(\bar{X}_n - 1/2)}{\sqrt{1/12}} < \frac{\sqrt{75}(0.55 - 1/2)}{\sqrt{1/12}}\right]$$

$$= P[-1.5 < 30(\bar{X} - 1/2) < 1.5] = 0.866.$$

Example. Suppose X_i, $i = 1, 2, \ldots, n$ is a random sample from a binomial distribution $B(n, p) = B(1, p)$ i.e., $\mu = np = p$ and $\sigma^2 = p(1 - p)$, and $M(t)$ exists $\forall t \in \Re$. If $Y_n = X_1 + X_2 + \cdots + X_n$, we know that Y_n is $B(n, p)$. We can use

$$\frac{Y_N - np}{\sqrt{np(1-p)}} = \frac{n(\bar{X}_n - p)}{\sqrt{np(1-p)}} = \frac{\sqrt{n}(\bar{X}_n - \mu)}{\sigma}$$

as a limiting distribution with mean 0 and variance 1. Suppose $n = 100$ and $p = 1/2$, and that we want to compute $P(Y = 48, 49, 50, 51, 52)$. Since Y is a discrete random variable, the events $Y = 48, 49, 50, 51, 52$ and $47.5 < Y < 52.5$ are equivalent (using the convention of taking 0.5 above and below the limiting discrete values). So instead we compute $P(47.5 < Y < 52.5)$. Thus, with $\mu = np = 50$ and $\sigma^2 = np(1 - p) = 25$, we have

$$P(47.5 < Y < 52.5) = P\left(\frac{47.5 - 50}{5} < \frac{Y - 50}{5} < \frac{52.5 - 50}{5}\right)$$

$$= P\left(-0.5 < \frac{Y - 50}{5} < 0.5\right) = 0.382$$

since $(Y - 50)/5$ has an approximately normal distribution.

There are many examples of stochastic variables whose values are determined by independent additive increments. The best known example of such a variable may be the momentum of a molecule in a dilute gas. At a given time, the x momentum mv_x is the vector sum of all momentum increments caused by past collisions with other molecules. If we suppose that the increments in mv_x are independent with zero mean, we may conclude from the central limit theorem that v_x is normally distributed, i.e.,

$$f(v_x) = \frac{1}{\sqrt{2\pi kT}} \exp\left(-\frac{v_x^2}{2kT}\right),$$

where k is Boltzmann's constant and T is the temperature. Similarly, we may argue that the y and z momenta are independent of the x momentum, and that the increments in the three directions are independent. Hence the y and z velocities also have normal or Gaussian distributions. The joint pdf $f(v_x, v_y, v_z)$ of the independent velocities is therefore the product

$$f(v_x, v_y, v_z) = f(v_x)f(v_y)f(v_z) = \left(\frac{1}{2\pi kT}\right)^{3/2} \exp\left(-\frac{v_x^2 + v_y^2 + v_z^2}{2kT}\right).$$

By introducing spherical coordinates in velocity space with $c^2 = v_x^2 + v_y^2 + v_z^2$, we obtain the Maxwell-Boltzmann distribution

$$f(c) = 4\pi \left(\frac{1}{2\pi kT}\right)^{3/2} c^2 \exp\left(-\frac{c^2}{2kT}\right).$$

The Maxwell-Boltzmann distribution for the gas is a consequence of the independence of the successive collisions experienced by a molecule. The three components are identically distributed because there is no preferred direction. This can be different for a magnetized flow in the presence of a large-scale or mean magnetic field. The distributions have a zero mean because the gas is at rest with respect to the chosen coordinate system. A mean flow will introduce an offset in the normal distribution.

Velocities in a turbulent fluid flow do not have a normal distribution because the momentum increments that a fluid parcel experiences at successive times are not necessarily independent. For example, eddies tend to be coherent and interact in sometimes complicated ways with other fluid particles. However, at some scales, the local motions may be nearly independent. Consequently, turbulent velocity fields are not Gaussian, although often not very different from having a Gaussian distribution. This difference is a fundamental property of the dynamics of turbulence. To analyze the dynamics of turbulence, non-Gaussian properties do need to be included typically, but if one is concerned primarily with the effects of turbulence, assuming a Gaussian distribution may be an adequate approximation.

Exercises

1. Compute an approximate probability that the mean of a random sample of size 15 from a distribution having pdf

$$f(x) = 3x^2, \quad 0 < x < 1,$$
$$= 0, \quad \text{elsewhere}$$

is between 3/5 and 4/5.

2. Let Y be $B(72, 1/3)$. Approximate $P(22 \leq Y \leq 28)$.

2.10 Relation Between Microscopic and Macroscopic Descriptions: Particles, the Gibbs Ensemble, and Liouville's Theorem

A gas of particles or plasma of charged particles are both characterized by a very large number of degrees of freedom. One can in principle describe the system in terms the spatial and momentum coordinates of each of the particles in the system. By contrast, a macroscopic description, such as a fluid mechanical model, may have as few as three variables (the density, the velocity, the pressure or the temperature), depending on the closure assumptions imposed. The statistical treatment of the same system may require as many as $6N$ variables, where N is the number of particles in the system. These variables are the $3N$ spatial coordinates $\mathbf{x}_1, \mathbf{x}_2, \cdots \mathbf{x}_N \equiv (\mathbf{x})$ and the $3N$ conjugate momenta $\mathbf{p}_1, \mathbf{p}_2, \cdots \mathbf{p}_N \equiv (\mathbf{p})$ of the constituent particles. This of course neglects effects specific to the particles themselves and treats the particles as point masses. A typical system can be described in terms of a Hamiltonian function $H[(\mathbf{x}), (\mathbf{p})]$, where

$$H[(\mathbf{x}), (\mathbf{p})] = E[(\mathbf{x}), (\mathbf{p})].$$

In the absence of external fields, $E[(\mathbf{x}), (\mathbf{p})]$ denotes the total energy, kinetic energy, and potential energy of the system. The equations of motion for the system are given by Hamilton's equations

$$\frac{d\mathbf{x}_i}{dt} = \dot{\mathbf{x}}_i = \frac{\partial H}{\partial \mathbf{p}_i}$$

$$\frac{d\mathbf{p}_i}{dt} = \dot{\mathbf{p}}_i = -\frac{\partial H}{\partial \mathbf{x}_i}, \qquad i = 1, 2, \cdots, N.$$

The state of the system at any time is given by a representative point in the $6N$ dimensional phase space (also called the Γ space) defined by the mutually orthogonal vectors $\mathbf{x}_1, \mathbf{x}_2, \cdots \mathbf{x}_N, \mathbf{p}_1, \mathbf{p}_2, \cdots \mathbf{p}_N$. Thus, for a given set of initial conditions, the trajectories of a particular system can be computed. Note that the Hamiltonian does not depend on time and so the equations of motion above are invariant under time reversal.

It is evident that a very large number of states of a gas corresponds to a particular macroscopic state of a gas e.g., a gas of a particular density contained in a box of fixed volume can be formed in an infinite number of ways according to the distribution of the particles in space.[1] However, macroscopically we cannot distinguish between one representative point or another i.e., between gases existing in different states. A gas that can be described by certain macroscopic conditions refers therefore to an infinite number of states and not to a single state. Thus, instead of considering a single system, we may consider a collection of systems that are identical in composition and macroscopic conditions but existing in different states. Such a collection of systems is called a Gibbs ensemble, and is the collection of systems that is microscopically equivalent to the system we are considering macroscopically. Each system in the ensemble can be represented by a point in phase space. As the number of systems becomes very large, the representative points become increasingly dense in phase space and we can describe their distribution in phase space by a density function. The density function is a continuous function of $(\mathbf{x})$ and $(\mathbf{p})$, which if normalized can be described by a probability density function $f_N(\mathbf{x}, \mathbf{p}, t)$ i.e., $f_N(\mathbf{x}, \mathbf{p}, t) d^{3N}\mathbf{p} d^{3N}\mathbf{x}$ is the number of representative points that at time t are in the infinitesimal volume $d^{3N}\mathbf{p} d^{3N}\mathbf{x}$ about the point $(\mathbf{x}, \mathbf{p})$ in phase space. Although f_N is a probability distribution function, it evolves in time in a completely deterministic manner, in principle though solving Hamilton's equations.

An ensemble average of the macroscopic property $M(\mathbf{x}, \mathbf{p})$ can be defined by

$$\langle M(t) \rangle = \int M(\mathbf{x}, \mathbf{p}) f_N(\mathbf{x}, \mathbf{p}, t) d^{3N}\mathbf{p} d^{3N}\mathbf{x};$$

i.e., the expectation of the property $M(\mathbf{x}, \mathbf{p})$. Another important and basic postulate of statistical mechanics is the so-called *ergodic statement*, which is that the time average $\bar{M}(\mathbf{x}, \mathbf{p})$

---

[1] Although it appears that this statement is self-evident, it is essentially a postulate. A basic postulate of both equilibrium and non-equilibrium statistical mechanics is that all macroscopic properties of a given system can be described in terms of the microscopic state of that system.

2.10 Particles, the Gibbs Ensemble, and Liouville's Theorem

$$\bar{M}(\mathbf{x}, \mathbf{p}) = \langle M(t) \rangle.$$

The ergodic statement asserts that we can consider ensemble averages rather than time averages as a basis for determining macroscopic properties from the microscopic description. Thus, we need to study the properties and behavior of the probability density function f_N.

The evolution of the pdf f_N is described by Liouville's theorem. The Hamilton equations determine how each ensemble member evolves in phase space. Consider the change df_N in the value of f_N at the point $(\mathbf{x}, \mathbf{p})$ at time t in phase space which results from an arbitrary, infinitesimal change in these variables. This yields

$$df_N = \frac{\partial f_n}{\partial t} dt + \sum_{i=1}^N \frac{\partial f_N}{\partial \mathbf{x}_i} \cdot d\mathbf{x}_i + \sum_{i=1}^N \frac{\partial f_N}{\partial \mathbf{p}_i} \cdot d\mathbf{p}_i$$

$$\Rightarrow \frac{df_N}{dt} = \frac{\partial f_N}{\partial t} + \sum_{i=1}^N \left[\frac{\partial f_N}{\partial \mathbf{x}_i} \cdot \dot{\mathbf{x}}_i + \frac{\partial f_N}{\partial \mathbf{p}_i} \cdot \dot{\mathbf{p}}_i \right]. \tag{2.1}$$

df_N/dt is the total change of f_N along the trajectory in the neighborhood of $(\mathbf{x})$ and $\partial f_N/\partial t$ is the local change in f_N, i.e., at the point $(\mathbf{x})$. Liouville's Theorem is the statement

$$\frac{df_N}{dt} = 0. \tag{2.2}$$

Liouville's theorem states that along the trajectory of any phase point, the probability density in the neighborhood of the point remains constant in time. Since Hamilton's equations have unique solutions, there can be no intersection of trajectories of separate ensemble members in phase space. Thus, an incremental volume about the point $(\mathbf{x})$ in phase space, defined by a specified surface of points in phase space, is also invariant in time, even though it may change its shape (points from inside the volume can never cross the surface since then they would intersect with the points defining the boundary). Since both f_N and the number of points inside the volume $d\mathbf{x}$ remain constant in time, the volume of $d\mathbf{x}$ is unchanged.

Note that since f_N is constant along a trajectory in phase space, so too is any function of f_N. Finally, Liouville's equation is reversible in the sense that the transformation $t \to -t$ leaves the form of the equation unaltered. Hence, if $f_N((\mathbf{x}(t)), (\mathbf{p}(t)), t)$ is a solution to Liouville's equation, then so is $f_N((\mathbf{x}(-t)), (\mathbf{p}(-t)), -t)$.

2.11 The Language of Fluid Turbulence

The mathematical description in the previous sections provides the statistical tools to understand fluid turbulence, for example, but like many areas of physics, one needs to translate the language of mathematics to that of physics. There will be a slight departure from some of the concepts introduced already in that we will use some of the tools of Fourier theory. As we have seen, an important physical concept is the notion of *ensemble average*, as it allows one to form averages for time-dependent processes.

An example of a random function in space and time in fluid dynamics is the velocity field of a turbulent jet or flow. The macroscopic boundary conditions for the flow field may be independent of time, but the velocity at a point varies in an unpredictable manner in time. The local time-average velocity is different in different locations, as are other averages, such as the square of the velocity departures from the mean $(\mathbf{v} - \mathbf{U})^2$ – the variance. For flows with gross boundary conditions that are constant, we can define time averages. For flows where the boundary conditions are temporal, time averages are not useful, and we need to use ensemble averages.

Consider an ensemble of macroscopically identical experiments, each of which produces as output a variable $u(t)$, where $t > 0$ is the time. The output from the jth experiment is the jth realization of $u(t)$, denoted by $^j u(t)$ say. The $^j u(t)$ may look like an oscillation with "noise" superimposed, for example, and each realization may be rather different from the others. The ensemble average of the values of $u(t)$ is defined as the limit

$$\langle u(t) \rangle = \lim_{n \to \infty} \frac{1}{n} \sum_{j=1}^{n} {}^j u(t),$$

or one can define the ensemble average of a function $g[u(t)]$ of $u(t)$ in the same way,

$$\langle g[u(t)] \rangle = \lim_{n \to \infty} \frac{1}{n} \sum_{j=1}^{n} g\left[{}^j u(t)\right].$$

Ensemble averages of powers of u are the moments, so the ensemble average of u^k is the kth moment of u at t, i.e.,

$$\langle u(t)^k \rangle = \lim_{n \to \infty} \frac{1}{n} \sum_{j=1}^{n} \left[{}^j u(t)\right]^k.$$

If we consider two distinct times t and t' and form the ensemble average of the product $u(t)u(t')$ for each realization, we then define the *covariance* R_{uu} by

2.11 The Language of Fluid Turbulence

$$R_{uu}(t,t') = \langle u(t)u(t')\rangle.$$

A random function is *stationary* if all its moments and joint moments are independent of the choice of time origin. For example, a flow becomes turbulent after passing through a grid at a starting time $t = 0$. After some time $T \gg 0$, the flow will have settled down and initial transients will have damped away. Then, for $t \gg T$, the values of velocities and other variables can be expected to be stationary random functions. Instead of using the above definitions, the stationary ensemble average is equivalent to a time average, e.g.,

$$\langle u \rangle = \lim_{T \to \infty} \frac{1}{T} \int_{-T/2}^{T/2} u(t)dt.$$

A stationary random variable is a significant simplification since then averages such as $\langle u(t) \rangle$ are independent if time, as are all expectations and moments of u.

Time covariances of stationary random functions should be independent of the choice of time origin but will depend on the time difference $\tau = t' - t$,

$$R_{uu}(\tau) = \langle u(t)u(t+\tau)\rangle.$$

The double subscript indicates that the covariance is the covariance of u with u, and the argument τ indicates that the velocities are measured at a time interval τ apart. We note too that the velocity is generally three dimensional and so one frequently expresses the tensor **R** as

$$R_{ij} = \langle u_i(t)u_j(t+\tau)\rangle.$$

Some properties of the covariance are easily derived.

1. $R_{uu}(\tau)$ is an even function. This can be seen from

$$R_{uu}(\tau) = \langle u(t)u(t+\tau)\rangle = \langle u(t'-\tau)u(t')\rangle = R_{uu}(-\tau).$$

2. The joint covariance of u and its time derivative $\dot{u}$ is the time derivative of R_{uu}. This follows from

$$R_{u\dot{u}}(\tau) = \langle u(t)\dot{u}(t+\tau)\rangle = \frac{\partial}{\partial \tau}\langle u(t)u(t+\tau)\rangle = \frac{\partial}{\partial \tau} R_{uu}(\tau).$$

Related results are derived easily.

Consider two joint random variables – that is, in each realization there are two results (such as the x and y components of velocity, or the velocity and the density at a point in the flow) – say, $u(t)$ and $v(t)$. The joint covariance is then given by

$$R_{uv}(\tau) = \langle u(t)v(t+\tau)\rangle,$$

assuming the process is stationary (since we express the joint covariance in terms of the relative time delay only). Some elementary properties are given in the Exercises.

The covariance and joint covariance functions are assumed to decay for large values of the lag or delay time τ, so that the functions are square integrable and possess Fourier transforms. The Fourier transform of the time *autocovariance* function $R_{uu}(\tau)$ is called the *power spectral density*, defined by

$$S_{uu}(\omega) = \frac{1}{2\pi} \int_{-\infty}^{\infty} e^{i\omega\tau} R_{uu}(\tau) d\tau.$$

For a function $f(t)$, Fourier's integral theorem yields the expression

$$f(t) = \frac{1}{2\pi} \int_{-\infty}^{\infty} e^{-i\omega t} \int_{-\infty}^{\infty} e^{i\omega t'} f(t') dt' d\omega.$$

Using this result in the power spectral density expression then yields

$$R_{uu}(\tau) = \int_{-\infty}^{\infty} e^{-i\omega\tau} S_{uu} d\omega.$$

The joint or cross-spectral density of the joint pair of random functions u and v is given by

$$S_{uv}(\omega) = \frac{1}{2\pi} \int_{-\infty}^{\infty} e^{i\omega\tau} R_{uv} d\tau = Co_{uv}(\omega) + i Qu_{uv}(\omega),$$

where the real part Co_{uv} is called the co-spectrum and the imaginary part Qu_{uv} is called the quadrature spectrum.

Let $u(\mathbf{x}, t)$ be a random function of position $\mathbf{x}$ and time t. The space-time covariance is expressed as

$$R_{uu}(\mathbf{x}, \mathbf{x}', t, t') = \langle u(\mathbf{x}, t) u(\mathbf{x}', t') \rangle.$$

If u is a stationary random function, then R_{uu} is independent of the choice of time origin. If R_{uu} is independent of the choice of spatial origin, and the same is true for other statistical measures, then $u(\mathbf{x}, t)$ is a *homogeneous* function of $\mathbf{x}$. Hence, for a stationary and homogeneous random function, the space-time covariance is

$$R_{uu}(\xi, \tau) = \langle u(\mathbf{x}, t) u(\mathbf{x} + \xi, t + \tau) \rangle.$$

We can Fourier transform the space-time covariance with respect to space and time. The power spectral density of u is the Fourier transform of the time autocovariance $R_{uu}(0, \tau)$, and is given by

$$S_{uu}(\omega) = \frac{1}{2\pi} \int_{-\infty}^{\infty} e^{i\omega\tau} R_{uu}(0, \tau) d\tau.$$

2.11 The Language of Fluid Turbulence

The *wave number spectrum* $\Phi(k)$ is defined by

$$\Phi(\mathbf{k}) = \left(\frac{1}{2\pi}\right)^3 \int_{-\infty}^{\infty}\int_{-\infty}^{\infty}\int_{-\infty}^{\infty} \exp[-i(\mathbf{k}\cdot\xi)] R_{uu}(\xi, 0) d\xi_1 d\xi_2 d\xi_3.$$

Finally, we may define the combined wave number-frequency spectrum as

$$\Phi(\mathbf{k}, \omega) = \left(\frac{1}{2\pi}\right)^4 \int_{-\infty}^{\infty}\int_{-\infty}^{\infty}\int_{-\infty}^{\infty}\int_{-\infty}^{\infty} \exp[-i(\mathbf{k}\cdot\xi - \omega\tau)] R_{uu}(\xi, \tau) d\xi_1 d\xi_2 d\xi_3 d\tau.$$

Example. Suppose that we have placed two surface gauges in the middle of the Pacific ocean, separated by a small distance ℓ, so that we can measure the cross-spectral density at each. With these gauges, we want to determine the phase velocity of the random surface waves – these are 2D with velocity (u, v) say. Assuming that the power spectrum is the same at both gauges, we have

$$S_{uv}(\omega, \ell) = [|S_{uu}||S_{vv}|]^{1/2}(\omega) Coh_{uv}(\omega) \exp[i\theta_{uv}(\omega)],$$

where the *coherence* $Coh_{uv}(\omega) = [(Co_{uv}^2 + Qu_{uv}^2)/|S_{vv}||S_{uu}|]^{1/2}$. The phase velocity of frequency Fourier component can be determined by a straightforward argument. The cross-spectral density is a complex function, so the phase is given by the argument of S_{uv}, i.e.,

$$\theta_{uv}(\omega) = \arg[S_{uv}(\omega, \ell)],$$

and this will give the phase velocity. The phase gives the time interval δt between arrivals of a wave, which is proportional to the phase difference times the wave period divided by 2π,

$$\delta t = \frac{(2\pi/\omega)\theta_{uv}}{2\pi} = \frac{\theta_{uv}}{\omega}.$$

The phase speed is simply the gauge separation distance divided by the time delay, so the phase velocity as a function of frequency is therefore

$$c(\omega) = \frac{\ell}{\delta t} = \frac{\ell\omega}{\theta_{uv}(\omega)}.$$

This expression also obviously gives the average wave number for the frequency as

$$k = \frac{\omega}{c(\omega)} = \frac{\theta_{uv}(\omega)}{\ell}.$$

For surface waves, it is also possible to derive the phase velocity from $S_{uu}(\omega)$. Since, if $u' = \partial u/\partial x$, then $S_{u'u'}(\omega) = k^2 S_{uu}(\omega)$ (Exercise), the mean square wave number is

$$\langle k^2 \rangle = \frac{S_{u'u'}(\omega)}{S_{uu}(\omega)}.$$

The phase speed is then obtained from

$$c_{slope}^2 = \frac{\omega^2}{\langle k^2 \rangle}.$$

Note that the phase speeds calculated from the two methods might well be different. For example, for a standing wave between the two gauges, the phase speed derived from the cross-spectral density would be zero while the slope method may yield a non-zero phase speed. For waves propagating in a single direction, the two methods should give similar results.

Example. Noise is often defined as having all its Fourier components as stochastic variables with zero mean i.e., $\langle u(\omega) \rangle = 0$. A signal therefore corresponds to a definite additive component. In the absence of periodic components, the correlation function or covariance tends to zero as $t \to \infty$. Suppose that the covariance decays exponentially, so that

$$\langle u(\tau)u(0) \rangle = Ce^{-\gamma|\tau|}.$$

The power spectrum is given by (Exercise)

$$S_{uu}(\omega) = \frac{1}{\pi} \frac{C\gamma}{\omega^2 + \gamma^2},$$

and is called the *Lorentz* distribution. Note that *white noise* corresponds to the limit $\gamma \to \infty$.

Consider now a periodic component to $u(t)$, say $u(t) = v(t) + Ae^{-i\omega_0 t}$, with $\langle v(t) \rangle = 0$ and $\langle v(\tau)v(0) \rangle = Ce^{-\gamma|\tau|}$. Hence,

$$\langle u(\tau)u(0) \rangle = Ce^{-\gamma|\tau|} + A^2 e^{-i\omega_0 \tau},$$

$$S_{uu}(\omega) = \frac{1}{\pi} \frac{\gamma C}{\omega^2 + 16} + A^2 \delta(\omega - \omega_0),$$

indicating that the periodic signal introduces a spike in the power density spectrum at $\omega = \omega_0$.

2.11 The Language of Fluid Turbulence

Exercises

1. Show that the joint covariance is not symmetric in the time lag τ, i.e., that

$$R_{uv}(\tau) = R_{vu}(-\tau).$$

2. Show that the joint covariance for u and a time derivative of v satisfies

$$R_{u\dot{v}} = \frac{\partial}{\partial \tau} R_{vu}(-\tau).$$

3. Show that the co-spectrum and quadrature spectrum may be expressed as integrals

$$Co_{uv}(\omega) = \frac{1}{2\pi} \int_0^\infty [R_{uv}(\tau) + R_{uv}(-\tau)] \cos(\omega\tau) d\tau;$$

$$Qu_{uv}(\omega) = \frac{1}{2\pi} \int_0^\infty [R_{uv}(\tau) - R_{uv}(-\tau)] \sin(\omega\tau) d\tau.$$

4. By introducing the *coherence* $Coh_{uv}(\omega) = [(Co_{uv}^2 + Qu_{uv}^2)/|S_{vv}||S_{uu}|]^{1/2}$ and the phase $\theta_{uv}(\omega) = \arg(S_{uv})$ (the argument of S_{uv}), show that the joint- or cross-spectral density can be expressed in terms its magnitude and argument,

$$S_{uv}(\omega) = [|S_{uu}(\omega)||S_{vv}(\omega)|]^{1/2} Coh_{uv}(\omega) \exp[i\theta_{uv}(\omega)].$$

5. Show that if $u' = \partial u/\partial x$, then $S_{u'u'} = k^2 S_{uu}(\omega)$.
6. Show that an exponentially decaying covariance $\langle u(\tau)u(0) \rangle = Ce^{-\gamma|\tau|}$ yields a Lorentz distribution for the power spectral density,

$$S_{uu}(\omega) = \frac{1}{\pi} \frac{\gamma C}{\omega^2 + \gamma^2}.$$

Sketch the covariance and the power spectral density.

7. Show that the autocovariance in the last example of the chapter is given by

$$\langle u(\tau)u(0) \rangle = Ce^{-\gamma|\tau|} + A^2 e^{-i\omega_0 \tau},$$

and that

$$S_{uu}(\omega) = \frac{1}{\pi} \frac{\gamma C}{\omega^2 + 16} + A^2 \delta(\omega - \omega_0).$$

Sketch the covariance and the power spectral density.

References

W. Feller, *An Introduction to Probability Theory and its Applications*, vol. 1, 3rd edn. (John Wiley, New York, 1968)

R.V. Hogg, A.T. Craig, *Introduction to Mathematical Statistics*, 4th edn. (McMillan, New York, 1978)

I.N. Gibra, *Probability and Statistical Inference for Scientists and Engineers*. (Prentice Hall, Englewood Cliffs, 1973)

Chapter 3
The Boltzmann Transport Equation

3.1 Derivation of the Boltzmann Transport Equation

We are not interested in the motion of each particle in detail but instead in the distribution function $f(\mathbf{x}, \mathbf{p}, t)$, which is defined so that

$$f(\mathbf{x}, \mathbf{p}, t) d^3 x\, d^3 p$$

is the number of particles dN in the phase space volume $(\mathbf{x}+d^3\mathbf{x}, \mathbf{p}+d^3\mathbf{p})$ about the point $(\mathbf{x}, \mathbf{p})$ at some time t. The space defined by $(\mathbf{x}, \mathbf{p})$ is a six-dimensional space in spatial volume $\mathbf{x}$ and momentum $\mathbf{p}$, and is called μ-space. Before proceeding to the derivation of the transport equation for the distribution function in phase space, we prove a useful result.

Invariance of phase space volume. Consider a frame comoving with a group of particles that occupy a phase space volume $(\mathbf{x} + d^3\mathbf{x}', \mathbf{p} + d^3\mathbf{p}')$ about the point $(\mathbf{x}, \mathbf{p})$ at some time t, all at the same energy. The particles therefore occupy a spatial volume element $d^3\mathbf{x}' = dx'dy'dz'$ and a momentum volume element $d^3\mathbf{p}' = dp'_x dp'_y dp'_z$. Consider now a frame K that is not comoving with the particles, say with velocity parameter $\beta \equiv v/c$ with respect to the comoving frame K'. In the spatial volume $d^3\mathbf{x}$ occupied by the particles as measured in frame K, perpendicular distances are unaffected by the motion, i.e., $dy = dy'$, $dz = dz'$, say, but there is a length contraction in the x direction (the assumed direction of motion), so that $dx = \gamma^{-1} dx'$, where $\gamma \equiv (1-\beta^2)^{-1/2}$, or

$$d^3\mathbf{x} = \gamma^{-1} d^3\mathbf{x}'.$$

Consider the momentum volume element $d^3\mathbf{p}$ measured in the frame K. Since there is no energy change from K to K', the x component of the momentum transforms according to $dp_x = \gamma dp'_x$ and the remaining momentum increments are unchanged. Thus,

$$d^3\mathbf{p} = \gamma d^3\mathbf{p}',$$

from which we find

$$d^3\mathbf{x}d^3\mathbf{p} = d^3\mathbf{x}'d^3\mathbf{p}'.$$

Hence, since the frames K and K' are arbitrary, $d^3\mathbf{x}d^3\mathbf{p}$ is a Lorentz invariant. Furthermore, since the number of particles dN within a phase space volume element is a countable quantity and therefore invariant, we have also established that the phase space density

$$f(\mathbf{x},\mathbf{p},t) = \frac{dN}{dV}, \quad dV \equiv d^3\mathbf{x}d^3\mathbf{p},$$

is an invariant.

Now suppose that the number density of points in phase space from volume element to volume element does not vary rapidly. This allows us to assume continuity of $f(\mathbf{x},\mathbf{p},t)$ over the entire μ-space and we can introduce an integral over the over the distribution function so that

$$\int f(\mathbf{x},\mathbf{p},t)d^3\mathbf{x}d^3\mathbf{p} = N,$$

where N is the total number of particles in a total volume V. If the particles are distributed uniformly in space, so that f is independent of $\mathbf{x}$, then

$$\int f(\mathbf{x},\mathbf{p},t)d^3\mathbf{p} = \frac{N}{V}.$$

Kinetic theory tries to find the distribution function f for particular forms of particle interaction in different physical settings.

To determine the equation of "motion" for the distribution function, suppose there are no particle collisions, so that particles at location $(\mathbf{x},\mathbf{p})$ at time t will find themselves at location $(\mathbf{x}+\mathbf{v}\delta t, \mathbf{p}+\mathbf{F}\delta t)$ at the time $t+\delta t$. Here δt is an infinitesimal change in time, $\mathbf{F}$ the force acting on the particle, and $\mathbf{v} = \mathbf{p}/m$ the velocity. Furthermore, these translated particles will find themselves in the new volume element $d^3\mathbf{x}'d^3\mathbf{p}'$ at time $t+\delta t$. Thus, in the absence of collisions, we have

$$f(\mathbf{x}+\mathbf{v}\delta t, \mathbf{p}+\mathbf{F}\delta t, t+\delta t)d^3\mathbf{x}'d^3\mathbf{p}' = f(\mathbf{x},\mathbf{p},t)d^3\mathbf{x}d^3\mathbf{p}$$
$$\implies f(\mathbf{x}+\mathbf{v}\delta t, \mathbf{p}+\mathbf{F}\delta t, t+\delta t) = f(\mathbf{x},\mathbf{p},t),$$

by the invariance of the phase space volume element.

In the presence of collisions, the above equality will be modified so that

$$f(\mathbf{x}+\mathbf{v}\delta t, \mathbf{p}+\mathbf{F}\delta t, t+\delta t) = f(\mathbf{x},\mathbf{p},t) + \left(\frac{\delta f}{\delta t}\right)_{coll}\delta t,$$

3.1 Derivation of the Boltzmann Transport Equation

which should be interpreted as a definition of the collisional term $(\delta f/\delta t)_{coll}$. By Taylor expanding the LHS about the time t, and retaining only linear terms, we obtain the equation of motion for the distribution function f as we let $\delta t \to 0$,

$$\frac{\partial f}{\partial t} + \frac{\mathbf{p}}{m} \cdot \nabla f + \mathbf{F} \cdot \nabla_p f = \left(\frac{\delta f}{\delta t}\right)_{coll},$$

where ∇ and ∇_p are the gradient operators in $\mathbf{x}$ and $\mathbf{p}$ respectively. The variables $\mathbf{x}$, $\mathbf{p}$, and t are independent variables. At this point, we will assume that the particles are non-relativistic and express the Boltzmann equation in terms of particle velocity $\mathbf{v}$,

$$\frac{\partial f}{\partial t} + \mathbf{v} \cdot \nabla f + \frac{\mathbf{F}}{m} \cdot \nabla_v f = \left(\frac{\delta f}{\delta t}\right)_{coll}. \quad (3.1)$$

Before deriving forms of the collisional term, it is often useful to express the Boltzmann equation (3.1) in terms of mixed variables related to convection in a background flow. One can separate the particle velocity into a random component and mean or a bulk fluid or gas component $\mathbf{u}(\mathbf{x}, t)$ according to

$$\mathbf{c}(\mathbf{x}, t) = \mathbf{v} - \mathbf{u}(\mathbf{x}, t),$$

where the bulk velocity is defined as a moment of the distribution function (more later)

$$\mathbf{u}(\mathbf{x}, t) = \frac{1}{\int_{-\infty}^{\infty} f(\mathbf{x}, \mathbf{v}, t) d^3\mathbf{v}} \int_{-\infty}^{\infty} \mathbf{v} f(\mathbf{x}, \mathbf{v}, t) d^3\mathbf{v}.$$

The bulk velocity is independent of $\mathbf{v}$, and the random velocity component $\mathbf{c}$, unlike that of $\mathbf{v}$, is a function of t. To express the Boltzmann equation (3.1) in terms of the random velocity $\mathbf{c}(\mathbf{x}, t)$ introduces a mixed phase space set of coordinates, giving (Exercise)

$$\frac{\partial f}{\partial t} + (u_i + c_i)\frac{\partial f}{\partial x_i} - \left(\frac{\partial u_i}{\partial t} + (u_j + c_j)\frac{\partial u_i}{\partial x_j} - \frac{F_i}{m}\right)\frac{\partial f}{\partial c_i} = \left(\frac{\delta f}{\delta t}\right)_{coll}, \quad (3.2)$$

and $f = f(\mathbf{x}, \mathbf{c}, t)$, $\mathbf{u} = \mathbf{u}(\mathbf{x}, t)$, $\mathbf{F} = \mathbf{F}(\mathbf{x}, \mathbf{v}, t)$. We shall see that this form of the Boltzmann equation is the starting point for studying energetic particle transport in a magnetically turbulent medium.

Exercises

1. Show that the Boltzmann equation (3.1) is invariant with respect to Galilean transformations.
2. Show that the Boltzmann equation (3.1) transforms into the mixed phase space coordinate form (3.2).

3. Find the general solution to the Boltzmann equation (3.1) in the absence of collisions i.e., $(\delta f/\delta t)_{coll} = 0$. Derive the general solution for the case that the force $\mathbf{F} = 0$.

3.2 The Boltzmann Collision Operator

Consider a suitably rarefied gas that satisfies the following conditions.

1. The gas is neutral;
2. The mean distance between particles is large in comparison to their size as expressed by their inter-particle forces;
3. The gas is sufficiently dilute that only binary collisions are important; and
4. Collisions conserve mass, momentum, and energy.

For simplicity, suppose we consider a gas of particles of a single species and let the mass be normalized to unity. Consider a two particle collision, where one particle has velocities in the range dv and the other in a range du before collision and in the ranges dv' and du' after. See Fig. 3.1. To keep the notation simple in this subsection, we do not use bold-face for the vector variables. As noted, collisions conserve momentum and energy, so that

$$v' + u' = v + u;$$
$$|v'|^2 + |u'|^2 = |v|^2 + |u|^2.$$

The total number of collisions per unit time per unit volume for a particle with velocities in the range dv may be expressed as (the number of particles per unit volume) times (the probability that a particle experiences a collision), or

$$f(x, v, t) dv \times P,$$

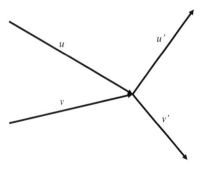

Fig. 3.1 Binary collision

3.2 The Boltzmann Collision Operator

where P is proportional to (the number of particles per unit volume) times $(dv' \times du')$ or $f(x, u, t) du \times dv' du'$. Thus the

total number of collisions/unit volume/unit time
$$= \Phi(u', v'; u, v) f(u) f(v) du\, dv\, du'\, dv'.$$

The function Φ is determined from the collision problem for the given physical system. The joint pdf for a pair of particles experiencing a collision with velocities u, v at (x, t) is proportional to the product $f(x, v, t) f(x, u, t)$, and was in fact hypothesized already by Maxwell. The hypothesis is called "molecular chaos" and in fact corresponds to an assumption of stochastic independence i.e., the joint pdf can be expressed as the product of the marginal pdfs.

The function Φ may be thought of as the transition matrix in quantum mechanics. Φ may be regarded as a symmetric function, i.e.,

$$\Phi(u', v'; u, v) = \Phi(u, v; u', v').$$

This follows from noting that in equilibrium, the number of collisions $(u, v) \mapsto (u', v')$ is equal to the number of collisions $(-u', -v') \mapsto (-u, -v)$. This is a consequence of the Newtonian equations being symmetric under time reversal. Thus, under such a mapping we expect to get

$$\Phi(u', v'; u, v) = \Phi(-u, -v; -u', -v'),$$

and hence it follows that Φ is a symmetric function.

For each collision of two particles, there is a transfer of velocity out of a particular range dv (losses). Given dv, the total number of collisions $(u, v) \mapsto (u', v')$ with all possible values of u, u', v' occurring in the volume element dx per unit time is

$$dx\, dv \int \Phi(u', v'; u, v) f(u) f(v) du\, du'\, dv'.$$

Consider the total number of collisions in the volume dx that bring particles into the range dv from outside the range per unit time (gains). Given v, these are collisions $(u', v') \mapsto (u, v)$ with all possible u, u', v' such that

the total number of such collisions in volume dx per unit time
$$= dx\, dv \int \Phi(u, v; u', v') f(u') f(v') du\, du'\, dv'.$$

Consequently, by virtue of the symmetry of Φ, the collisional term can be expressed as

$$\left(\frac{\delta f}{\delta t}\right)_{coll} \equiv \mathcal{C}(f) = \int \Phi(u, v; u', v') \left[f(u') f(v') - f(u) f(v)\right] du\, du'\, dv'.$$

The *differential cross section* $d\sigma$ is frequently introduced,

$$d\sigma = \frac{\Phi du' dv'}{|v-u|}.$$

The differential cross section contains delta functions that express conservation of momentum and energy,

$$\delta(u' + v' - u - v) \cdot \delta\left(|v'|^2 + |u'|^2 - |v|^2 - |u|^2\right).$$

If the delta functions are removed, then $d\sigma$ is the scattering cross section, expressed in terms of the solid angle

$$d\Omega = \sin\theta d\theta d\phi,$$

and relative velocity $|u - v|$, such that

$$d\sigma = \xi(\Omega, |u-v|) d\Omega.$$

Hence,

$$C(f) = \int_{\Re^3} \int_{|\Omega|=1} \xi(\Omega, |u-v|)[f(u')f(v') - f(u)f(v)] d\Omega du.$$

This is the Boltzmann collision integral, and is integrated over five independent variables. The Boltzmann transport equation (3.1) is therefore a nonlinear integro-differential equation

$$\frac{\partial f}{\partial t} + \mathbf{v} \cdot \nabla f + \mathbf{F} \cdot \nabla_v f = \int_{\Re^3} \int_{|\Omega|=1} \xi(\Omega, |u-v|)[f(u')f(v') - f(u)f(v)] d\Omega du. \tag{3.3}$$

3.2.1 Collision Dynamics

Consider first the simplest possible case of hard sphere scattering. In this case, we have

$$\Phi(u, v; u', v') = \frac{b_0}{2} \delta(u + v - u' - v') \delta\left(|v|^2 + |u|^2 - |v'|^2 - |u'|^2\right),$$

where b_0 is related to the size of the spheres and the factor of 2 is due to energy conservation. Then $C(f)$ can be expressed as

3.2 The Boltzmann Collision Operator

Fig. 3.2 Geometry for hard sphere scattering of a particle

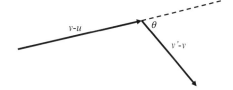

$$C(f) = \frac{b_0}{2} \int\int \left[f(v')f(u+v-v') - f(u)f(v) \right] \times$$
$$\delta\left(|v|^2 + |u|^2 - |v'|^2 - |u+v-v'|^2\right) du\, dv'$$
$$= \frac{b_0}{2} \int\int \left[f(v-\lambda)f(u+\lambda) - f(u)f(v) \right] \times$$
$$\delta\left(|v|^2 + |u|^2 - |v-\lambda|^2 - |u+\lambda|^2\right) du\, d\lambda,$$

where $\lambda = v - v'$. Write $d\lambda = \mu^2 d\mu\, d\Omega = \mu^2 d\mu \sin\theta\, d\theta\, d\phi$ using spherical coordinates (μ, θ, ϕ) with the polar axis along $v - u$ so that $|\mu| \leq |v - u|$ and (Fig. 3.2)

$$\cos\theta = \frac{\lambda \cdot (v-u)}{|\lambda||v-u|} = \frac{(v-v') \cdot (v-u)}{|v'-v||v-u|}.$$

We can express the argument of the energy delta function above as

$$|v|^2 + |u|^2 - |v-\lambda|^2 - |u+\lambda|^2$$
$$= |v|^2 + |u|^2 - (|v|^2 - 2v\cdot\lambda + |\lambda|^2) - (|u|^2 + 2u\cdot\lambda + |\lambda|^2)$$
$$= -2|\lambda|^2 + 2(v-u)\cdot\lambda = 2\mu|v-u|\cos\theta - 2\mu^2.$$

To evaluate the delta function, we need the following result (see the Exercises),

$$\delta((x-a)(x-b)) = \frac{1}{|a-b|}\left[\delta(x-a) + \delta(x-b)\right], \quad a \neq b.$$

Hence,

$$\frac{1}{2}\delta\left(|v|^2 + |u|^2 - |v'|^2 - |u+v-v'|^2\right) = \delta\left(\mu^2 - \mu|v-u|\cos\theta\right)$$
$$= \delta(\mu(\mu - |v-u|\cos\theta))$$
$$= \frac{1}{|u-v||\cos\theta|} \times$$
$$\left[\delta(\mu) + \delta(\mu - |v-u|\cos\theta)\right].$$

The $\delta(\mu)$ term in the integrand yields 0 because of the $\mu^2 d\mu$ term, whereas the second term contributes

$$\frac{1}{|u-v||\cos\theta|}|v-u|^2 \cos^2\theta = |u-v|\cos\theta.$$

In the integrand for $C(f)$, we have

$$f(v-\lambda) \equiv f(v'), \quad f(u+\lambda) \equiv f(u'),$$

so that

$$C = b_0 \int_{|\Omega|=1} \int_{\Re^3} |u-v||\cos\theta|[f(u')f(v') - f(u)f(v)]du\,d\Omega$$

for the scattering of hard spheres i.e., for hard sphere scattering, $\xi(\Omega,|u-v|) = b_0|u-v||\cos\theta|$, where b_0 is an impact parameter that reflects the size of the scattering spheres and $0 < \theta < \pi/2$.

A more general analysis can be developed for particles with a potential energy $V(r)$. Consider the relative motion of particles P_1 and P_2 moving in each other's field of force. The particles have position vectors $\mathbf{r}_1$ and $\mathbf{r}_2$, masses m_1 and m_2, and subject to central forces $\mathbf{F}_1$ and $\mathbf{F}_2$. Of course, $\mathbf{F}_1 = -\mathbf{F}_2$, and are parallel to $\mathbf{r} = \mathbf{r}_1 - \mathbf{r}_2$ and depend only on $r = |\mathbf{r}_1 - \mathbf{r}_2|$. Thus, $m_1\ddot{\mathbf{r}}_1 = \mathbf{F}_1, m_2\ddot{\mathbf{r}}_2 = \mathbf{F}_2$ so that $m_1 m_2 (\ddot{\mathbf{r}}_1 - \ddot{\mathbf{r}}_2) = m_2\mathbf{F}_1 - m_1\mathbf{F}_2$, or

$$M\ddot{\mathbf{r}} = \frac{m_1 m_2}{m_1 + m_2}\ddot{\mathbf{r}} = \mathbf{F}_1 = \mathbf{F},$$

where M is the reduced mass.

Introduce polar coordinates r and θ,

$$x = r\cos\theta \qquad y = r\sin\theta,$$

to describe particle motion subject to a central force (Fig. 3.3). To obtain the particle velocity and acceleration vector, differentiating with respect to time yields

$$\dot{x} = \dot{r}\cos\theta - r\dot{\theta}\sin\theta;$$
$$\dot{y} = \dot{r}\sin\theta + r\dot{\theta}\cos\theta;$$
$$\ddot{x} = \ddot{r}\cos\theta - 2\dot{r}\dot{\theta}\sin\theta - r\dot{\theta}^2\cos\theta - r\ddot{\theta}\sin\theta;$$
$$\ddot{y} = \ddot{r}\sin\theta + 2\dot{r}\dot{\theta}\cos\theta - r\dot{\theta}^2\sin\theta + r\ddot{\theta}\cos\theta, \qquad (3.4)$$

where as usual the dot denotes a time derivative. The velocity and acceleration vectors may be expressed in terms of the orthonormal vectors $\hat{\mathbf{r}}$ and $\hat{\boldsymbol{\theta}}$ using $\mathbf{v} = v_r\hat{\mathbf{r}} + v_\theta\hat{\boldsymbol{\theta}}$ or $\mathbf{a} = a_r\hat{\mathbf{r}} + a_\theta\hat{\boldsymbol{\theta}}$. Furthermore, the velocity and acceleration

3.2 The Boltzmann Collision Operator

Fig. 3.3 Scattering of a particle by a central force

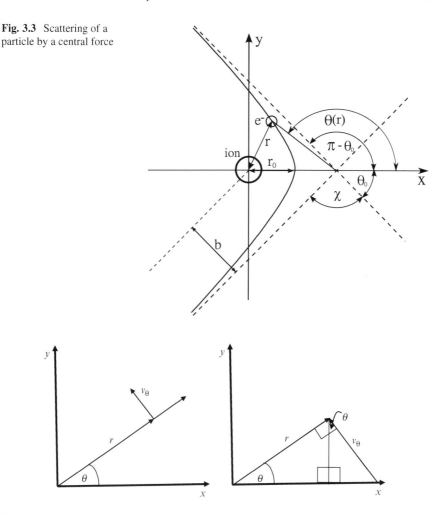

Fig. 3.4 Particle motion subject to a central force

components of the Cartesian coordinates of the particle can be expressed using the time derivatives of the Cartesian coordinates. One obtains geometrically from Fig. 3.4 the following relations

$$\begin{aligned} v_r &= \dot{x}\cos\theta + \dot{y}\sin\theta; \\ v_\theta &= -\dot{x}\sin\theta + \dot{y}\cos\theta; \\ a_r &= \ddot{x}\cos\theta + \ddot{y}\sin\theta; \\ a_\theta &= -\ddot{x}\sin\theta + \ddot{y}\cos\theta. \end{aligned} \qquad (3.5)$$

By using (3.5) in (3.4), we easily obtain

$$v_r = \dot{r}; \qquad v_\theta = r\dot{\theta}; \tag{3.6}$$

$$a_r = \ddot{r} - r\dot{\theta}^2; \qquad a_\theta = 2\dot{r}\dot{\theta} + r\ddot{\theta}. \tag{3.7}$$

The terms in (3.7) correspond to the centrifugal ($r\dot{\theta}^2$) and Coriolis ($2\dot{r}\dot{\theta}$) forces respectively.

On returning to the reduced mass equation of motion, we may use the above results to obtain

$$M(\ddot{r} - r\dot{\theta}^2)\hat{\mathbf{r}} + M(r\ddot{\theta} + 2\dot{r}\dot{\theta})\hat{\boldsymbol{\theta}} = -\frac{\partial V}{\partial r}\hat{\mathbf{r}},$$

where $V(r)$ is the potential energy ($V(\infty) = 0$) of the force **F**. The conservation laws for angular momentum and energy are therefore

$$r^2\dot{\theta} = const. = gb;$$

$$\frac{1}{2}M(\dot{r}^2 + r^2\dot{\theta}^2) + V(r) = const. = \frac{1}{2}Mg^2,$$

where b is a distance, and is known as the "impact parameter", and g is the constant relative velocity. It is then straightforwardly established that

$$\frac{g^2 b^2}{2r^4}\left[\left(\frac{dr}{d\theta}\right)^2 + r^2\right] = \frac{1}{2}g^2 - \frac{V(r)}{M},$$

from which we find

$$\frac{d\theta}{dr} = \pm\frac{b}{r^2}\left[1 - \frac{b^2}{r^2} - \frac{2V(r)}{Mg^2}\right]^{-1/2}.$$

The path of particle P_1 in a force field centered on P_2 exhibits two asymptotes, one along the initial direction of approach and the other along the final direction of motion as the particle recedes to "infinity." The *"scattering angle"* χ is the angle between the asymptotes (see Fig. 3.3). The reference frame is chosen so that prior to the interaction of the particles, P_1 is at $\theta = 0$, and its final position after the interaction is at $\theta = \pi - \chi$. The trajectory is symmetric about the apse line i.e., the line from P_1 to P_2 at closest approach (r_0, θ_0). At this point, $dr/d\theta = 0$. Hence,

$$r_0^2 - b^2 - \frac{2r_0^2}{Mg^2}V(r_0) = 0.$$

Evidently, $\chi = \pi - 2\theta_0$. To determine θ_0, we integrate $d\theta/dr$ over (r_0, ∞), choosing the negative root since the slope is negative along the incoming trajectory. Hence,

$$\chi = \pi - 2b\int_{r_0}^{\infty}\frac{dr}{r^2\sqrt{1 - b^2/r^2 - 2V(r)/(Mg^2)}}.$$

3.2 The Boltzmann Collision Operator

As discussed above, the differential cross section can be expressed as

$$d\sigma = \xi(\Omega, |u-v|)d\Omega.$$

We can use the results above for the scattering angle to determine ξ. Suppose that P_2 is incident on an annulus of inner radius b and outer radius $b + db$ with fractional area $2\pi b\, db\, d\rho$, where $d\rho$ is the angular width of the area. Let $\xi d\Omega$ be the probability that P_1 is deflected into the solid angle $d\Omega = \sin\chi\, d\chi\, d\rho$. Of N incident particles per unit area per second, $|b\, db\, d\rho| N$ are scattered into solid angle $d\Omega$, which by definition is $\xi d\Omega N$. Hence, we obtain

$$\xi = \frac{b}{\sin\chi}\left|\frac{db}{d\chi}\right|,$$

where $\chi = \chi(b, g)$ can be computed if the potential $V(r)$ is known. An important example is the Coulomb or Rutherford scattering cross section.

Exercises

1. Show that for $a \neq b$,

$$\delta((x-a)(x-b)) = \frac{1}{|a-b|}[\delta(x-a) + \delta(x-b)].$$

2. Complete the steps in the derivation of

$$\frac{d\theta}{dr} = \pm\frac{b}{r^2}\left[1 - \frac{b^2}{r^2} - \frac{2V(r)}{Mg^2}\right]^{-1/2},$$

starting from

$$M(\ddot{r} - r\dot{\theta}^2)\hat{\mathbf{r}} + M\left(r\ddot{\theta} + 2\dot{r}\dot{\theta}\right)\hat{\boldsymbol{\theta}} = -\frac{\partial V}{\partial r}\hat{\mathbf{r}}.$$

3. Consider the scattering of an electron (charge $Z_1 = -1$) in the Coulomb field of an ion of charge eZ_2,

$$\mathbf{E} = \frac{eZ_2}{4\pi\varepsilon_0}\frac{\mathbf{r}}{r^3},$$

where ε_0 is the permittivity of free space. Show that

$$\frac{b^2}{rb_0} = 1 + \epsilon\cos\theta, \quad b_0 \equiv |Z_1 Z_2|e^2/(4\pi\varepsilon_0 Mg^2),$$

and $\epsilon \equiv \sqrt{1+(b/b_0)^2}$. Show that at the point of closest approach, $\tan\theta_0 = b/b_0$ and hence that

$$\sigma = \frac{b_0^2}{4\sin^4\frac{1}{2}\chi} = \left(\frac{Z_1 Z_2 e^2}{8\pi\varepsilon_0 M g \sin^2\frac{1}{2}\chi}\right)^2.$$

This is the Coulomb or Rutherford scattering cross section.

3.3 Conservation Laws, the H-Theorem, and the Maxwell-Boltzmann Distribution Function

The collisional integral is frequently expressed as the operator

$$Q(f,f)(v) = \int_{\Re^3}\int_{|\Omega|=1} \xi(\Omega,|u-v|)[f(u')f(v') - f(u)f(v)]du\,d\Omega.$$

A related quadratic form can be introduced,

$$Q^*(f,g)(v) = \frac{1}{2}\int_{\Re^3}\int_{|\Omega|=1} \xi[f(v')g(u') + f(u')g(v') - f(u)g(v) - f(v)g(u)]d\Omega\,du.$$

Q^* is symmetric and $Q^*(f,f) = Q(f,f)$.

We can obtain explicit forms of u' and v' from the momentum and energy conservation laws since they impose four constraints on the six variables. This implies two degrees of freedom. We can write

$$u' = u + a(u,v,\Omega)\Omega$$
$$v' = v - a(u,v,\Omega)\Omega,$$

where a is a scalar function and $|\Omega| = 1$. Evidently, momentum is automatically conserved. Considering energy conservation implies

$$|u'|^2 + |v'|^2 = |u|^2 + a^2 + 2a\Omega\cdot u + |v|^2 + a^2 - 2a\Omega\cdot v = |u|^2 + |v|^2.$$

Hence,

$$a^2 = a(\Omega\cdot v - \Omega\cdot u),$$

or, provided $a \neq 0$,

$$a(u,v,\Omega) = \Omega\cdot(v-u).$$

3.3 Conservation Laws, the H-Theorem, and the Maxwell-Boltzmann Distribution Function

Introduce a smooth function $\eta(v)$, and consider $\int Q^*(f, g)\eta(v)dv$, and exchange u and v in the integral. Then, since

$$u' = u + [\Omega \cdot (v - u)]\Omega;$$
$$v' = v - [\Omega \cdot (v - u)]\Omega,$$

swapping u and v yields

$$v - [\Omega \cdot (v - u)]\Omega = v';$$
$$u + [\Omega \cdot (u - v)]\Omega = u'.$$

Consequently, we obtain the following equivalent expression,

$$\int Q^*(f,g)\eta(v)dv = \frac{1}{2}\int\int\int \xi \left[f(v')g(u') + f(u')g(v') \right.$$
$$\left. - f(u)g(v) - f(v)g(u) \right] \eta(v) d\Omega du dv$$
$$= \frac{1}{2}\int\int\int \xi \left[f(v')g(u') + f(u')g(v') \right.$$
$$\left. - f(u)g(v) - f(v)g(u) \right] \eta(u) d\Omega du dv$$

since ξ is invariant.

Now consider the change in variables $(u, v) \to (u', v')$ so that the integral becomes

$$\int Q^*(f,g)\eta(v)dv$$
$$= \frac{1}{2}\int\int\int \xi(\Omega, |v' - u'|)\eta(v(u', v'))|J| \left[f(v')g(u') + f(u')g(v') \right.$$
$$\left. - f(u(u', v'))g(v(u', v')) - f(v(u', v'))g(u(u', v')) \right] d\Omega du' dv'.$$

It is a tedious if straightforward exercise to show that the Jacobian $J = -1$. Since

$$|u - v|^2 = 2|u|^2 + 2|v|^2 - |u + v|^2 = |u' - v'|^2,$$

(from conservation of energy and momentum), ξ is invariant. Also, since

$$v' - u' = v - u - 2(\Omega \cdot (v - u))\Omega,$$
$$\Omega \cdot (v' - u') = \Omega \cdot (v - u) - 2(\Omega \cdot (v - u)) = -\Omega \cdot (v - u).$$

These equations can be inverted to yield

$$v = v' + [\Omega \cdot (v - u)]\Omega = v' - [\Omega \cdot (v' - u')]\Omega \equiv v(u', v');$$

$$u = u' - [\Omega \cdot (v-u)]\Omega = u' + [\Omega \cdot (v'-u')]\Omega \equiv u(u',v').$$

On renaming (u',v') to (u,v), we get

$$\int Q^*(f,g)\eta(v)dv = \frac{1}{2}\int\int\int \xi(\Omega,|v-u|)\left[f(v)g(u) + f(u)g(v)\right.$$
$$-f(u - \Omega \cdot (v-u)\Omega)g(v - \Omega \cdot (v-u)\Omega)$$
$$\left.-f(v - \Omega \cdot (v-u)\Omega)g(u + \Omega \cdot (v-u)\Omega)\right]\eta(v - \Omega \cdot (v-u)\Omega)d\Omega du dv$$
$$= \frac{1}{2}\int\int\int \xi(\Omega,|v-u|)\left[f(v)g(u) + f(u)g(v) - f(u')g(v')\right.$$
$$\left.-f(v')g(u')\right]\eta(v')d\Omega du dv.$$

Finally, if we switch u and v above and again let $u' \to v'$, $v' \to u'$ as above, we obtain the last equivalent expression,

$$\int Q^*(f,g)\eta(v)dv$$
$$= \frac{1}{2}\int\int\int \xi\left[f(v)g(u) + f(u)g(v) - f(u')g(v') - f(v')g(u')\right]\eta(u')d\Omega du dv.$$

Let $f = g$ in $Q^*(f,g)(v)$ and the four equivalent forms to obtain

$$\int Q(f,f)\eta(v)dv = \frac{1}{4}\int\int\int \xi\left[f(u')f(v') - f(u)f(v)\right]$$
$$\times \left[\eta(v) + \eta(u) - \eta(v') - \eta(u')\right]d\Omega du dv. \qquad (3.8)$$

Hence it follows that

$$\int Q(f,f)\eta(v)dv = 0 \text{ if } \eta(v) + \eta(u) = \eta(v') + \eta(u').$$

These are called collisional invariants. In particular, if we choose

$$\eta(v) = 1; \quad \eta(v) = v_j, \quad (j = 1,2,3); \quad \eta(v) = |v|^2,$$

it follows immediately (conservation of momentum and energy) that the following moments are zero:

$$\int Q(f,f)dv = 0; \quad \int Q(f,f)v_j dv = 0; \quad \int Q(f,f)|v|^2 dv = 0,$$

for $j = 1,2,3$. Consequently, if we choose $\eta(v)$ to be one of the three moments and multiply the Boltzmann equation by $\eta(v)$, and integrate, we obtain the conservation laws,

3.3 Conservation Laws, the H-Theorem, and the Maxwell-Boltzmann Distribution Function

$$\int\int f \, dv \, dx = \text{const.} \quad \text{mass conservation;}$$

$$\int\int v_j f \, dv \, dx = \text{const.} \quad \text{momentum conservation;}$$

$$\int\int |v|^2 f \, dv \, dx = \text{const.} \quad \text{energy conservation.}$$

We now come to a fundamental result for the collisional Boltzmann equation, viz. the H-theorem. This theorem shows that Boltzmann's equation possesses the irreversibility associated with dissipation – the left-hand side has negative parity under time reversal while the right-hand side has positive parity. Negative parity means that its sign changes when all the velocities and the time are reversed in sign; with positive parity, there is no change in sign.

Theorem. *If* f *satisfies the Boltzmann equation, then*

$$\frac{dH}{dt} \leq 0, \qquad H(t) \equiv \int\int f \ln f \, dv \, dx.$$

The expression $-f \ln f$ *is the entropy density.*

Proof. Let $\eta = 1 + \ln f$ and use Eq. (3.8) to obtain

$$4 \int Q(f,f)(1 + \ln f) dv = \int\int\int \xi \left[f(u')f(v') - f(u)f(v) \right] \times$$

$$\left[\ln f(v) + \ln f(u) - \ln f(v') - \ln f(u') \right] d\Omega \, du \, dv$$

$$= \int\int\int \xi \left[f(u')f(v') - f(u)f(v) \right] \left[\ln(f(u)f(v)) - \ln(f(u')f(v')) \right] d\Omega \, du \, dv$$

$$= \int\int\int \xi \left[f(u')f(v') - f(u)f(v) \right] \ln \frac{f(u)f(v)}{f(u')f(v')} d\Omega \, du \, dv$$

$$= \int\int\int \xi f(u')f(v')(1 - \mu) \ln \mu \, d\Omega \, du \, dv,$$

where

$$\mu = \frac{f(u)f(v)}{f(u')f(v')}.$$

Now clearly $(1 - \mu) \ln \mu \leq 0$ for all $\mu > 0$, implying that

$$\int Q(f,f)\ln f\,dv \le 0,$$

for $f > 0$. Therefore,

$$\frac{d}{dt}\int\int f\ln f\,dv\,dx = \int\int f_t(1+\ln f)\,dv\,dx$$

$$= \int\int (-v\cdot\nabla f + Q(f,f))(1+\ln f)\,dv\,dx$$

$$= \int\int Q(f,f)\ln f\,dv\,dx \le 0,$$

thus establishing the H-theorem.

To conclude this section, consider $\int Q(f,f)\ln f\,dv = 0$. It follows that

$$\ln f(v) + \ln f(u) = \ln f(v') + \ln f(u'),$$

i.e., $\eta = \ln f$ satisfies $\eta(v)+\eta(u) = \eta(v')+\eta(u')$. Thus, conservation of momentum and energy imply immediately that

$$\eta(v) = a + b\cdot v + c|v|^2,$$

and thus f is a Gaussian distribution, $f(v) = \exp(a + b\cdot v + c|v|^2)$, $c < 0$. The precise values of a, b, and c are determined from the constraints,

$$n = \int f\,dv, \quad n\mathbf{u} = \int \mathbf{v}f\,dv, \quad \frac{3}{2}nkT = \frac{m}{2}\int (\mathbf{v}-\mathbf{u})^2 f\,dv.$$

We can rewrite the equilibrium distribution $f(v)$ as

$$f(\mathbf{v}) = \exp\left[-\alpha - \beta(\mathbf{v}-\mathbf{u})^2\right],$$

where α, β, and $\mathbf{u}$ are determined from the above constraints. For simplicity, suppose we move into the translational frame so that $\mathbf{c} = \mathbf{v} - \mathbf{u}$. The first constraint yields

$$n = e^{-\alpha}\int_{-\infty}^{\infty}\int_{-\infty}^{\infty}\int_{-\infty}^{\infty}\exp\left[-\beta(c_x^2+c_y^2+c_z^2)\right]dc_x dc_y dc_z,$$

and since

$$\sqrt{\frac{\pi}{\beta}} = \int_{-\infty}^{\infty}\exp(-\beta x^2)\,dx,$$

we have

$$\exp(-\alpha) = n\left(\frac{\beta}{\pi}\right)^{3/2}.$$

The third constraint yields

$$\frac{3}{2}nkT = e^{-\alpha} \int_{-\infty}^{\infty}\int_{-\infty}^{\infty}\int_{-\infty}^{\infty} \frac{1}{2}m\left(c_x^2 + c_y^2 + c_z^2\right)$$
$$\exp\left[-\beta\left(c_x^2 + c_y^2 + c_z^2\right)\right] dc_x dc_y dc_z.$$

The expectation of c_x^2 is given by

$$E[c_x^2] = \frac{1}{n}\int c_x^2 f d\mathbf{c}$$
$$= \left(\frac{\beta}{\pi}\right)^{3/2} \int_{-\infty}^{\infty}\int_{-\infty}^{\infty}\int_{-\infty}^{\infty} c_x^2 \exp\left[-\beta\left(c_x^2 + c_y^2 + c_z^2\right)\right] dc_x dc_y dc_z.$$

Since

$$\frac{\sqrt{\pi}}{2\beta^{3/2}} = \int_{-\infty}^{\infty} x^2 \exp(-\beta x^2) dx,$$

$E[c_x^2] = 1/2\beta = E[c_y^2] = E[c_z^2]$. Hence, $kT = m/(2\beta)$ or $\beta = m/(2kT)$. This yields the familiar Maxwell-Boltzmann distribution

$$f(\mathbf{x},\mathbf{v},t) = n\left(\frac{m}{2\pi kT}\right)^{3/2} \exp\left[-m\frac{(\mathbf{v}-\mathbf{u})^2}{2kT}\right],$$

where k is Boltzmann's constant, $\mathbf{u}$, n, and T are the bulk velocity, number density, and temperature of the gas.

3.4 The Boltzmann Equation and the Fluid Equations

Suppose $f(\mathbf{x},\mathbf{v},t)$ is a solution to the Boltzmann equation in the absence of forces,

$$\frac{\partial f}{\partial t} + \mathbf{v}\cdot\nabla f = Q(f,f). \tag{3.9}$$

For each species, we define the *number density* and the *hydrodynamic (Eulerian) velocity* by

$$n(\mathbf{x},t) = \int f(\mathbf{x},\mathbf{v},t) d^3v,$$

and

$$\mathbf{u}(\mathbf{x},t) = \frac{1}{n}\int \mathbf{v} f(\mathbf{x},\mathbf{v},t) d^3v.$$

On multiplying (3.9) by 1 and integrating with respect to **v** yields

$$\frac{\partial}{\partial t} \int f d^3v + \nabla \cdot \int \mathbf{v} f d^3v = 0$$

$$\Longrightarrow \frac{\partial n}{\partial t} + \nabla \cdot (n\mathbf{u}) = 0,$$

where we have exploited the vanishing of the velocity integrated collision integral. This is the continuity equation in physical space, or the conservation of mass (sometimes charge) equation.

We define the *pressure tensor* p_{ij} by

$$p_{ij}(\mathbf{x}, t) = \int m(v_i - u_i)(v_j - u_j) f(\mathbf{x}, \mathbf{v}, t) d^3v$$

$$= m \int v_i v_j f d^3v - mn u_i u_j,$$

which is the negative of the *stress tensor* as generally defined in fluid mechanics. Multiply (3.9) by v_i and integrate over velocity space. The first term is $\partial(nu_i)/\partial t$, and the second is

$$\frac{\partial}{\partial x_j} \int v_i v_j f d^3v = \frac{\partial}{\partial x_j} \left(\frac{1}{m} p_{ij} + n u_i u_j \right).$$

This then yields the conservation form of the equation of momentum,

$$\frac{\partial}{\partial t}(n u_i) + \frac{\partial}{\partial x_j}\left(\frac{1}{m} p_{ij} + n u_i u_j\right) = 0,$$

which, if we use the continuity equation, reduces to

$$mn\left(\frac{\partial u_i}{\partial t} + u_j \frac{\partial u_i}{\partial x_j}\right) = -\frac{\partial p_{ij}}{\partial x_j},$$

for smooth flows i.e., flows without discontinuities such as shock waves or contact discontinuities.

The continuity and momentum equations admit the further unknown p_{ij}, so we can take the next moment of equation (3.9) to determine the evolution equation of the pressure tensor. However, this introduces further unknowns – this is referred to as the "closure problem." Nonetheless, by introducing the following definitions,

$$\varepsilon(\mathbf{x}, t) = \frac{1}{2mn} \sum_i p_{ii},$$

3.4 The Boltzmann Equation and the Fluid Equations

for the *internal energy*, (and note that $\rho \equiv mn$) and

$$q_i(\mathbf{x}, t) = \frac{1}{2} m \int (v_i - u_i)(\mathbf{v} - \mathbf{u}) \cdot (\mathbf{v} - \mathbf{u}) f d^3 v,$$

for the heat flux vector, we can derive an energy conservation equation. On multiplying (3.9) by $|\mathbf{v}|^2$ and integrating, we obtain

$$\frac{\partial}{\partial t} \int |v|^2 f d^3 v + \nabla \cdot \int |v|^2 \mathbf{v} f d^3 v = 0.$$

The internal energy can be expressed as

$$\varepsilon(\mathbf{x}, t) = \frac{1}{2mn} \sum_i p_{ii} = \frac{1}{2n} \sum_i \int |v_i - u_i|^2 f d^3 v$$

$$= \frac{1}{2n} \int (|u|^2 + |v|^2 - 2\mathbf{u} \cdot \mathbf{v}) f d^3 v$$

$$= \frac{1}{2n} \left(n|u|^2 + \int |v|^2 f d^3 v - 2n\mathbf{u} \cdot \mathbf{u} \right)$$

$$= \frac{1}{2n} \int |v|^2 f d^3 v - \frac{1}{2} u^2.$$

It therefore follows that

$$\frac{m}{2} \frac{\partial}{\partial t} \int |v|^2 f d^3 v = \frac{\partial}{\partial t} (mn\varepsilon + \frac{1}{2} mnu^2).$$

To compute $\int v^2 v_i f d^3 v$, we need to use the heat flux vector,

$$q_i = \frac{1}{2} m \int (v_i - u_i) |\mathbf{v} - \mathbf{u}|^2 f d^3 v$$

$$= \frac{1}{2} m \int (v_i - u_i) \left[u^2 + v^2 - 2 \sum_j v_j u_j \right] f d^3 v$$

$$= -\frac{1}{2} u_i \left[mnu^2 + (2mn\varepsilon + mnu^2) - 2mn\mathbf{u} \cdot \mathbf{u} \right]$$

$$+ \frac{1}{2} m u^2 \int v_i f d^3 v + \frac{1}{2} m \int v_i v^2 f d^3 v - m \sum_j u_j \int v_i v_j f d^3 v$$

$$= -mn\varepsilon u_i + \frac{1}{2} mnu^2 u_i + \frac{1}{2} m \int v_i v^2 f d^3 v - \sum_j u_j (p_{ij} + mn u_i u_j)$$

$$= -mn\varepsilon u_i - \frac{1}{2} mn u_i u^2 - \sum_j u_j p_{ij} + \frac{1}{2} m \int v_i v^2 f d^3 v.$$

Combining these results yields the conservation of energy equation,

$$\frac{\partial}{\partial t}\left[mn\left(\varepsilon + \frac{1}{2}u^2\right)\right] + \sum_i \frac{\partial}{\partial x_i}\left[mnu_i\left(\varepsilon + \frac{1}{2}u^2\right) + \sum_j u_j p_{ij} + q_i\right] = 0.$$

The conservation of mass, momentum, and energy are the five equations that are the basis of continuum mechanics – fluid mechanics or hydrodynamics, magnetohydrodynamics (including Maxwell's equations), etc. As noted above, the five equations possess more than five unknowns. We have the two scalars, n and e, the three vectors $\mathbf{u}$ and q, giving six unknowns, and finally the 3×3 pressure tensor p_{ij}, which is symmetric and yields a further six unknowns. However, the trace of the pressure tensor is related to the internal energy e through $\sum_i p_{ii} = 2\rho e$, which reduces the unknowns by one. Hence we are left with five equations for 13 unknowns. We therefore need to impose constitutive equations or relations to relate p_{ij}, q_i to $n, \mathbf{u}, e$.

Finally, we may define a "local temperature" $T(\mathbf{x}, t)$ by

$$3nkT = p_{ii} = m \int |\mathbf{v} - \mathbf{u}|^2 f d^3 v.$$

The energy density of the random translational motion is thus $\frac{3}{2}nkT$.

Example. The *Euler equations* result from assuming that

$$p_{ij} = p(\mathbf{x}, t)\delta_{ij}, \quad q_i = 0,$$

and $p(\mathbf{x}, t)$ is the scalar pressure. We have

$$\frac{\partial n}{\partial t} + \nabla \cdot (n\mathbf{u}) = 0;$$

$$nm\left(\frac{\partial \mathbf{u}}{\partial t} + \mathbf{u} \cdot \nabla \mathbf{u}\right) = -\nabla p;$$

$$\frac{\partial p}{\partial t} + \mathbf{u} \cdot \nabla p + \frac{5}{3}p\nabla \cdot \mathbf{u} = 0.$$

Example. The *Navier-Stokes equations* for viscous fluid result from assuming that there exist viscosity coefficients μ and λ such that

$$p_{ij} = p(\mathbf{x}, t)\delta_{ij} - \mu\left(\frac{\partial u_i}{\partial x_j} + \frac{\partial u_j}{\partial x_i}\right) - \lambda \sum_k \frac{\partial u_k}{\partial x_k}\delta_{ij};$$

$$q_i = -\kappa \frac{\partial T}{\partial x_i}.$$

Note that the non-diagonal terms in the pressure tensor p_{ij} i.e., excluding the scalar pressure $p\delta_{ij}$, comprise the rate-of-strain tensor.

Exercises

1. By using the conservation equations for mass, momentum and energy, derive the evolution equation for p_{ij} assuming the flow is smooth.
2. Use the Boltzmanns-Maxwellian distribution to show that the definitions for the number density n, velocity **u**, pressure tensor p_{ij}, and temperature T do indeed yield these quantities. Show too that the heat flux **q** vanishes and that the pressure tensor can be expressed as $p_{ij} = p(\mathbf{x}, t)\delta_{ij}$.
3. Using the above results, derive the Euler equations.
4. Linearize the 1D Euler equations about the constant state $\Psi_0 = (n_0, u_0, p_0)$ i.e., consider perturbations $\delta\Psi$ such that $\Psi = \Psi_0 + \delta\Psi$. Derive a linear wave equation in terms of a single variable, say δn. Seek solutions to the linear wave equation in the form $\exp i(\omega t - kx)$, and show that the Euler equations admit a non-propagating zero-frequency wave and forward and backward propagating acoustic modes satisfying the dispersion relation $\omega' \equiv \omega - u_0 k = \pm C_s k$ where C_s is a suitably defined sound speed.

3.5 The Relaxation Time Approximation

The form of the Boltzmann collision operator suggests that we may approximate

$$Q(f, f)(v) = -\frac{f - f_0}{\tau} = -\nu(f - f_0),$$

for a relaxation time parameter τ or scattering frequency ν (Bhatnagar et al. 1954). Here, f_0 is the Maxwellian equilibrium distribution. The relaxation time or BGK operator must vanish under the appropriate moments, as with $Q(f, f)(v)$. The relaxation time approximation describes the rate of loss of particles $-\nu f$ due to collisions from a small element of phase space while νf_0 represents the corresponding gain of particles as the result of collisions. The detailed dynamics and statistics of the collisions are neglected, as is the fact that the velocity after a collision is correlated with that before. The relaxation operator is purely local and simulates the effect of close binary collisions in which there is a substantial change of velocity. The collisions can be thought of as a Poisson process, occurring with probability νdt in the time interval $[t, t+dt]$, and the relaxation operator establishes a Maxwellian or normal distribution in a time of the order of a few ν^{-1}.

3.6 The Chapman-Enskog Expansion

We have seen that taking moments of the Boltzmann equation leads to a closure problem. A formal approach to solving the Boltzmann equation and closing the moment expansion was developed independently by D. Enskog and S. Chapman.[1] The general approach will be outlined, and then by way of example, the Navier-Stokes equations will be derived using the relaxation time approximation.

We consider an expansion of the distribution function about the equilibrium or Maxwellian distribution f_0 in the form

$$f = f_0 + \varepsilon f_1 + \varepsilon^2 f_2 + \cdots,$$

where $f_1, f_2, \ldots$ are successive corrections to f. The right-hand collisional term is to be regarded as providing the fastest time scale in the problem and an order larger than the left, allowing us to write the equation as

$$\frac{\partial f}{\partial t} + v_i \frac{\partial f}{\partial x_i} = \frac{1}{\varepsilon} Q(f, f)(v).$$

To capture the fast time scale behavior, we need to introduce a multiple scales expansion of the time derivative,

$$\frac{\partial}{\partial t} = \frac{\partial^0}{\partial t} + \varepsilon \frac{\partial^1}{\partial t} + \varepsilon^2 \frac{\partial^2}{\partial t} + \cdots.$$

On using these expansions in the Boltzmann equation, we obtain

$$0 = Q^0(f_0, f_0);$$

$$\frac{\partial^0 f_0}{\partial t} + v_i \frac{\partial f_0}{\partial x_i} = Q^1(f_0, f_1);$$

$$\frac{\partial^0 f_1}{\partial t} + \frac{\partial^1 f_0}{\partial t} + v_i \frac{\partial f_1}{\partial x_i} = Q^2(f_0, f_1, f_2),$$

where $Q^1, Q^2, \ldots$ are appropriate functionals, assumed known. The first of these equations is satisfied automatically by our choice of the Maxwellian distribution. The moment equations applied to the lowest order then give $\partial n_0/\partial t$, $\partial \mathbf{u}_0/\partial t$, and $\partial T_0/\partial t$. Since we therefore know $\partial f_0/\partial t$, the second equation can in principle be solved for f_1. Since f_1 must not contribute to n_0, $\mathbf{u}_0$, and T_0, it is also subject to the constraints

$$\int f_1 d^3v = 0, \quad \int \mathbf{c} f_1 d^3v = 0, \quad \int c^2 f_1 d^3v = 0.$$

---

[1] A wonderful reference to much of this section is the classic monograph by Chapman and Cowling (1970).

3.6 The Chapman-Enskog Expansion

It transpires that this system can be solved uniquely and terms describing viscosity and heat conduction can be derived. This can be laborious for the full collision integral.

Consider, by way of illustration, the Chapman-Enskog approach applied to the Boltzmann equation with a relaxation form of the collisional operator. Suppose then that we consider a gas of particles governed by

$$\frac{\partial f}{\partial t} + v_k \frac{\partial f}{\partial x_k} = Q(f, f) = -\nu(f - f_0),$$

where $\nu = O(1/\varepsilon)$ and the equilibrium distribution is the Boltzmann-Maxwellian distribution,

$$f_0 = n \left(\frac{m}{2\pi kT}\right)^{3/2} \exp\left[-\frac{m(\mathbf{v} - \mathbf{u})^2}{2kT}\right] \quad (3.10)$$

$$= n \left(\frac{\beta}{\pi}\right)^{3/2} e^{-\beta c^2}, \quad (3.11)$$

where $\mathbf{c} = \mathbf{v} - \mathbf{u}$ and $\beta = m/(2kT)$. We expand as before $f = f_0 + \varepsilon f_1 + \varepsilon^2 f_2 + \cdots$ to obtain

$$\frac{\partial f_0}{\partial t} + v_k \frac{\partial f_0}{\partial x_k} = -\nu f_1.$$

Hence, for the relaxation time operator, solving for f_1 is straightforward. The left-hand side of the reduced Boltzmann equation can be evaluated since f_0 is the Maxwellian distribution. Thus,

$$\frac{\partial f_0}{\partial t} = f_0 \left[\frac{1}{n}\frac{\partial n}{\partial t} - \frac{3}{2}\frac{1}{T}\frac{\partial T}{\partial t} + \frac{m}{kT}(\mathbf{v} - \mathbf{u}) \cdot \frac{\partial \mathbf{u}}{\partial t} + \frac{m}{2kT}(\mathbf{v} - \mathbf{u})^2 \frac{1}{T}\frac{\partial T}{\partial t}\right].$$

To evaluate the time derivatives, we employ the zeroth-order or Euler form of the fluid equations (i.e., that are solved exactly by the Maxwell-Boltzmann distribution). We therefore have

$$\frac{\partial n}{\partial t} + \frac{\partial}{\partial x_k}(nu_k) = 0;$$

$$n\left(\frac{\partial u_i}{\partial t} + u_k \frac{\partial u_i}{\partial x_k}\right) = -\frac{1}{m}\frac{\partial p_{ik}}{\partial x_k}; \quad (3.12)$$

$$\frac{\partial}{\partial t}(nkT) + u_k \frac{\partial}{\partial x_k}(nkT) + \frac{5}{3}nkT\frac{\partial u_i}{\partial x_i} = 0,$$

from which we can express the time derivatives in terms of spatial derivatives, i.e.,

$$\frac{1}{n}\frac{\partial n}{\partial t} = -\frac{\partial u_k}{\partial x_k} - \frac{u_k}{n}\frac{\partial n}{\partial x_k};$$

$$\frac{1}{T}\frac{\partial T}{\partial t} = -\frac{2}{3}\frac{\partial u_k}{\partial x_k} - \frac{u_k}{T}\frac{\partial T}{\partial x_k};$$

$$\frac{\partial u_i}{\partial t} = -u_k\frac{\partial u_i}{\partial x_k} - \frac{1}{mn}\frac{\partial p}{\partial x_i}.$$

This then yields, assuming $p = nkT$,

$$\frac{1}{f_0}\frac{\partial f_0}{\partial t} = -u_k\left(\frac{mc^2}{2kT} - \frac{3}{2}\right)\frac{1}{T}\frac{\partial T}{\partial x_k} - \frac{c_k}{T}\frac{\partial T}{\partial x_k}$$
$$- \frac{m}{kT}\left(c_i u_k + \frac{1}{3}c^2\delta_{ki}\right)\frac{\partial u_i}{\partial x_k} - \frac{u_k}{n}\frac{\partial n}{\partial x_k} - \frac{c_i}{n}\frac{\partial n}{\partial x_i}.$$

By taking the spatial derivative of f_0, we can derive after a little algebra,

$$v_k \frac{\partial f_0}{\partial x_k} = f_0\left[\frac{v_k}{n}\frac{\partial n}{\partial x_k} + v_k\left(\frac{mc^2}{2kT} - \frac{3}{2}\right)\frac{1}{T}\frac{\partial T}{\partial x_k} + \frac{m}{kT}v_k c_i \frac{\partial u_i}{\partial x_k}\right].$$

Combining these expression yields

$$-vf_1 = f_0\left[\frac{m}{kT}\left(c_i c_k - \frac{1}{3}c^2\delta_{ki}\right)\frac{\partial u_i}{\partial x_k} + c_k\left(\frac{mc^2}{2kT} - \frac{5}{2}\right)\frac{1}{T}\frac{\partial T}{\partial x_k}\right].$$

It is easily established that f_1 does not introduce collisional source terms.

For the Euler equations, we have the relations (which is seen by taking moments of the Maxwell-Boltzmann equation)

$$P_{ij} = p\delta_{ij} \text{ and } q_i = 0,$$

for the pressure tensor and the heat flux vector. To determine the corrections to the Euler equations, we need to evaluate

$$p_{ij}^1 = m\int c_i c_j f_1 d^3v, \qquad q_i^1 = \frac{m}{2}\int c_i c^2 f_1 d^3v,$$

since $P_{ij} = p\delta_{ij} + p_{ij}^1$ and $q_i = 0 + q_i^1$. Consider the contributions term-by-term and use $dv_1 dv_2 dv_3 = dc_1 dc_2 dc_3$. The first term is

$$\frac{m^2}{kT}\int_{-\infty}^{\infty}\int_{-\infty}^{\infty}\int_{-\infty}^{\infty} c_i c_j c_k c_l f_0 dc_1 dc_2 dc_3 \frac{\partial u_l}{\partial x_k}.$$

For the moment, we suspend the convention that repeated indices implies summation.

3.6 The Chapman-Enskog Expansion

Case 1: $i = j, i \neq k \neq l, p_{ii}^1$:

$$\frac{m^2}{kT} \int_{-\infty}^{\infty}\int_{-\infty}^{\infty}\int_{-\infty}^{\infty} c_i^2 c_k c_l f_0 dc_1 dc_2 dc_3 \frac{\partial u_l}{\partial x_k}$$

$$= \frac{m^2}{kT} \int_{-\infty}^{\infty}\int_{-\infty}^{\infty} c_i^2 c_k n \left(\frac{m}{2\pi kT}\right)^{3/2} e^{-\beta(c_i^2+c_k^2)} \frac{-1}{2\beta} e^{-\beta c_l^2}\Big|_{-\infty}^{\infty} dc_i dc_k \frac{\partial u_l}{\partial x_k}$$

$$= 0.$$

Case 2: $i = j, i \neq k = l, p_{ii}^1$:

$$\frac{m^2}{kT} \int_{-\infty}^{\infty}\int_{-\infty}^{\infty}\int_{-\infty}^{\infty} c_i^2 c_k^2 f_0 dc_1 dc_2 dc_3 \frac{\partial u_k}{\partial x_k}$$

$$= \frac{m^2}{kT} n \left(\frac{m}{2\pi kT}\right)^{3/2} \frac{\sqrt{\pi}}{2\beta^{3/2}} \frac{\sqrt{\pi}}{2\beta^{3/2}} \sqrt{\frac{\pi}{\beta}} \frac{\partial u_k}{\partial x_k} = nkT \frac{\partial u_k}{\partial x_k}.$$

Case 3: $i = k, j = l, p_{ij}^1$:

$$\frac{m^2}{kT} \int_{-\infty}^{\infty}\int_{-\infty}^{\infty}\int_{-\infty}^{\infty} c_i^2 c_j^2 f_0 dc_1 dc_2 dc_3 \frac{\partial u_j}{\partial x_i} = nkT \frac{\partial u_j}{\partial x_i}.$$

Case 4: $i = l, j = j, p_{ij}^1$:

$$\frac{m^2}{kT} \int_{-\infty}^{\infty}\int_{-\infty}^{\infty}\int_{-\infty}^{\infty} c_i^2 c_j^2 f_0 dc_1 dc_2 dc_3 \frac{\partial u_j}{\partial x_i} = nkT \frac{\partial u_j}{\partial x_i}.$$

Case 5: $i \neq j, k = l, p_{ij}^1$:

$$\frac{m^2}{kT} \int_{-\infty}^{\infty}\int_{-\infty}^{\infty}\int_{-\infty}^{\infty} c_i c_j c_k^2 f_0 dc_1 dc_2 dc_3 \frac{\partial u_j}{\partial x_i} = 0.$$

Consider the term $\frac{1}{3} c^2 \delta_{kl} \partial u_l / \partial x_k$. The pressure moment then yields

$$\frac{1}{3} \frac{m^2}{kT} \int_{-\infty}^{\infty}\int_{-\infty}^{\infty}\int_{-\infty}^{\infty} c_i c_j c^2 f_0 dc_1 dc_2 dc_3 \frac{\partial u_k}{\partial x_k}.$$

Case 1: $i \neq j, p_{ij}^1$:

$$\frac{1}{3} \frac{m^2}{kT} \int_{-\infty}^{\infty}\int_{-\infty}^{\infty}\int_{-\infty}^{\infty} c_i c_j (c_1^2 + c_2^2 + c_3^2) f_0 dc_1 dc_2 dc_3 \frac{\partial u_k}{\partial x_k} = 0.$$

Case 2: $i = j$, p_{ii}^1:

$$\frac{1}{3}\frac{m^2}{kT}\int_{-\infty}^{\infty}\int_{-\infty}^{\infty}\int_{-\infty}^{\infty} c_i^2(c_1^2 + c_2^2 + c_3^2) f_0 dc_1 dc_2 dc_3 \frac{\partial u_k}{\partial x_k}$$

$$= \frac{2}{3} nkT \frac{\partial u_k}{\partial x_k} + \frac{1}{3}\frac{m^2}{kT}\int_{-\infty}^{\infty}\int_{-\infty}^{\infty}\int_{-\infty}^{\infty} c_i^4 f_0 dc_1 dc_2 dc_3 \frac{\partial u_k}{\partial x_k}$$

$$= \frac{2}{3} nkT \frac{\partial u_k}{\partial x_k} + nkT \frac{\partial u_k}{\partial x_k}.$$

Hence, collecting terms $\propto \partial u_l / \partial x_k$ yields

$$\frac{m}{kT}\int_{-\infty}^{\infty}\int_{-\infty}^{\infty}\int_{-\infty}^{\infty} c_i c_j \left(c_k c_l - \frac{1}{3} c^2 \delta_{kl} \right) f_0 d^3 c \frac{\partial u_l}{\partial x_k}$$

$$= nkT \left(\frac{\partial u_i}{\partial x_j} + \frac{\partial u_j}{\partial x_i} - \frac{2}{3}\frac{\partial u_i}{\partial x_i} \right).$$

Consider now terms $\propto (1/T)\partial T/\partial x_k$, so we need evaluate terms like

$$\frac{m}{2kT}\int_{-\infty}^{\infty}\int_{-\infty}^{\infty}\int_{-\infty}^{\infty} c_i c_j c_k c^2 f_0 d^3 c;$$

$$\frac{5}{2}\int_{-\infty}^{\infty}\int_{-\infty}^{\infty}\int_{-\infty}^{\infty} c_i c_j c_k f_0 d^3 c.$$

It is easily checked that these terms are all zero. Consequently, the pressure can be expressed as the sum of the zeroth-order and first-order terms,

$$p_{ij} = p\delta_{ij} - \mu \left(\frac{\partial u_i}{\partial x_j} + \frac{\partial u_j}{\partial x_i} - \frac{2}{3}\delta_{ij}\frac{\partial u_k}{\partial x_k} \right),$$

where

$$\mu = \frac{mnkT}{\nu},$$

is the coefficient of viscosity. The term in brackets is the rate-of-strain tensor. It is left as an exercise to show that the heat flux vector can be written as

$$q_i = -\lambda \frac{\partial T}{\partial x_i}, \quad \lambda = \frac{5}{2}\frac{nk^2 T}{m\nu},$$

and λ is the coefficient of heat conduction.

3.6 The Chapman-Enskog Expansion

The Navier-Stokes equations can therefore be expressed as

$$\frac{\partial n}{\partial t} + \frac{\partial}{\partial x_i}(nu_i) = 0;$$

$$n\left(\frac{\partial u_i}{\partial t} + u_k \frac{\partial u_i}{\partial x_k}\right) = -\frac{1}{m}\frac{\partial p}{\partial x_i} + \frac{\partial}{\partial x_j}\left[\mu\left(\frac{\partial u_i}{\partial x_j} + \frac{\partial u_j}{\partial x_i} - \frac{2}{3}\delta_{ij}\frac{\partial u_i}{\partial x_i}\right)\right];$$

$$n\left(\frac{\partial e}{\partial t} + u_k \frac{\partial e}{\partial x_k}\right) = \nabla \cdot (\lambda \nabla T) - p\nabla \cdot \mathbf{u} + \Phi, \qquad (3.13)$$

where Φ is the viscous flux.

Exercises

1. Complete the details for the derivation of the expressions above for $\partial f_0/\partial t$ and $\partial f_0/\partial x_k$. Use these results to complete the derivation of the expression for f_1.
2. Consider the 1D pdf

$$f(x) = \sqrt{\frac{\beta}{\pi}} e^{-\beta x^2}. \qquad -\infty < x < \infty$$

Show that the moment generating function is given by $M(t) = \exp(t^2/4\beta)$. Derive the expectations $E(X)$, $E(X^2)$, $E(X^3)$, $E(X^4)$, $E(X^5)$, and $E(X^6)$. Hence show that the integrals

$$\frac{\sqrt{\pi}}{2\beta^{3/2}} = \int_{-\infty}^{\infty} x^2 e^{-\beta x^2} dx, \qquad \frac{3\sqrt{\pi}}{4\beta^{5/2}} = \int_{-\infty}^{\infty} x^4 e^{-\beta x^2} dx,$$

$$\frac{15\sqrt{\pi}}{8\beta^{7/2}} = \int_{-\infty}^{\infty} x^6 e^{-\beta x^2} dx.$$

3. Show that the Chapman-Enskog expression for f_1 satisfies the constraints

$$\int f_1 d^3v = 0, \quad \int \mathbf{c} f_1 d^3v = 0, \quad \int c^2 f_1 d^3v = 0.$$

4. Show that the terms $\propto (1/T)\partial T/\partial x_k$ in the pressure moment term vanish identically.
5. Show that the heat flux vector is given by

$$q_i = -\lambda \frac{\partial T}{\partial x_i}, \quad \lambda = \frac{5}{2}\frac{nk^2 T}{mv},$$

where λ is the coefficient of heat conduction.

3.7 Application 1: Structure of Weak Shock Waves

By way of application, consider the structure of weak shock waves derived as solutions to first, the 1D Euler equations, and then the 1D Navier-Stokes equations. The 1D Euler equations for an arbitrary adiabatic index γ are given by

$$\frac{\partial \rho}{\partial t} + \frac{\partial}{\partial x}(\rho u) = 0; \tag{3.14}$$

$$\rho\left(\frac{\partial u}{\partial t} + u\frac{\partial u}{\partial x}\right) = -\frac{\partial p}{\partial x}; \tag{3.15}$$

$$\frac{\partial p}{\partial t} + u\frac{\partial p}{\partial x} + \gamma p\frac{\partial u}{\partial x} = 0, \tag{3.16}$$

where ρ denotes the mass density, u the flow velocity, and p the adiabatic pressure. Consider a characteristic time T and length scale L such that a characteristic phase velocity V_p, to be identified, satisfies the relation

$$\frac{V_p T}{L} = 1.$$

This allows the Euler equations (3.14)–(3.16) to be expressed in dimensionless form using the variables

$$\bar{x} = x/L \Rightarrow \frac{\partial}{\partial x} = \frac{1}{L}\frac{\partial}{\partial \bar{x}} \quad \bar{t} = t/T \Rightarrow \frac{\partial}{\partial t} = \frac{1}{T}\frac{\partial}{\partial \bar{t}}$$

$$\bar{\rho} = \rho/\rho_0; \quad \bar{u} = u/V_p; \quad \bar{p} = p/p_0,$$

where ρ_0 and p_0 are equilibrium values far upstream of any shock transition. We then have

$$\frac{\partial \bar{\rho}}{\partial \bar{t}} + \frac{\partial}{\partial \bar{x}}(\bar{\rho}\bar{u}) = 0;$$

$$\bar{\rho}\left(\frac{\partial \bar{u}}{\partial \bar{t}} + \bar{u}\frac{\partial \bar{u}}{\partial \bar{x}}\right) = -\frac{a_{c0}^2}{\gamma V_p^2}\frac{\partial \bar{p}}{\partial \bar{x}};$$

$$\frac{\partial \bar{p}}{\partial \bar{t}} + \bar{u}\frac{\partial \bar{p}}{\partial \bar{x}} + \gamma \bar{p}\frac{\partial \bar{u}}{\partial \bar{x}} = 0,$$

where the square of the background sound speed $a_{c0}^2 \equiv \gamma p_0/\rho_0$ has been introduced. If we make no assumptions about the magnitude of the normalized sound speed, then there are no natural time or length scales in the system. Nonetheless, we expect that the quadratically nonlinear terms in (3.14)–(3.16) will lead to the "steepening" of wave forms over a long time. Introduce therefore the *fast* and *slow* variables

3.7 Application 1: Structure of Weak Shock Waves

$$\xi = \bar{x} - \bar{t}; \quad \tau = \varepsilon \bar{t} \Rightarrow \frac{\partial}{\partial \bar{x}} = \frac{\partial \xi}{\partial \bar{x}} \frac{\partial}{\partial \xi} = \frac{\partial}{\partial \xi}; \quad \frac{\partial}{\partial \bar{t}} = \varepsilon \frac{\partial}{\partial \tau} - \frac{\partial}{\partial \xi},$$

and expand the flow variables about a uniform far-upstream background,

$$\bar{\rho} = 1 + \varepsilon \rho_1 + \varepsilon^2 \rho_2 \cdots ;$$
$$\bar{u} = \varepsilon u_1 + \varepsilon^2 u_2 + \cdots ;$$
$$\bar{p} = 1 + \varepsilon p_1 + \varepsilon^2 p_2 + \cdots .$$

On substituting the new time and spatial coordinates τ and ξ and expanding the state variables, we get, for example,

$$\left(\varepsilon \frac{\partial}{\partial \tau} - \frac{\partial}{\partial \xi}\right) \left(1 + \varepsilon \rho_1 + \varepsilon^2 \rho_2 \cdots\right) +$$
$$\frac{\partial}{\partial \xi} \left[\left(1 + \varepsilon \rho_1 + \varepsilon^2 \rho_2 \cdots\right) \left(\varepsilon u_1 + \varepsilon^2 u_2 + \cdots\right)\right] = 0.$$

Collecting terms of different orders of ε for each of the normalized Euler equations (3.14)–(3.16), we find (Exercise)

$$O(\varepsilon): \quad -\frac{\partial \rho_1}{\partial \xi} + \frac{\partial u_1}{\partial \xi} = 0;$$

$$-\frac{\partial u_1}{\partial \xi} + \frac{\bar{a}_{c0}^2}{\gamma} \frac{\partial p_1}{\partial \xi} = 0;$$

$$-\frac{\partial p_1}{\partial \xi} + \gamma \frac{\partial u_1}{\partial \xi} = 0,$$

which, for uniform upstream conditions, yields the relations

$$p_1 = \gamma u_1; \quad \rho_1 = u_1; \quad u_1 = \frac{\bar{a}_{c0}^2}{\gamma} p_1.$$

The last of the relations imposes a compatibility condition on the sound and phase speed since

$$u_1 = \frac{\bar{a}_{c0}^2}{\gamma} \gamma u_1 \Leftrightarrow a_{c0}^2 = V_p^2,$$

corresponding to the dispersion relation of the system. The expansion is therefore following the propagation of a sound mode in the fluid. To determine the nonlinear evolution of the wave, consider the expansion at order ε^2,

$$O(\varepsilon^2): \quad -\frac{\partial p_2}{\partial \xi} + \frac{\partial u_2}{\partial \xi} = -\frac{\partial \rho_1}{\partial \tau} - \frac{\partial}{\partial \xi}(\rho_1 u_1);$$

$$-\frac{\partial u_2}{\partial \xi} + \frac{\bar{a}_{c0}^2}{\gamma}\frac{\partial p_2}{\partial \xi} = -\frac{\partial u_1}{\partial \tau} + \rho_1 \frac{\partial u_1}{\partial \xi} - u_1 \frac{\partial u_1}{\partial \xi};$$

$$-\frac{\partial p_2}{\partial \xi} + \gamma \frac{\partial u_2}{\partial \xi} = -\frac{\partial p_1}{\partial \tau} - u_1 \frac{\partial p_1}{\partial \xi} - \gamma p_1 \frac{\partial u_1}{\partial \xi}.$$

By substituting for ρ_1 and p_1 using the $O(\varepsilon)$ relations, we may combine the $O(\varepsilon^2)$ equations as a single nonlinear partial differential equation in u_1 after eliminating the second terms with the dispersion relation (Exercise)

$$\frac{\partial u_1}{\partial \tau} + \frac{\gamma+1}{2} u_1 \frac{\partial u_1}{\partial \xi} = 0. \qquad (3.17)$$

This is the simplest quasilinear nonlinear wave equation and is sometimes called the *inviscid* form of Burgers' equation. Suppose we have initial data

$$u_1(\xi, 0) = f(\xi),$$

where $f(\xi)$ is a given smooth function. To solve the initial value problem generally, we parameterize the initial curve as

$$\xi = \eta, \quad \tau = 0, \quad u_1 = f(\eta).$$

The characteristic equations are

$$\frac{d\xi}{ds} = \frac{\gamma+1}{2} u_1, \quad \frac{d\tau}{ds} = 1, \quad \frac{du_1}{ds} = 0,$$

with initial conditions at $s = 0$ given by the initial curve. Evidently, the last relation shows that $u_1(s, \eta)$ is constant along the characteristic curves. Therefore,

$$u_1(s, \eta) = u_1(0, \eta) = f(\eta).$$

We can therefore immediately solve for ξ and τ, finding that

$$\xi(s, \eta) = \eta + s\frac{\gamma+1}{2} f(\eta), \quad \tau(s, \eta) = s.$$

This system can be inverted generally to give $s = \tau$ and $\eta = \eta(\xi, \tau)$ and

$$u_1 = f[\eta(\xi, \tau)],$$

3.7 Application 1: Structure of Weak Shock Waves

or since $s = \tau$ and $\eta = \xi - s(\gamma + 1)f(\eta)/2 = \xi - (\gamma + 1)/2 \cdot u_1\tau$, we have the implicit solution

$$u_1 = f\left(\xi - (\gamma + 1)/2 \cdot u_1\tau\right).$$

An obvious but important point about the solution $u_1 = u_1(\xi, \tau)$ is that the two characteristic equations imply

$$\frac{d\xi}{d\tau} = \frac{\gamma + 1}{2} u_1.$$

Thus, the greater the amplitude $|u_1(\xi, \tau)|$ of the wave, the greater the speed of the corresponding point x on the wave. Assuming $u_1 > 0$, points ξ where $u_1(\xi, \tau)$ has larger values and the wave is higher move more rapidly to the right than points ξ where $u_1(\xi, \tau)$ has smaller values and the wave is of lower amplitude. If, initially, there are portions of the wave form located to the left or the rear of the lower portions, the higher points may eventually catch up and pass the lower points, at which time the wave is said to *break*. At the time of breaking, the wave is multi-valued and is no longer a valid solution of the inviscid Burgers' equation. To avoid wave breaking and the introduction of multi-valued solutions, we can try to introduce higher-order derivatives that can act to smooth the discontinuity. This is discussed below on the basis of the Navier-Stokes equations.

Alternatively, we can introduce a discontinuous solution – a *shock wave* – that extends the validity of the solution beyond the breaking time. To determine the breaking time, we use $u_1(s, \eta) = f(\eta)$ and $\xi = \eta + (\gamma + 1)/2 \cdot u_1\tau$ and implicit differentiation to find the slope of the wave

$$\frac{\partial u_1}{\partial \xi} = \frac{f'(\eta)}{1 + [(\gamma + 1)/2]\tau f'(\eta)}.$$

Hence for $f'(\eta) < 0$, the slope u_{1x} becomes infinite and the wave begins to break when

$$\tau = -\frac{2}{\gamma + 1} \frac{1}{f'(\eta)},$$

has the smallest non-negative value τ_0. This condition corresponds to intersection of the characteristics.

To extend the solution beyond the breaking time, we need to define *weak solutions* which exist beyond the time τ_0. We can write the wave equation (3.17) in conservation form

$$(u_1)_\tau + \left(\frac{1}{2}u_1^2\right)_\chi = 0,$$

where we introduced the normalization $\chi = \xi/[(\gamma + 1)/2]$ for convenience.

3.7.1 Weak Solutions and the Rankine-Hugoniot Relations

In view of the inviscid form of Burgers' equation above, consider more generally nonlinear equations expressed in the conservation form

$$u_t + (f(u))_x = 0. \tag{3.18}$$

For notational convenience, let $\mathbf{F} \equiv (f(u), u)$ and note that

$$\text{Div}\mathbf{F} = 0,$$

is equivalent to (3.18) provided we introduce the spacetime divergence $\text{Div}(f_1, f_2) \equiv (f_1)_x + (f_2)_t$. Let ϕ be a smooth function with compact support in the (x, t) plane, meaning that ϕ is zero outside a compact set. We may therefore express (3.18) in the weak form

$$\int \phi \cdot \text{Div}\mathbf{F} \, dx \, dt = 0,$$

for all ϕ. On integrating by parts, and using the compact support of ϕ, yields at once

$$\int \text{Grad}\phi \cdot \mathbf{F} \, dx \, dt = 0. \tag{3.19}$$

If u is smooth, then the *differential form* (3.18) and the integral form (3.19) are equivalent. However, if u is not smooth, then the latter expression (3.19) remains valid unlike the differential expression (3.18). A *weak solution* of the differential equation (3.18) is one that satisfies (3.19) for all smooth ϕ with compact support.

Besides the differential (3.18) and weak form (3.19), we can also express (3.18) as an *integral form*. Suppose we consider the interval $[a, b]$ on the x-axis, so that from (3.18)

$$\frac{d}{dt} \int_a^b u \, dx = \int_a^b u_t \, dx = -\int_a^b (f(u))_x \, dx = -f(u)\big|_a^b.$$

A weak solution does not have to be differentiable, and neither does a function that satisfies the integral form of the equations. A weak solution does satisfy the integral form of the equation, and therefore weak solutions are the objects that we seek when a flow is discontinuous. This result is straightforward to show in general but we omit the details here.

Let us consider the conservation law (3.18) and the weak formulation (3.19) in the presence of a jump discontinuity. Suppose u is a weak solution with a jump discontinuity across a smooth curve Σ in the (x, t) plane, and let ϕ be a smooth function with compact support on the closed region S. This is illustrated in Fig. 3.5. The closed region can be expressed as $S = S_1 \cup S_2$. Then

3.7 Application 1: Structure of Weak Shock Waves

Fig. 3.5 Spacetime plot illustrating a region S in which ϕ is a smooth function that vanishes outside S. The jump discontinuity is denoted by Σ

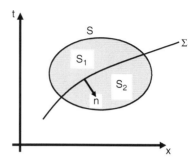

$$0 = \int \operatorname{Grad}\phi \cdot \mathbf{F} \, dx\, dt = \int_{S_1} \operatorname{Grad}\phi \cdot \mathbf{F} \, dx\, dt + \int_{S_2} \operatorname{Grad}\phi \cdot \mathbf{F} \, dx\, dt.$$

For u smooth in the regions S_1 and S_2, we have

$$\int_{S_1} \operatorname{Grad}\phi \cdot \mathbf{F} \, dx\, dt = \int_{S_1} \operatorname{Div}(\phi\mathbf{F}) \, dx\, dt - \int_{S_1} \phi \operatorname{Div}\mathbf{F} \, dx\, dt$$

$$= \int_{\Sigma} \phi \mathbf{F} \cdot \mathbf{n} \, ds - \int_{S_1} \phi \operatorname{Div}\mathbf{F} \, dx\, dt.$$

The last term above is zero since in the region S_1 where u is smooth, $\operatorname{Div}\mathbf{F} = 0$, thus yielding

$$\int_{S_1} \operatorname{Grad}\phi \cdot \mathbf{F} \, dx\, dt = \int_{\Sigma} \phi \mathbf{F}_1 \cdot \mathbf{n} \, ds,$$

where the notation $\mathbf{F}_1$ means that u is evaluated by taking the limit from the region S_1. In similar fashion, one has

$$\int_{S_2} \operatorname{Grad}\phi \cdot \mathbf{F} \, dx\, dt = -\int_{\Sigma} \phi \mathbf{F}_2 \cdot \mathbf{n} \, ds,$$

and the negative sign is because the outward normal $\mathbf{n}$ for S_1 is the inward normal for S_2. We therefore immediately obtain

$$\int_{\Sigma} \phi \left(\mathbf{F}_1 - \mathbf{F}_2 \right) \cdot \mathbf{n} \, ds = 0,$$

which is valid for all ϕ. It therefore follows that the jump condition

$$[\mathbf{F} \cdot \mathbf{n}] = 0$$

holds generally on the curve Σ, and $[\mathbf{F} \cdot \mathbf{n}] \equiv \mathbf{F}_1 \cdot \mathbf{n} - \mathbf{F}_2 \cdot \mathbf{n}$ denotes the jump in $\mathbf{F} \cdot \mathbf{n}$ across Σ.

Suppose we parameterize Σ by $x = x(t)$ so that the speed of the discontinuity is $s \equiv dx/dt$. The normal $\mathbf{n}$ may then be expressed as

$$\mathbf{n} = \frac{(1, -s)}{\sqrt{s^2 + 1}},$$

and with $\mathbf{F} = (f(u), u)$, we can express the jump condition as

$$-s[u] + [f(u)] = 0 \tag{3.20}$$

on Σ. Equation (3.20) is the constraint that the weak form of the equations (3.19) imposes on the values of u on both sides of the discontinuity. Notice that for the inviscid form of Burgers' equation, we have

$$-s[u] + \frac{1}{2}[u^2] = 0. \tag{3.21}$$

Hence,

$$s = \frac{u_0 + u_1}{2}, \tag{3.22}$$

is the shock jump relation for the inviscid Burgers' equation, connecting the speed of propagation s of the discontinuity with the amounts by which the velocity u jumps. The subscripts 0 (front) and 1 (back) denote the different sides of the discontinuity.

The above analysis holds exactly for systems of conservation laws where we use the vector unknown $\mathbf{u}$. The Euler equations are an example, and in this case satisfy the one-dimensional *Rankine-Hugoniot* conditions

$$\begin{aligned} s[\rho] &= [\rho u] \equiv [m]; \\ s[\rho u] &= [\rho u^2 + P]; \\ s[e] &= [(e + p)u], \end{aligned} \tag{3.23}$$

where we have introduced the total energy

$$e \equiv \frac{1}{2}\rho u^2 + \frac{P}{\gamma - 1} = \frac{1}{2}\rho u^2 + \rho \varepsilon.$$

Here, $\varepsilon = P/(\rho(\gamma - 1))$ is the expression for the internal energy.

Since the Euler equations are Galilean invariant, we may transform the Rankine-Hugoniot conditions into a coordinate system moving with a uniform velocity such that the speed of the discontinuity is 0. The steady-state Rankine-Hugoniot conditions can then be written as

$$\rho_0 u_0 = \rho_1 u_1; \tag{3.24}$$

$$\rho_0 u_0^2 + p_0 = \rho_1 u_1^2 + p_1; \tag{3.25}$$

$$(e_0 + p_0) u_0 = (e_1 + p_1) u_1. \tag{3.26}$$

3.7 Application 1: Structure of Weak Shock Waves

As defined above, if we let $m = \rho_0 u_0 = \rho_1 u_1$, we can distinguish between two classes of discontinuity. If $m = 0$, the discontinuity is a *contact discontinuity* or *slip line*. Since $u_0 = u_1 = 0$, these discontinuities convect with the fluid. From (3.25), we observe that $p_0 = p_1$ across a contact discontinuity but in general $\rho_0 \neq \rho_1$. By contrast, if $m \neq 0$, then the discontinuity is called a *shock wave*. Since $u_0 \neq 0$ and $u_1 \neq 0$, the gas crosses the shock, or equivalently, the shock propagates through the fluid. The side of the shock that comprises gas that has not been shocked is the front or upstream of the shock, and the shocked gas is the back or downstream of the shock. A detailed discussion of the properties of gas shocks based on the Euler equations can be found in Landau and Lifshitz (2000).

Exercises

1. Explicitly derive the $O(\varepsilon)$ and $O(\varepsilon^2)$ expansions of the Euler equations.
2. As outlined in the text, derive the nonlinear wave equation (3.17).
3. Solve the linear wave equation

$$u_t + c u_x = 0, \quad c = \text{const.}$$

with $u(x, t = 0) = f(x)$. Write down the solution if $f(x) = \sin kx$.

4. Consider the initial data

$$U(x, 0) = \begin{cases} 0, & x \geq 0, \\ 1 & x < 0 \end{cases}$$

for the partial differential equation written in conservative form

$$U_t + \left(\frac{1}{2}U^2\right)_x = 0.$$

Sketch the characteristics. What is the shock propagation speed necessary to prevent the characteristics from crossing?

5. From the stationary Rankine-Hugoniot conditions (3.24)–(3.26), show that

$$m^2 = \frac{p_0 - p_1}{\tau_0 - \tau_1},$$

where $\tau \equiv 1/\rho$. Show that

$$e_0 \tau_0 - e_1 \tau_1 = p_1 \tau_1 - p_0 \tau_0,$$

and hence that

$$\varepsilon_1 - \varepsilon_0 + \frac{p_0 + p_1}{2}(\tau_1 - \tau_0) = 0,$$

(the *Hugoniot* equation for the shock) where $\varepsilon \equiv \frac{1}{\gamma - 1} p \tau$.

We have seen that the weak shock limit of the Euler equations leads to an equation that required the introduction of discontinuous shock jumps to connect solutions when the wave breaking occurs. The inviscid Burgers' equation might be thought of as the limit in which the coefficient of a higher order derivative vanishes. This suggests that we should consider higher-order corrections to the Euler equations that admit second-order and possibly higher order derivatives. As we have seen, the simplest example is the Navier-Stokes equations. Consider the 1D form of the equations. Then

$$\sum_i p_{ii} = 3p - \sum_i \mu \left(\frac{\partial u_i}{\partial x_i} + \frac{\partial u_i}{\partial x_i} - \frac{2}{3} \frac{\partial u_k}{\partial x_k} \delta_{ii} \right) = 3p.$$

The internal energy is therefore simply

$$\varepsilon = \frac{3}{2} \frac{p}{n}.$$

Finally, we need

$$\sum_j u_j p_{ij} = u p_{11} = up - u \frac{4}{3} \mu \frac{\partial u}{\partial x},$$

from which we obtain the total energy equation as

$$\frac{\partial}{\partial t} \left[n \left(\varepsilon + \frac{1}{2} u^2 \right) \right]$$
$$+ \frac{\partial}{\partial x} \left[nu \left(\varepsilon + \frac{1}{2} u^2 \right) + up - u \frac{4}{3} \mu \frac{\partial u}{\partial x} - \lambda \frac{\partial T}{\partial x} \right] = 0.$$

The corresponding momentum equation can be expressed as

$$\rho \left(\frac{\partial u}{\partial t} + u \frac{\partial u}{\partial x} \right) = -\frac{\partial p}{\partial x} + \frac{\partial}{\partial x} \left(\frac{4}{3} \mu \frac{\partial u}{\partial x} \right).$$

On expanding the energy equation and using the continuity equation and the momentum equation and the result

$$n \frac{De}{Dt} = \frac{3}{2} \frac{Dp}{Dt} + \frac{3}{2} p \frac{\partial u}{\partial x},$$

where $D/Dt = \partial/\partial t + u \partial/\partial x$ is the convective derivative, we find

$$\frac{3}{2} \frac{Dp}{Dt} + \frac{5}{2} p \frac{\partial u}{\partial x} = \frac{4}{3} \mu \left(\frac{\partial u}{\partial x} \right)^2 + \frac{\partial}{\partial x} \left(\lambda \frac{\partial T}{\partial x} \right).$$

3.7 Application 1: Structure of Weak Shock Waves

To summarize, the 1D Navier-Stokes equations are therefore given by the higher-order system of equations

$$\frac{\partial \rho}{\partial t} + \frac{\partial}{\partial x}(\rho u) = 0; \tag{3.27}$$

$$\rho\left(\frac{\partial u}{\partial t} + u\frac{\partial u}{\partial x}\right) = -\frac{\partial p}{\partial x} + \frac{\partial}{\partial x}\left(\frac{4}{3}\mu \frac{\partial u}{\partial x}\right); \tag{3.28}$$

$$\frac{\partial p}{\partial t} + u\frac{\partial p}{\partial x} + \gamma p \frac{\partial u}{\partial x} = \frac{8}{9}\mu\left(\frac{\partial u}{\partial x}\right)^2 + \frac{\partial}{\partial x}\left(\frac{2}{3}\lambda \frac{\partial T}{\partial x}\right), \tag{3.29}$$

and $p = nkT$.

We normalize the Navier-Stokes equations in the same way that we normalized the Euler equations except that we introduce $\bar{p} = p/\rho_0 V_p^2$. This yields

$$\frac{\partial \rho}{\partial t} + \frac{\partial}{\partial x}(\rho u) = 0;$$

$$\rho\left(\frac{\partial u}{\partial t} + u\frac{\partial u}{\partial x}\right) = -\frac{\partial p}{\partial x} + \frac{\partial}{\partial x}\left(\frac{4}{3}\frac{1}{Re}\frac{\partial u}{\partial x}\right);$$

$$\frac{\partial p}{\partial t} + u\frac{\partial p}{\partial x} + \gamma p \frac{\partial u}{\partial x} = \frac{8}{9}\frac{1}{Re}\left(\frac{\partial u}{\partial x}\right)^2 + \frac{\partial}{\partial x}\left(\frac{5}{3}\chi \frac{\partial T}{\partial x}\right),$$

where for convenience we have omitted the bars. We have also used the normalization $T_0 = mV_p^2/k$, where k is Boltzmann's constant. In the normalized equations, we have introduced the dimensionless Reynolds number

$$Re \equiv \frac{\rho_0 V_p L}{\mu},$$

which is therefore a measure of the relative importance of a fluids inertia and viscosity. Thus, if viscous forces dominate, as in the case of a flow very near a body, then Re is small. If on the other hand, inertial effects dominate, then Re is large. Problems for which Re are typically large include turbulent flows, inviscid flows, potential flows, and flows well removed from boundaries. Cases where Re is small include laminar flows, bubble flows, and flows close to a boundary. The remaining nondimensional number is related to the Prandtl number, and is given by

$$\chi \equiv \frac{\lambda}{V_p L \rho_0} \frac{2m}{5k},$$

and $2m/5k$ is the specific heat of a monatomic gas at constant pressure. The Prandtl number is essentially the ratio of the heat conductivity and the viscosity,

$$Pr = \frac{\lambda}{\mu}\frac{2m}{5k},$$

so that
$$\chi^{-1} = \frac{1}{\mu Pr} \frac{5}{2} \rho_0 V_p L.$$

To derive the weak shock equation from the Navier-Stokes equations, we use again the multiple scales method. An important step in the multiple scales analysis is the relative ordering of the dissipative or dispersive terms, both of which are assumed to be small. To isolate the richest possible evolution equation from the Navier-Stokes equations, we impose the most general scaling on Re^{-1} and χ, which is that both are $O(\varepsilon)$. As before, we use the slow and fast scales and expand the flow variables about a uniform far-upstream background,

$$\bar{\rho} = 1 + \varepsilon \rho_1 + \varepsilon^2 \rho_2 \cdots ;$$
$$\bar{u} = \varepsilon u_1 + \varepsilon^2 u_2 + \cdots ;$$
$$\bar{p} = p_0 + \varepsilon p_1 + \varepsilon^2 p_2 + \cdots ;$$
$$kT = \frac{p}{n} = p_0 + \varepsilon(p_1 - p_0 n_1) + \varepsilon^2(p_2 - n_1 p_1 + p_0 n_1^2) + \cdots$$
$$= k(T_0 + \varepsilon T_1 + \cdots).$$

On expanding to $O(\varepsilon)$, we obtain

$$\frac{\partial \rho_1}{\partial \xi} = \frac{\partial u_1}{\partial \xi}; \quad \frac{\partial u_1}{\partial \xi} = \frac{\partial p_1}{\partial \xi}; \quad \frac{\partial p_1}{\partial \xi} = \gamma p_0 \frac{\partial u_1}{\partial \xi}.$$

These are the same as for the Euler equation case discussed above. This yields the eigenrelations and acoustic dispersion equation of before (although normalized slightly differently)

$$\rho_1 = u_1; \quad p_1 = u_1; \quad p_1 = \gamma p_0 u_1 \Rightarrow \gamma p_0 = 1 \Leftrightarrow V_p^2 = a_{c0}^2.$$

At the $O(\varepsilon^2)$ expansion, we find

$$-\frac{\partial \rho_2}{\partial \xi} + \frac{\partial u_2}{\partial \xi} = -\frac{\partial u_1}{\partial \tau} - \frac{\partial u_1^2}{\partial \xi};$$
$$-\frac{\partial u_2}{\partial \xi} + \frac{\partial p_2}{\partial \xi} = -\frac{\partial u_1}{\partial \tau} - u_1 \frac{\partial u_1}{\partial \xi} - \rho_1 \frac{\partial u_1}{\partial \xi} + \frac{\partial}{\partial \xi}\left(\frac{4}{3}\frac{1}{Re}\frac{\partial u_1}{\partial \xi}\right);$$
$$-\frac{\partial p_2}{\partial \xi} + \gamma p_0 \frac{\partial u_2}{\partial \xi} = -\frac{\partial u_1}{\partial \tau} - (\gamma + 1)u_1 \frac{\partial u_1}{\partial \xi} + \frac{\partial}{\partial \xi}\left(\frac{5}{3}\chi \frac{\partial T_1}{\partial \xi}\right),$$

3.7 Application 1: Structure of Weak Shock Waves

where $kT_1 = p_1 - n_1 = u_1(m-1)/m$, and m is the mass of a gas molecule. Use of the normalized acoustic dispersion relation $1 = \gamma p_0$ yields a general form of Burgers' equation

$$\frac{\partial u_1}{\partial \tau} + \frac{\gamma+1}{2} u_1 \frac{\partial u_1}{\partial \xi} = \frac{\partial}{\partial \xi}\left(\frac{4}{3}\frac{1}{Re}\frac{\partial u_1}{\partial \xi}\right) + \frac{\partial}{\partial \xi}\left(\frac{5}{3}\chi \frac{\partial u_1}{\partial \xi}\right), \qquad (3.30)$$

which resembles the inviscid nonlinear wave equation derived above except that dissipation terms in the form of the second-order spatial derivatives are now present. Physically, the importance of the dissipative terms will depend on the value of the Prandtl number Pr. If $Pr \gg 1$, then the heat conduction term dominates and determines the characteristic length scale for a shock transition.

For notational convenience, consider instead the canonical form of Burgers' equation

$$u_t + uu_x = \kappa u_{xx}, \qquad (3.31)$$

which is straightforwardly obtained from (3.30). The steady-state form of the shock structure equation (3.31) satisfies

$$-V_p u_X + uu_X = \kappa u_{XX}, \quad X \equiv x - V_p t.$$

Hence

$$\frac{1}{2}u^2 - V_p u + C = \kappa u_X.$$

If $u \to u_1, u_2$ as $X \to \pm\infty$ and $u_X(\pm\infty) = 0$, then

$$V_p = \frac{u_1 + u_2}{2}, \quad C = \frac{1}{2}u_1 u_2,$$

and the equation may be written

$$(u - u_1)(u - u_2) = -2\kappa u_X.$$

The solution is simply

$$X = \frac{2\kappa}{u_2 - u_1} \log\left|\frac{u - u_1}{u - u_2}\right|, \quad u_1 \neq u_2,$$

which yields

$$u = u_1 + \frac{u_2 - u_1}{1 - \exp\left[\frac{u_2 - u_1}{2\kappa}(x - V_p t)\right]}, \quad V_p = \frac{u_1 + u_2}{2}.$$

This solution illustrates that the discontinuity of the shock is smoothed by the dissipative term in Burgers' equation and the length scale of the transition is determined by diffusion coefficient κ/V_p. The steady-state Burgers' equation solution can be expressed in terms of a hyperbolic tan (tanh) function (Exercise).

Exercises

1. Show that the Cole-Hopf transformation

$$u = -2\kappa \frac{\phi_x}{\phi},$$

removes the nonlinear term in the Burgers' equation

$$u_t + u u_x = \kappa u_{xx},$$

and yields the heat equation as the transformed equation. For the initial problem $u(x,t=0) = F(x)$, show that this transforms to the initial problem

$$\phi = \Phi(x) = \exp\left[-\frac{1}{2\kappa}\int_0^x F(\eta)d\eta\right], \quad t = 0,$$

for the heat equation. Show that the solution for u is

$$u(x,t) = \frac{\int_{-\infty}^{\infty} \frac{x-\eta}{t} e^{-G/2\kappa} d\eta}{\int_{-\infty}^{\infty} e^{-G/2\kappa} d\eta},$$

where

$$G(\eta; x, t) = \int_0^\eta F(\eta')d\eta' + \frac{(x-\eta)^2}{2t}.$$

2. Show that the characteristic form of the steady Burgers' equation admits a solution that can be expressed as a hyperbolic tan $tanh$ profile given $u(-\infty) = u_0$ and $u(\infty) = u_1$.

3.8 Application 2: The Diffusion and Telegrapher Equations

A basic problem in space physics and astrophysics is the transport of charged particles in the presence of a magnetic field that is ordered on some large scale and highly random and temporal on other smaller scales. This will be considered in more detail later. Here we discuss a simplified form of the Fokker-Planck transport equation that describes particle transport via particle scattering in pitch-angle in a magnetically turbulent medium since it resembles closely the basic Boltzmann

3.8 Application 2: The Diffusion and Telegrapher Equations

equation. In the absence of both focusing and adiabatic energy changes, the BGK form of the Boltzmann equation reduces to the simplest possible integro-differential equation,

$$\frac{\partial f}{\partial t} + \mu v \frac{\partial f}{\partial r} = \frac{\langle f \rangle - f}{\tau}, \tag{3.32}$$

where $f(r, t, \mu, v)$ is a gyrophase averaged velocity distribution function at position r and time t for particles of speed v and pitch-angle cosine $\mu = \cos\theta$. Observe that $\mu \in [-1, 1]$. Here, we are no longer restricting ourselves to the equilibrium Boltzmann distribution but instead define

$$\langle f \rangle \equiv \frac{1}{2} \int_{-1}^{1} f d\mu,$$

as the mean or isotropic distribution function averaged over μ. Finally, as before, τ is the collision time. We follow the approach of Zank et al. (2000).

To solve (3.32), we may exploit $\mu \in [-1, 1]$ and expand $f(r, t, \mu, v)$ in an infinite series of Legendre polynomials $P_n(\mu)$,

$$f = \frac{1}{2} \sum_{n=0}^{\infty} (2n + 1) P_n(\mu) f_n(r, t, v),$$

where $f_n(r, t, v)$ is the nth harmonic of the scattered distribution function

$$f_n(r, t, v) = \int\int f(r, t, \mu, v) P_n(\mu) d\Omega = 2\pi \int_{-1}^{1} P_n(\mu) f(r, t, \mu, v) d\mu,$$

and

$$P_0(\mu) = 1; \quad P_1(\mu) = \mu; \quad P_2(\mu) = \frac{1}{2}(3\mu^2 - 1); \quad \ldots,$$

are the first three Legendre polynomials. The Legendre polynomials are mutually orthogonal, satisfying the orthogonality condition,

$$\int_{-1}^{1} P_m(\mu) P_n(\mu) d\mu = \begin{cases} 0 & m \neq n \\ \frac{2}{2n+1} & m = n \end{cases},$$

and form an infinite basis set about which to expand the distribution function.

By means of the recurrence relation

$$(n + 1) P_{n+1} + n P_{n-1} = (2n + 1) \mu P_n, \quad n = 1, 2, 3, \ldots,$$

we can use the polynomial expansion and the orthogonality of the Legendre polynomials to reduce the BGK Boltzmann equation (3.32) to an infinite set of partial differential equations

$$(2n+1)\frac{\partial f_n}{\partial t} + (n+1)v\frac{\partial f_{n+1}}{\partial r} + nv\frac{\partial f_{n-1}}{\partial r} + (2n+1)\frac{f_n}{\tau} = (2n+1)\frac{f_0}{\tau}\delta_{n0},$$

$$n = 0, 1, 2, \ldots, \quad (3.33)$$

where $\delta_{ij} = 0$ ($i \neq j$) or 1 ($i = j$).

The problem, of course, that one is faced with in solving the infinite set of equations (3.33) corresponds to the standard closure problem, this time expressed as a decision about the order at which to truncate the series. This is commonly addressed by simply truncating the infinite set of equations at some arbitrary order with the hope that this does not introduce any unphysical character into the reduced model. Typically, truncations are made at the lowest order possible. For the f_1 approximation (i.e. assume $f_n = 0 \ \forall \ n \geq 2$), we have

$$\frac{\partial f_0}{\partial t} + v\frac{\partial f_1}{\partial r} = 0; \quad (3.34)$$

$$\frac{\partial f_1}{\partial t} + \frac{v}{3}\frac{\partial f_0}{\partial r} = -\frac{f_1}{\tau}, \quad (3.35)$$

which can be combined to yield the homogeneous telegrapher equation

$$\tau\frac{\partial^2 f_0}{\partial t^2} + \frac{\partial f_0}{\partial t} - \kappa\frac{\partial^2 f_0}{\partial r^2} = 0, \quad (3.36)$$

where the spatial diffusion coefficient $\kappa = \frac{1}{3}v^2\tau$ has been introduced. Before considering the properties of the telegrapher equation, notice that if, in Eq. (3.35), we assume that the "inertial" term $\partial f_1/\partial t \simeq 0$ compared to the remaining more rapidly varying terms (cf., the discussion related to the Chapman-Enskog expansion), then combining the f_1 approximation equations yields

$$f_1 = -\frac{v\tau}{3}\frac{\partial f_0}{\partial r} \Rightarrow \frac{\partial f_0}{\partial t} - \kappa\frac{\partial^2 f_0}{\partial r^2} = 0, \quad (3.37)$$

which is the classical diffusion equation.

Consider the properties of the transport equation expressed by the telegrapher equation. The telegrapher equation can be expressed in matrix form

$$\frac{\partial}{\partial t}\begin{pmatrix} f_0 \\ f_1 \end{pmatrix} + \begin{pmatrix} 0 & v \\ v/3 & 0 \end{pmatrix}\frac{\partial}{\partial r}\begin{pmatrix} f_0 \\ f_1 \end{pmatrix} = \begin{pmatrix} 0 \\ -f_1/\tau \end{pmatrix}.$$

3.8 Application 2: The Diffusion and Telegrapher Equations

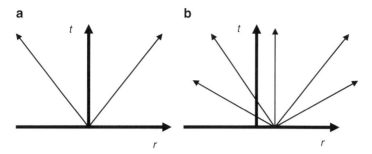

Fig. 3.6 The space-time plane showing the characteristics $r = \pm vt/\sqrt{3}$ (**a**) for the f_1 truncation, and (**b**) the offset characteristics for the f_4 truncation. Note the presence in (**b**) of two forward and two backward characteristics and a stationary or non-propagating characteristic

This is a hyperbolic system and the characteristics are determined by solving the characteristic equation $|A - \lambda I| = 0$ for λ, where A is the matrix above and I is the identity matrix. Thus

$$|A - \lambda I| = \begin{vmatrix} -\lambda & v \\ v/3 & -\lambda \end{vmatrix} = \lambda^2 - \frac{v^2}{3} = 0,$$

yields the characteristic or "sound" speeds for the system

$$\lambda = \pm \frac{v}{\sqrt{3}} \quad \text{and characteristic equations} \quad \left.\frac{dr}{dt}\right|_{\pm} = \pm \frac{v}{\sqrt{3}}.$$

Consequently, particles released at some initial time in a 1D scattering medium propagate in opposite directions at the fixed speed $\pm v/\sqrt{3}$ in the limit that their transport is described by the telegrapher equation. This is illustrated in the space-time diagram Fig. 3.6 that shows the characteristics $\pm vt/\sqrt{3}$ along which information propagates.

Exact solutions to the telegrapher equation for initial data can be derived and can be expressed in terms of the modified Bessel function of order 0. A somewhat more revealing approach is to use asymptotic expansions since this clarifies the late-time evolution of the solution to the telegrapher equation. For convenience, let us normalize the telegrapher equation (3.36) by introducing new time and space scales according to $\bar{t} = t/\tau$, $\bar{r} = x/L$, and ensuring that $L^2/\tau = \kappa$. This then yields the canonical form

$$\frac{\partial^2 f}{\partial t^2} + \frac{\partial f}{\partial t} - \frac{\partial^2 f}{\partial r^2} = 0, \qquad (3.38)$$

after dropping the bars and the 0 subscript for convenience.

Since the telegrapher equation is a second-order (dissipative) equation, the Cauchy problem requires that we specify smooth initial data $f(r, 0)$ and $f_t(r, 0)$

at $t = 0$, and we seek solutions for $t > 0$, $-\infty < r < \infty$. Let $F(\omega, t)$ be the Fourier transform of $f(r, t)$,

$$F(\omega, t) = \frac{1}{\sqrt{2\pi}} \int_{-\infty}^{\infty} e^{i\omega r} f(r, t) dr,$$

so that the transformed equation becomes

$$\frac{\partial^2 F}{\partial t^2} + \frac{\partial F}{\partial t} + \omega^2 F = 0, \quad t > 0.$$

The initial values are similarly transformed but we are not particularly concerned about the specific data since we seek to characterize the solution in general terms. The solution to the transformed equation is simply

$$F(\omega, t) = F_+(\omega) \exp\left[\left(-\frac{1}{2} + \frac{1}{2}\sqrt{1 - 4\omega^2}\right) t\right]$$
$$+ F_-(\omega) \exp\left[\left(-\frac{1}{2} - \frac{1}{2}\sqrt{1 - 4\omega^2}\right) t\right],$$

and the inverse transform therefore yields

$$f(r, t) = \frac{1}{\sqrt{2\pi}} \int_{-\infty}^{\infty} F_+(\omega) \exp\left[\left(-\frac{1}{2} + \frac{1}{2}\sqrt{1 - 4\omega^2}\right) t - i\omega r\right] d\omega$$
$$+ \frac{1}{\sqrt{2\pi}} \int_{-\infty}^{\infty} F_-(\omega) \exp\left[\left(-\frac{1}{2} - \frac{1}{2}\sqrt{1 - 4\omega^2}\right) t - i\omega r\right] d\omega,$$

where $F_\pm(\omega)$ is specified by the initial data.

To examine the behavior in the limit of large time t, it is evident that for $\omega^2 > \frac{1}{4}$, we have $\sqrt{1 - 4\omega^2} = i\sqrt{4\omega^2 - 1}$, so that

$$\exp\left[\left(-\frac{1}{2} \pm \frac{1}{2}\sqrt{1 - 4\omega^2}\right) t\right] = e^{-t/2} \exp\left[\pm\frac{i}{2}\sqrt{4\omega^2 - 1}\, t\right], \quad \omega^2 > \frac{1}{4}.$$

We then obtain

$$\left| \frac{1}{\sqrt{2\pi}} \int_{\omega^2 > \frac{1}{4}} F_\pm(\omega) e^{-t/2} \exp\left[\pm\frac{i}{2}\sqrt{4\omega^2 - 1}\, t - i\omega r\right] d\omega \right|$$
$$\leq \frac{e^{-t/2}}{\sqrt{2\pi}} \int_{-\infty}^{\infty} |F_\pm|(\omega) d\omega \leq \frac{M e^{-t/2}}{\sqrt{2\pi}},$$

for constant M. Thus, the contributions to both integrals are exponentially small whenever $\omega^2 > \frac{1}{4}$ for large t. For values $\omega^2 \leq \frac{1}{4}$, we have

3.8 Application 2: The Diffusion and Telegrapher Equations

$$\exp\left[\left(-\frac{1}{2}+\frac{1}{2}\sqrt{1-4\omega^2}\right)t\right] \le e^0 = 1, \quad \omega^2 \le \frac{1}{4};$$

$$\exp\left[\left(-\frac{1}{2}-\frac{1}{2}\sqrt{1-4\omega^2}\right)t\right] \le e^{-t/2}, \quad \omega^2 \le \frac{1}{4}.$$

The latter exponential is maximum for $\omega = \pm\frac{1}{2}$ and the first when $\omega = 0$. Hence, the entire $F_-(\omega)$ integral is bounded over the full interval of integration since

$$\left|\frac{1}{\sqrt{2\pi}}\int_{-\infty}^{\infty} F_-(\omega)\exp\left[\left(-\frac{1}{2}-\frac{1}{2}\sqrt{1-4\omega^2}\right)t - i\omega r\right]d\omega\right|$$

$$\le \frac{e^{-t/2}}{\sqrt{2\pi}}\int_{-\infty}^{\infty} |F_-(\omega)|\,d\omega.$$

Consequently, the entire $F_-(\omega)$ integral is bounded by $(M/\sqrt{2\pi})e^{-t/2}$ and decays exponentially and uniformly in r as $t \to \infty$. Thus, we expect the main contribution to come from the $F_+(\omega)$ integral as $t \to \infty$. Since the F_+ integral decays for all $\omega \ne 0$, the major contribution to the solution $f(r,t)$ must come from the neighborhood of $\omega = 0$. The asymptotic evaluation of the $F_+(\omega)$ integral can be accomplished using the method of Sirovich rather than the stationary phase method.[2] This is because the integral that we are interested in decays exponentially

---

[2] The Sirovich method proceeds as follows. Consider an integral of the form

$$I(r,t) = \int_{-\infty}^{\infty} F(\omega)\exp[-g(\omega)t - i\omega r]d\omega,$$

where the conditions (i) $\int_{-\infty}^{\infty} |F(\omega)|d\omega < M < \infty$; (ii) $\max|F(\omega)| < M \; \forall \; \omega$; (iii) $\Re\,[g(\omega)] \ge 0$; (iv) $g(\omega) = 0 \leftrightarrow \omega = 0$; (v) $g(\omega) = i\alpha\omega + \beta\omega^2 + O(|\omega|^3)$, for α, β real, $\beta > 0$, $\omega \to 0$, and (vi) $g(\omega)$ is continuous in ω, are all satisfied. Condition (v) is essentially a Taylor expansion of $g(\omega)$ about $\omega = 0$, recognizing that the maximum contribution comes from the neighborhood of $\omega = 0$. Then as $t \to \infty$, we have

$$I(r,t) = \int_{-\infty}^{\infty} F(\omega)\exp\left[-\beta\omega^2 t - i\alpha\omega t - i\omega r\right]d\omega + O\left[\frac{1}{t^{1-\delta}}\right],$$

where δ is small positive constant. Recall that the notation $O[\cdots]$ means that if $K(x) = O[G(x)]$, then $|K(x)/G(x)| \to C$ for some constant C as $x \to \zeta$. Thus, $O[1/t^{1-\delta}]$ implies that as $t \to \infty$ the error term above decays as $C/t^{1-\delta}$. Now expand $F(\omega)$ about $\omega = 0$ and retain only the leading term $F(0)$, to obtain

$$I(r,t) \approx \int_{-\infty}^{\infty} F(0)\exp\left[-\beta\omega^2 t - i\alpha\omega t - i\omega r\right]d\omega.$$

This integral can be evaluated exactly, yielding

$$I(r,t) \approx \sqrt{\frac{\pi}{\beta t}}F(0)\exp\left[-\frac{(r+\alpha t)^2}{4\beta t}\right], \quad t \to \infty.$$

as the relevant parameter becomes large whereas in the stationary phase method, the integral oscillates rapidly and the main contributions are due to the "slow" oscillations.

As described in the footnote, we can use the Sirovich approximation to evaluate the $F_+(\omega)$ integral. Identify $F_+(\omega)/\sqrt{2\pi}$ with $F(\omega)$ and the term $\frac{1}{2} - \frac{1}{2}\sqrt{1 - 4\omega^2}$ with $g(\omega)$. We assume that $F_+(\omega)$ satisfies conditions (i) and (ii), and we have (iii) $\Re\left[\frac{1}{2} - \frac{1}{2}\sqrt{1 - 4\omega^2}\right] \geq 0$ for $-\infty < \omega < \infty$; (iv) $\frac{1}{2} - \frac{1}{2}\sqrt{1 - 4\omega^2} = 0$ at $\omega = 0$ only, and (v) $\frac{1}{2} - \frac{1}{2}\sqrt{1 - 4\omega^2} = \frac{1}{2} - \frac{1}{2}[1 - 2\omega^2 + O(|\omega|^3)] = \omega^2 + O(|\omega|^3)$ as $\omega \to 0$ on using the binomial expansion. Finally, since $g(\omega)$ is continuous for all ω, all the conditions on $g(\omega)$ are met. We therefore have

$$\frac{1}{\sqrt{2\pi}} \int_{-\infty}^{\infty} F_+(\omega) \exp\left[\left(-\frac{1}{2} + \frac{1}{2}\sqrt{1 - 4\omega^2}\right)t - i\omega r\right] d\omega \approx \frac{F_+(0)}{\sqrt{2t}} \exp\left[-\frac{r^2}{4t}\right],$$

in the limit $t \to \infty$. Hence, we find that asymptotically the solution for $f(r,t)$ is given by

$$f(r,t) \approx \frac{F_+(0)}{\sqrt{2t}} \exp\left[-\frac{r^2}{4t}\right] = \sqrt{2\pi} F_+(0) \times \frac{1}{\sqrt{4\pi t}} \exp\left[-\frac{r^2}{4t}\right], \quad t \to \infty, \tag{3.39}$$

where we have expressed the solution in terms of the fundamental solution of the heat or diffusion equation (3.37). On rewriting Eq. (3.39) in non-normalized variables, we obtain

$$f(r,t) \approx \frac{F_+(0)}{\sqrt{2t/\tau}} \exp\left[-\frac{r^2}{4\kappa t}\right].$$

Evidently, $f(r,t)$ is nonzero essentially only in the parabolic region defined by $r^2/(4t\kappa) = O(1)$ since otherwise the exponential term is vanishingly small. This solution behavior is illustrated in Fig. 3.7. The solution to the telegrapher equation at long times is essentially diffusive, occurring in the wake of the wave fronts or characteristics $r = vt/\sqrt{3}$ and yields the major contribution to the solution because dissipation has damped everything else away. The decay of the "heat equation"-like solution also occurs as $t \to \infty$ but only algebraically like $t^{-1/2}$ for $r^2 \leq 4t\kappa$ rather than exponentially as for all other values of r. It is worth noting that the asymptotic solution to the telegrapher equation (and many others) is often more useful than the exact solution in extracting the physical character of the solution.

The expanded system of partial differential equations (3.33) in the f_n truncation forms a linear hyperbolic system of pdes,

$$\mathbf{\Psi}_t + v\mathbf{A}\mathbf{\Psi}_r = \mathbf{C},$$

3.8 Application 2: The Diffusion and Telegrapher Equations

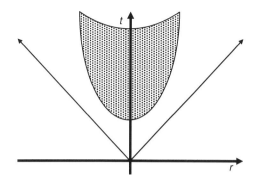

Fig. 3.7 The space-time plane showing the characteristics $r = \pm vt/\sqrt{3}$ for the telegrapher equation. The characteristics bound the solution space, and the shaded region identifies the long-time diffusion regime into which solutions of the telegrapher equation evolve

with a discrete spectrum of characteristic speeds. Here $\Psi \equiv (f_0, f_1, f_2, \ldots f_n)^t$, where t denotes transpose, $\mathbf{A}$ is the tridiagonal matrix

$$\mathbf{A} = \begin{pmatrix} 0 & 1 & 0 & 0 & 0 & \cdots & & \cdots & 0 \\ \frac{1}{3} & 0 & \frac{2}{3} & 0 & 0 & \cdots & & \cdots & 0 \\ 0 & \frac{2}{5} & 0 & \frac{3}{5} & 0 & \cdots & & \cdots & 0 \\ 0 & 0 & \frac{3}{7} & 0 & \frac{4}{7} & \cdots & & \cdots & 0 \\ \vdots & & & & & \ddots & \ddots & & \vdots \\ 0 & 0 & 0 & 0 & \cdots & \frac{n-1}{2n-1} & 0 & & \frac{n}{2n-1} \\ 0 & 0 & 0 & 0 & \cdots & & 0 & \frac{n}{2n+1} & 0 \end{pmatrix},$$

and

$$\mathbf{C} = \tau^{-1} (0, f_1, f_2, \ldots, f_n)^t.$$

The characteristic equation $|\mathbf{A} - \lambda \mathbf{I}| = 0$ yields the $n+1$ characteristics of the linear hyperbolic system, all of which are distinct. When the truncation of (3.33) is even, i.e. when n is even, the number of characteristics is odd, and consists of $n/2$ propagating information forward, $n/2$ propagating information backward, and one that is stationary. For example, the f_2 characteristics are

$$\left. \frac{dr}{dt} \right)_{0,\pm} = 0, \pm\sqrt{\frac{3}{5}}v \Rightarrow r_{0,\pm} = 0, \pm\sqrt{\frac{3}{5}}vt.$$

At the f_1 (telegrapher equation) level of truncation, all scattered particles propagate along $\pm vt/\sqrt{3}$ characteristics, whereas the f_2 truncation is more refined, substituting $0, \pm\sqrt{\frac{3}{5}}v$ for the speeds of the scattered particles. When n is odd, the number of characteristics is even, with $n/2$ propagating information forward and $n/2$ backward. No stationary or zero characteristic exists for the f_n (n odd) truncation, revealing that the even and odd closures are fundamentally different. By

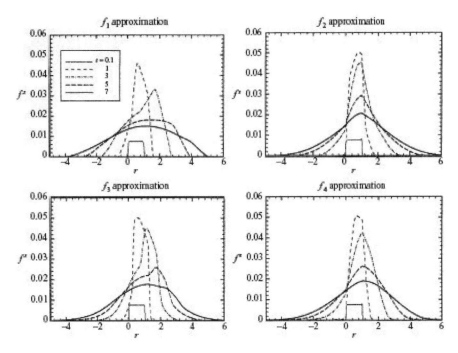

Fig. 3.8 Numerical comparison of solutions obtained using f_1, f_2, f_3, and f_4 truncations. The scattered distribution f_s is plotted for times ranging from early ($t = 0.1$ in normalized units) until later

increasing the number of equations in the truncation (i.e. increasing n), the accuracy with which information is propagated forwards and backwards is increased (see Fig. 3.6). However, the odd truncation can never capture the non-propagating mode, and is therefore always intrinsically less accurate than the even expansion, even when the even truncation is of lower order. This is illustrated in Fig. 3.8 where the method of characteristics is implemented numerically for the f_1, f_2, f_3, and f_4 truncations.

The telegrapher equation, and its higher-order truncations, does not capture the early phases of particle propagation when particles have experienced little or no scattering and the distribution is typically quite anisotropic. However, a simple extension of the approach described here can capture the so-called flash phase corresponding to early times of particle propagation (Zank et al. 2000).

Exercises

1. Legendre polynomials $P_n(\mu)$ and $P_m(\mu)$ satisfy Legendre's differential equation

$$(1 - \mu^2)y'' - 2\mu y' + n(n+1)y = 0, \quad n = 0, 1, 2, \ldots.$$

Hence show that for $n \neq m$, the orthogonality condition

$$\int_{-1}^{1} P_m(\mu) P_n(\mu) d\mu = 0, \quad n \neq m.$$

2. The generating function for the Legendre polynomials is given by

$$L(\mu, t) = (1 - 2\mu t + t^2)^{-1/2} = \sum_{n=0}^{\infty} P_n(\mu) t^n, \quad |t| < 1.$$

By differentiating the generating function with respect to t and equating coefficients, derive the recursion relation

$$(n+1) P_{n+1} + n P_{n-1} = (2n+1) \mu P_n, \quad n = 1, 2, 3, \ldots.$$

3. By using the generating function and Problem 1 above, show that

$$\int_{-1}^{1} P_n^2(\mu) d\mu = \frac{2}{2n+1}.$$

4. Complete the steps to derive the infinite set of partial differential equations (3.33).
5. Show that the integral

$$\int_{-\infty}^{\infty} \exp\left[-\beta \omega^2 t - i\alpha \omega t - i\omega r\right] d\omega = \sqrt{\frac{\pi}{\beta t}} \exp\left[-\frac{(r + \alpha t)^2}{4\beta t}\right].$$

References

P.L. Bhatnagar, E.P. Gross, M. Krook, A model for collision processes in gases. I. Small amplitude processes in charged and neutral one-component systems. Phys. Rev. **94**(3), 511–525 (1954)

S. Chapman, T.G. Cowling, *The Mathematical Theory of Non-uniform Gases*, 3rd edn. (Cambridge University Press, Cambridge, 1970)

L.D. Landau, E.M. Lifshitz, *Fluid Mechanics*, 2nd edn. (Butterworth-Heinemann, Oxford, 2000)

G.P. Zank, J.Y. Lu, W.K.M. Rice, G.M. Webb, Transport of energetic charged particles in a radial magnetic field. Part 1. Large-angle scattering. J. Plasma Phys. **64**(4), 507–541 (2000)

Chapter 4
Charged Particle Transport in a Collisional Magnetized Plasma

4.1 The Kinetic Equation and Moments for a Magnetized Plasma

A plasma is an admixture of charged particles – electrons, and possibly many different populations of ions, including protons – each of which can be characterized by their position $\mathbf{x}$ and velocity $\mathbf{v}$, and, as with a gas, can be represented by a point in the 6-dimensional phase space $(\mathbf{x}, \mathbf{v})$. The distribution function $f_a(\mathbf{x}, \mathbf{v}, t)$ for a particle species a is the number of particles of that species per unit volume in phase space near the point $(\mathbf{x}, \mathbf{v})$ at time t as before, i.e.,

$$f_a(\mathbf{x}, \mathbf{v}, t) d^3 x d^3 v$$

is the number of particles in the volume element $d^3 x d^3 v$ about the point $(\mathbf{x}, \mathbf{v})$. The zeroth moment gives the number density $n_a(\mathbf{x}, t)$ of the a particles in real space. In a plasma, each particle moves according to

$$\dot{\mathbf{x}} = \mathbf{v};$$
$$\dot{\mathbf{v}} = \frac{q_a}{m_a}(\mathbf{E} + \mathbf{v} \times \mathbf{B}),$$

where q_a denotes the electric charge of particle species a, m_a the mass, $\mathbf{E}$ the total electric field, and $\mathbf{B}$ the total magnetic field. Like the Boltzmann equation, the distribution function obeys a conservation equation (since the number density is conserved in phase space), so that if $\boldsymbol{\xi} \equiv (\mathbf{x}, \mathbf{v})$, then $\dot{\boldsymbol{\xi}} = (\dot{\mathbf{x}}, \dot{\mathbf{v}})$ or

$$\frac{\partial f_a}{\partial t} + \frac{\partial}{\partial \boldsymbol{\xi}}\left(\dot{\boldsymbol{\xi}} f_a\right) = 0.$$

This of course corresponds to the magnetized form of the Boltzmann equation or the Vlasov equation

$$\frac{\partial f_a}{\partial t} + \mathbf{v} \cdot \nabla f_a + \frac{q_a}{m_a}(\mathbf{E} + \mathbf{v} \times \mathbf{B}) \cdot \nabla_v f_a = 0, \tag{4.1}$$

where we used

$$\nabla_v \cdot (\mathbf{E} + \mathbf{v} \times \mathbf{B}) = 0.$$

In the neighborhood of each discrete charged particle, the fields can be large and dominate the macroscopic large-scale fields. Thus, $\mathbf{E}$ and $\mathbf{B}$ fluctuate strongly on short length scales compared to the Debye length (i.e., $\lambda_D = \sqrt{\varepsilon_0 kT/nq^2}$, which is the distance over which charged carriers are screened). We take $\mathbf{E}$ and $\mathbf{B}$ to be the average of the actual electric and magnetic fields over many Debye lengths, and the effects of the short-range electromagnetic fluctuations or collisions will be included through a collision operator

$$C_a(f_a) = \left.\frac{\delta f_a}{\delta t}\right|_{coll},$$

and (4.1) is now modified to read

$$\frac{\partial f_a}{\partial t} + \mathbf{v} \cdot \nabla f_a + \frac{q_a}{m_a}(\mathbf{E} + \mathbf{v} \times \mathbf{B}) \cdot \nabla_v f_a = C_a(f_a), \tag{4.2}$$

where $\mathbf{E}$ and $\mathbf{B}$ are the averaged fields. For plasmas, rather than using the Boltzmann collision operator as we did before, we instead use a *Fokker-Planck* operator C_a, and the corresponding kinetic equation is called the Fokker-Planck equation. In the absence of collisions, it becomes the Vlasov equation.

The collision operator

$$C_a = \sum_b C_{ab}(f_a, f_b)$$

is a sum of the contributions from collisions with each particle species b, including self-collisions $a = b$. Like the Boltzmann collision operator, the number density, momentum, and energy moments of the Fokker-Planck collisional operator must satisfy

$$\int C_{ab}(f_a) d^3v = 0;$$

$$\int m_a \mathbf{v} C_{ab}(f_a) d^3v = -\int m_b \mathbf{v} C_{ba}(f_b) d^3v;$$

$$\int \frac{1}{2} m_a v^2 C_{ab}(f_a) d^3v = -\int \frac{1}{2} m_b v^2 C_{ba}(f_b) d^3v,$$

4.1 The Kinetic Equation and Moments for a Magnetized Plasma

since the force a species a exerts on a species b must be equal and opposite to that which b exerts on a, so that no net momentum or energy change results from collisions. For $b = a$, we have

$$\int C_{aa}(f_a)d^3v = 0;$$

$$\int m_a \mathbf{v} C_{aa}(f_a)d^3v = 0;$$

$$\int \frac{1}{2} m_a v^2 C_{aa}(f_a)d^3v = 0.$$

Any model collision operator has to satisfy these properties.

Finally, another important property of the collision operator is that it induces the distribution function to relax to a local thermodynamical equilibrium, i.e., to a velocity shifted Maxwellian distribution for each species,

$$f_a^0 = n_a(\mathbf{x},t) \left(\frac{m_a}{2\pi k T_a(\mathbf{x},t)} \right)^{3/2} \exp\left[\frac{-m_a [\mathbf{v} - \mathbf{u}_a(\mathbf{x},t)]^2}{2k T_a(\mathbf{x},t)} \right],$$

where u_a is the mean velocity of species a, and the temperature is $T_a(\mathbf{x},t)$. The collision operator vanishes if and only if all species have the same mean velocity and temperature. For a single species plasma, $C_{aa}(f_a^0) = 0$ for an arbitrary mean velocity $\mathbf{u}_a$, indicating that the collision operator is Galilean invariant.

As with a simple non-magnetized gas, we can introduce moments for each particle species,

$$\langle M \rangle \equiv \frac{1}{n_a} \int M f_a d^3v,$$

where M is any polynomial function of the components of $\mathbf{v}$. For example, the zeroth-order moment, $M = 1$, yields the plasma species number density through

$$n_a(\mathbf{x},t) = \int f_a(\mathbf{x},\mathbf{v},t) d^3v.$$

The higher-order moment $\langle \mathbf{v} \rangle$ yields the average velocity of all particles of species a at some point in phase space, i.e., $\mathbf{u}_a(\mathbf{x},t) = \langle \mathbf{v} \rangle$. As before, let $\mathbf{c}_a = \mathbf{v} - \mathbf{u}_a$. The mean value of $\mathbf{c}_a$ is zero, but higher-order moments are generally non-zero and are related to the thermal momentum and energy flux, and hence to the pressure and heat flux. Define the temperature T_a so that $3k T_a/2$ is the average kinetic energy associated with the random velocities,

$$\frac{3}{2} k T_a = \left\langle \frac{1}{2} m_a c_a^2 \right\rangle.$$

This of course is quite independent of whether the plasma is in local thermodynamical equilibrium or not. Observe that

$$\frac{1}{2}m_a n_a \langle v^2 \rangle = \frac{1}{2}m_a n_a u_a^2 + \frac{3}{2}n_a k T_a,$$

showing that the total energy is the sum of the kinetic energy associated with the mean flow and the thermal energy.

Following the procedure described in Chap. 2, we can derive fluid equations by taking moments of the kinetic equation, obtaining the conservation laws

$$\frac{\partial n}{\partial t} + \nabla \cdot (n\mathbf{u}) = 0; \tag{4.3}$$

$$\frac{\partial}{\partial t}(mn\mathbf{u}) + \nabla \cdot \mathbf{P} = ne\,(\mathbf{E} + \mathbf{u} \times \mathbf{B}) + \int m\mathbf{v} C(f) d^3 v; \tag{4.4}$$

$$\frac{\partial}{\partial t}\left(\frac{3}{2}nkT + \frac{1}{2}mnu^2\right) + \nabla \cdot \mathbf{Q} = en\mathbf{E} \cdot \mathbf{u} + \int \frac{1}{2}mv^2 C(f) d^3 v, \tag{4.5}$$

where $\mathbf{P}$ is the momentum flux tensor,

$$\mathbf{P}_{ij} \equiv \langle mn v_i v_j \rangle,$$

and

$$\mathbf{Q} \equiv \frac{1}{2}mn\langle v^2 \mathbf{v} \rangle,$$

is the energy flux vector. For convenience, we neglect the subscript a until we need to distinguish between particle species again and we have used the notation $(\nabla \cdot \mathbf{P})_i = \partial P_{ij}/\partial x_j$. The conservation laws are the same as those of gas dynamics except for the inclusion of the electric and magnetic fields.

We may separate the tensor $\mathbf{P}$ into two tensors related to the thermal properties of the plasma and another related to the kinetic energy,

$$P_{ij} = p\delta_{ij} + \pi_{ij} + mn u_i u_j,$$

where we introduce the scalar pressure p (for an isotropic plasma) through

$$p = \frac{1}{3}nm\langle c^2 \rangle = nkT,$$

and a traceless tensor, the viscosity tensor,

$$\pi_{ij} \equiv mn\langle c_i c_j \rangle - p\delta_{ij}.$$

4.1 The Kinetic Equation and Moments for a Magnetized Plasma

Finally, the heat flux vector **q** that describes the flux of thermal energy relative to the flow velocity is defined by

$$\mathbf{q} \equiv n \langle \tfrac{1}{2} m c^2 \mathbf{c} \rangle.$$

The momentum and energy flux in the conservation equations (4.4) and (4.5) can therefore be expressed as

$$P_{ij} = p\delta_{ij} + \pi_{ij} + mnu_i u_j; \tag{4.6}$$

$$Q_i = q_i + \tfrac{5}{2} u_i p + \pi_{ij} u_j + \tfrac{1}{2} mnu^2 u_i. \tag{4.7}$$

The energy flux **Q** therefore comprises a heat flux **q**, a convective flux $5/2 \mathbf{u} p$, the viscous transport of energy, and the convection of kinetic energy. Note that a heat flux tensor

$$Q_{ijk} \equiv m \langle c_i c_j c_k \rangle$$

is sometimes introduced, which is then related to the heat flux vector via

$$q_i(\mathbf{x}, t) = \tfrac{1}{2} Q_{ijj}(\mathbf{x}, t).$$

The right-hand sides of the fluid equations contain the rate of change of momentum and energy due to the electromagnetic fields and the collisional transfer of momentum and energy via collisions to and from other species, and may be expressed as

$$\int m\mathbf{v} C(f) d^3 v = \mathbf{R};$$

$$\int \tfrac{1}{2} m v^2 C(f) d^3 v = Q + \mathbf{R} \cdot \mathbf{u},$$

where

$$Q \equiv \int \tfrac{1}{2} m c^2 C(f) d^3 v.$$

This Q should not be confused with the heat flux vector or tensor, but unfortunately the notation is standard and so we use it here. **R** is the rate of transfer of momentum to the particle species of interest due to collisions with other species in the plasma, and Q describes the corresponding rate of thermal energy transfer. The work $\mathbf{R} \cdot \mathbf{u}$

performed by the force $\mathbf{R}$ plus Q gives the total energy transfer rate. Evidently, for a plasma comprised of protons (p) and electrons (e) only, the collisional relations

$$\mathbf{R}_p = -\mathbf{R}_e;$$
$$Q_p = -Q_e - (\mathbf{u}_e - \mathbf{u}_p) \cdot \mathbf{R}_e,$$

follow from the collisional conservation relations.

In general, the effects of collisions with different charged particle species are additive (recall the collision operator is a sum of contributions from collisions with each particle species b including self-collisions) so

$$\mathbf{R}_a = \sum_b \mathbf{R}_{ab}$$
$$Q_a = \sum_b Q_{ab}$$

where each $\mathbf{R}_{ab}$ or Q_{ab} represents the interaction between particle species a and b.

It is straightforward to rewrite the conservation laws (4.3)–(4.5) as equations for the density, flow velocity, and temperature (Exercise)

$$\frac{\partial n_a}{\partial t} + \nabla \cdot (n_a \mathbf{u}_a) = 0; \tag{4.8}$$

$$m_a n_a \left(\frac{\partial \mathbf{u}_a}{\partial t} + \mathbf{u}_a \cdot \nabla \mathbf{u}_a \right) = -\nabla p - \nabla \cdot \boldsymbol{\pi}_a + q_a n_a (\mathbf{E} + \mathbf{u}_a \times \mathbf{B}) + \mathbf{R}; \tag{4.9}$$

$$\frac{3}{2} n_a \left(\frac{\partial (kT_a)}{\partial t} + \mathbf{u}_a \cdot \nabla (kT_a) \right) + p \nabla \cdot \mathbf{u}_a = -\nabla \cdot \mathbf{q}_a - \boldsymbol{\pi}_a : \nabla \mathbf{u}_a + Q_a, \tag{4.10}$$

where

$$\boldsymbol{\pi}_a : \nabla \mathbf{u}_a = \pi_{jk} \frac{\partial u_k}{\partial x_j}.$$

Note too that the convective derivative describes the rate of change experienced by the fluid element itself, and is sometimes denoted

$$\frac{d}{dt} = \frac{\partial}{\partial t} + \mathbf{u} \cdot \nabla.$$

The momentum equation is similar to the Navier-Stokes equation with the addition of the Lorentz force and the frictional term $\mathbf{R}_a$. A form of the energy equation that is often useful follows from rewriting the conservation of mass relation (4.8) as

$$\frac{d}{dt} \ln n_a = -\nabla \cdot \mathbf{u}_a.$$

On using this expression in the energy equation (4.10) with $p_a = n_a k T_a$, we find

$$\frac{3}{2} nk \frac{dT_a}{dt} - p_a \frac{d}{dt} \ln n_a = \frac{3}{2} \frac{dp_a}{dt} - \frac{5}{2} p_a \frac{d}{dt} \ln n_a$$

$$= \frac{3}{2} p_a \frac{d}{dt} \ln \left(\frac{p_a}{n_a^{5/3}} \right),$$

from which it follows that

$$p \frac{dS_a}{dt} = -\nabla \cdot \mathbf{q}_a - \pi_a : \nabla \mathbf{u}_a + Q_a.$$

Here we have introduced the entropy per particle $S_a \equiv \frac{3}{2} \ln \left(p_a / n_a^{5/3} \right)$. Thus, $n_a S_a$ is the entropy density per unit volume of species a and the right-hand-side describes the production of entropy.

Exercises

1. By taking moments of the Fokker-Planck equation, derive the fluid equations (4.3)–(4.5).
2. Derive the momentum equation (4.9) and energy equation (4.10) from the conservation laws (4.3)–(4.5).

4.2 Markov Processes, the Chapman-Kolmogorov Equation, and the Fokker-Planck Equation

The analysis of the previous section provides little insight into the underlying description of the collisional term, beyond identifying that it satisfies certain conservation laws. Here we derive a basic formalism that provides a structure for describing the slow change in time of a particle probability distribution function in response to an enormous number of small rapid changes. These rapidly occurring small changes may be due to small angle particle-particle collisions, or the pitch-angle scattering of charged particles by turbulent magnetic field fluctuations. Typically, the small rapid changes (particle scattering) can be regarded as independent in some sense. As will be shown here, the Fokker-Planck equation is a very general equation that can be used to describe any phenomena that in some approximate sense can be described as a *Markov process*. A Markov process is one whose value at the next measurement depends only on its present measurement and not on any previous measurements from earlier times. Thus, if $X(t)$ is a random process, and $X_n \equiv X(t_n)$ with $t_n > t_{n-1} > \cdots > t_1 > t_0$, then a Markov process has a probability density function such that

$$f(X_n | x_{n-1} x_{n-2} \ldots x_1 x_0) = f(X_n | x_{n-1}).$$

Markov processes can be both discrete and continuous.

Fig. 4.1 A smooth function representing a physical process

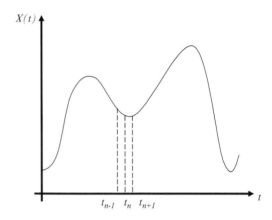

Example. A simple and obvious example of a discrete Markov process is the flipping of a coin where for each toss, the random variable $X(t_n) = x_n = +1$ for a head and -1 for a tail. Clearly, X is a Markov process since $f(x_n) = \frac{1}{2}\delta(x_n - 1) + \frac{1}{2}\delta(x_n + 1)$ is independent of x_{n-1} and all preceding values of x_n.

Example. Suppose

$$X(t_n) \equiv X_n = \sum_{i=1}^{n} x_i,$$

where x_i are as in the previous example i.e., $x_n = +1$ for a head and -1 for a tail. Evidently, X is a Markov process as the value of X_n depends on the value of X_{n-1} only and on no previous values.

An example of a continuous Markov process is not really possible to provide since a continuous Markov process cannot exist in nature. Suppose, for example, that a random function can be expressed as a smooth curve (Fig. 4.1). From elementary calculus, the value of x_{n+1} depends not only on x_n but also on x_{n-1} since we need to know the slope of the function x_n,

$$\left.\frac{dx(t)}{dt}\right|_{t=t_n} \approx \frac{x_n - x_{n-1}}{\Delta t},$$

for Δt small. Consequently, neither this function, nor any other that can be expressed as a smooth curve, can be a Markov process. Nonetheless, a Markov process can be a good approximation to a physical process. A particle experiencing collisions fluctuates rapidly in position and slows down due to a net frictional force. On a time scale much longer than the collisional time scale, the particle performs a random walk in velocity space and soon "forgets" the details of its orbit near $t = 0$, but it does typically remember its initial velocity $v(t = 0)$. The process has essentially three time scales: the collisional time τ_c; the time Δt after which we

4.2 Markov Processes, the Chapman-Kolmogorov Equation...

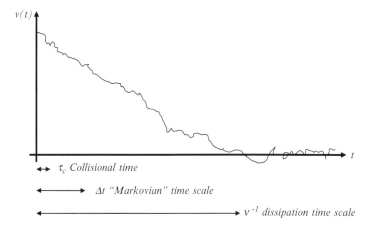

Fig. 4.2 A realization of the velocity of a particle experiencing scattering. The three time scales distinguishing the different physical regimes are identified

may assume that to a good approximation, the process is Markovian, and finally the dissipation time ν^{-1} at which the mean particle velocity is approximately zero. We require $\Delta t \gg \tau_c$, and will generally assume that $\Delta t \ll \nu^{-1}$. See Fig. 4.2.

Using Δt as described, we would expect to derive the particle distribution function $f(\mathbf{v}, t + \Delta t)$ governing the probability of occurrence of $\mathbf{v}$ at time $t + \Delta t$ from the distribution function $f(\mathbf{v}, t)$ at time t if we know the transition probability $\Psi(\mathbf{v}, \Delta \mathbf{v})$ that $\mathbf{v}$ changes by $\Delta \mathbf{v}$ in time Δt. This suggests that the relation

$$f(\mathbf{v}, t + \Delta t) = \int f(\mathbf{v} - \Delta \mathbf{v}, t) \Psi(\mathbf{v} - \Delta \mathbf{v}, \Delta \mathbf{v}) d(\Delta \mathbf{v}), \qquad (4.11)$$

holds. This is essentially the Chapman-Kolmogorov equation and is used as the starting point for deriving the Fokker-Planck equation, which may be thought of as a generalization of Liouville's theorem to include random motions. We follow the derivation given by Chandrasekhar (1943)[1] in deriving the Fokker-Planck equation.

We note that sometimes specific forms of the transition probability can be identified, such as for Brownian motion. Holding $\Psi(\mathbf{v}, \Delta \mathbf{v})$ general, we may Taylor expand $f(\mathbf{v}, t + \Delta t)$, $f(\mathbf{v} - \Delta \mathbf{v}, t)$, and $\Psi(\mathbf{v} - \Delta \mathbf{v}, \Delta \mathbf{v})$, to obtain

$$f(\mathbf{v}, t + \Delta t) = f(\mathbf{v}, t) + \frac{\partial f}{\partial t} \Delta t + O(\Delta t^2)$$

$$= \int_{-\infty}^{\infty} \int_{-\infty}^{\infty} \int_{-\infty}^{\infty} \left[f(\mathbf{v}, t) - \sum_i \frac{\partial f}{\partial v_i} \Delta v_i + \frac{1}{2} \sum_i \frac{\partial^2 f}{\partial v_i^2} (\Delta v_i)^2 \right.$$

---

[1] The classic paper on this topic is that by Chandrasekhar (1943).

$$+ \sum_{i<j} \frac{\partial^2 f}{\partial v_i \partial v_j} \Delta v_i \Delta v_j + \cdots \Bigg] \Bigg[\Psi(\mathbf{v}, \Delta \mathbf{v}) - \sum_i \frac{\partial \Psi}{\partial v_i} \Delta v_i + \frac{1}{2} \sum_i \frac{\partial^2 f}{\partial v_i^2} (\Delta v_i)^2$$

$$+ \sum_{i<j} \frac{\partial^2 \Psi}{\partial v_i \partial v_j} \Delta v_i \Delta v_j + \cdots \Bigg] d(\Delta v_1) d(\Delta v_2) d(\Delta v_3).$$

Let us introduce moments of the velocity increments as

$$\langle \Delta v_i \rangle = \int_{-\infty}^{\infty} \Delta v_i \Psi(\mathbf{v}, \Delta \mathbf{v}) d(\Delta \mathbf{v});$$

$$\langle \Delta v_i^2 \rangle = \int_{-\infty}^{\infty} (\Delta v_i)^2 \Psi(\mathbf{v}, \Delta \mathbf{v}) d(\Delta \mathbf{v});$$

$$\langle \Delta v_i \Delta v_j \rangle = \int_{-\infty}^{\infty} \Delta v_i \Delta v_j \Psi(\mathbf{v}, \Delta \mathbf{v}) d(\Delta \mathbf{v}),$$

which allows us to rewrite the Taylor expanded integral equation as

$$\frac{\partial f}{\partial t} \Delta t + O(\Delta t^2) = -\sum_i \frac{\partial f}{\partial v_i} \langle \Delta v_i \rangle + \frac{1}{2} \sum_i \frac{\partial^2 f}{\partial v_i^2} \langle \Delta v_i^2 \rangle$$

$$+ \sum_{i<j} \frac{\partial^2 f}{\partial v_i \partial v_j} \langle \Delta v_i \Delta v_j \rangle - \sum_i f \frac{\partial}{\partial v_i} \langle \Delta v_i \rangle + \sum_i \frac{\partial}{\partial v_i} \langle \Delta v_i^2 \rangle \frac{\partial f}{\partial v_i}$$

$$+ \sum_{i \neq j} \frac{\partial}{\partial v_j} \langle \Delta v_i \Delta v_j \rangle \frac{\partial f}{\partial v_i} + \frac{1}{2} \sum_i \frac{\partial^2}{\partial v_i^2} \langle \Delta v_i^2 \rangle f + \sum_{i<j} f \frac{\partial^2}{\partial v_i \partial v_j} \langle \Delta v_i \Delta v_j \rangle$$

$$+ O\left(\langle \Delta v_i \Delta v_j \Delta v_k \rangle\right).$$

This equation is written more conveniently as

$$\frac{\partial f}{\partial t} + \frac{O(\Delta t^2)}{\Delta t} = \sum_i \frac{\partial f}{\partial v_i} \left(f \frac{\langle \Delta v_i \rangle}{\Delta t} \right)$$

$$+ \frac{1}{2} \sum_i \frac{\partial^2 f}{\partial v_i^2} \left(f \frac{\langle \Delta v_i^2 \rangle}{\Delta t} \right) + \sum_{i<j} \frac{\partial^2 f}{\partial v_i \partial v_j} \left(f \frac{\langle \Delta v_i \Delta v_j \rangle}{\Delta t} \right) + \frac{O\left(\langle \Delta v_i \Delta v_j \Delta v_k \rangle\right)}{\Delta t},$$

(4.12)

which is the most general form of the Fokker-Planck equation.

4.2.1 A More Formal Derivation of the Chapman-Kolmogorov Equation and the Fokker-Planck Equation

More formally, consider the probability of a sequence of values of the random function $X(t)$ such that
$f(x_n, x_{n-1}, \ldots, x_1, x_0) \equiv$ Probability that at time t_0, the process $X(t)$ has value x_0 and at time t_1 has value x_1, and ... and at time t_n, $x(t)$ has value x_n where $t_n > t_{n-1} > \cdots > t_1 > t_0$.

Hence, by applying repeatedly the definition of a conditional probability for a Markov process,

$$f(x_n, x_{n-1}, \ldots, x_1, x_0) = f(x_n|x_{n-1}, \ldots, x_1, x_0) f(x_{n-1}, \ldots, x_1, x_0)$$
$$= f(x_n|x_{n-1}) f(x_{n-1}, \ldots, x_1, x_0)$$
$$= f(x_n|x_{n-1}) f(x_{n-1}|x_{n-2}) f(x_{n-2}, x_{n-3}, \ldots, x_1, x_0)$$
$$= \cdots$$
$$= f(x_n|x_{n-1}) f(x_{n-1}|x_{n-2}) f(x_{n-2}|x_{n-3}) \ldots f(x_2|x_1) f(x_1|x_0) f(x_0).$$

We have also seen that

$$f(x_n, x_{n-1}, \ldots, x_1, x_0) = f(x_n, x_{n-1}, \ldots, x_1|x_0) f(x_0),$$

so that

$$f(x_n, x_{n-1}, \ldots, x_1|x_0)$$
$$= f(x_n|x_{n-1}) f(x_{n-1}|x_{n-2}) f(x_{n-2}|x_{n-3}) \ldots f(x_2|x_1) f(x_1|x_0).$$

Hence, choosing $n = 2$ yields

$$f(x_2, x_1|x_0) = f(x_2|x_1) f(x_1|x_0).$$

Upon integrating over all possible values of x_1, we obtain

$$f(x_2|x_0) = \int f(x_2, x_1|x_0) dx_1 \quad \text{or}$$
$$f(x_2|x_0) = \int f(x_2|x_1) f(x_1|x_0) dx_1, \quad (4.13)$$

which is called either the *Chapman-Kolmogorov equation* or the *Smoluchowsky equation*.

In the formal definition of the Chapman-Kolmogorov equation, identify x_1 with the time t and let $x_2 = x(t + \Delta t)$. Furthermore, suppose that

$$f(x_0) \equiv f(x, t = t_0) = \delta(x - x_0).$$

This allows us to neglect x_0 in (4.13) and express

$$f(x_2|x_0) = f(x, t + \Delta t),$$

where x_2 is now denoted simply by x and

$$f(x_1|x_0) = f(x_1, t).$$

The introduction of the definition

$$\Delta x \equiv x - x_1,$$

allows us to use the following notation for $f(x_2|x_1)$,

$$f(x_2|x_1) = f(x, t + \Delta t | x - \Delta x, t)$$
$$= \Psi(\Delta x, t + \Delta t | x - \Delta x, t).$$

The transition probability Ψ gives the probability that at time $t + \Delta t$ the random process has made a jump of Δx from its previous value of $x - \Delta x$ at time t. The Chapman-Kolmogorov equation then becomes

$$f(x, t + \Delta t) = \int \Psi(\Delta x, t + \Delta t | x - \Delta x, t) f(x - \Delta x, t) d(\Delta x). \quad (4.14)$$

The value of x appears only in combination with incremental changes Δx through $x - \Delta x$. In plasma (and a gas in general), most Coulomb collisions cause only a small change in the velocity of a particle. So we may assume in general that all of the important physical changes will happen on the small Δx scale, and we therefore introduce a Taylor expansion on the right-hand-side of (4.14), obtaining

$$f(x, t + \Delta t) = \int \sum_{k=0}^{\infty} \frac{(-\Delta x)^k}{k!}$$

$$\times \left[\frac{\partial^k}{\partial x^k} [\Psi(\Delta x, t + \Delta t | x - \Delta x, t) f(x - \Delta x, t)]_{x - \Delta x = x} d(\Delta x) \right],$$

or

$$f(x, t + \Delta t) = \int \sum_{k=0}^{\infty} \frac{(-\Delta x)^k}{k!} \frac{\partial^k}{\partial x^k}$$
$$\times \Psi(\Delta x, t + \Delta t | x, t) f(x, t) d(\Delta x)$$

4.2 Markov Processes, the Chapman-Kolmogorov Equation...

$$= \sum_{k=0}^{\infty} \frac{(-1)^k}{k!} \frac{\partial^k}{\partial x^k} f(x,t) \int (\Delta x)^k \Psi(\Delta x, t + \Delta t | x, t) d(\Delta x)$$

$$= \sum_{k=0}^{\infty} \frac{(-1)^k}{k!} \frac{\partial^k}{\partial x^k} \left[f(x,t) \langle (\Delta x)^k \rangle (x,t) \right],$$

after assuming that the infinite sum is convergent and interchanging the summation and integration. The last expression was evaluated by introducing the kth moment or kth expectation

$$\langle (\Delta x)^k \rangle = \int (\Delta x)^k \Psi(\Delta x, t + \Delta t | x, t) d(\Delta x),$$

and is a function of (x,t). On rewriting the expanded Chapman-Kolmogorov equation as

$$\frac{f(x, t + \Delta t) - f(x, t)}{\Delta t} = \sum_{k=1}^{\infty} \frac{(-1)^k}{k! \Delta t} \frac{\partial^k}{\partial x^k} \left[f(x,t) \langle (\Delta x)^k \rangle (x,t) \right],$$

and letting $\Delta t \to 0$, implies the left-hand-side becomes

$$\lim_{\Delta t \to 0} \frac{f(x, t + \Delta t) - f(x, t)}{\Delta t} = \frac{\partial f}{\partial t}.$$

There is some subtlety in the limit which is not immediately apparent. The time Δt really refers to the Markovian time scale (Fig. 4.2) which is certainly smaller than the macroscopic dissipation time scale. However, to ensure that the system is Markovian requires the collisional time scale $\tau_c \ll \Delta t$, and so in this sense Δt cannot tend to zero! We conveniently overlook this technical point and express the limiting form of the Chapman-Kolmogorov equation as

$$\frac{\partial f}{\partial t} = \sum_{k=1}^{\infty} (-1)^k \frac{\partial^k}{\partial x^k} \left[\lim_{\Delta t \to 0} \frac{\langle (\Delta x)^k \rangle}{k! \Delta t} f(x,t) \right].$$

By defining diffusion coefficients

$$D^{(k)}(x,t) \equiv \lim_{\Delta t \to 0} \frac{\langle (\Delta x)^k \rangle}{k! \Delta t},$$

we derive the infinite order "diffusion" equation

$$\frac{\partial f}{\partial t} = \sum_{k=1}^{\infty} (-1)^k \frac{\partial^k}{\partial x^k} \left[D^{(k)}(x,t) f(x,t) \right].$$

If we truncate the expansion after $k = 2$, we have the general form of the *Fokker-Planck equation*,

$$\frac{\partial f}{\partial t} = -\frac{\partial}{\partial x}\left[D^{(1)}(x,t)f(x,t)\right] + \frac{\partial^2}{\partial x^2}\left[D^{(2)}(x,t)f(x,t)\right], \quad (4.15)$$

in the variable x on which incremental changes occur.

For particle motion in which the velocity experiences incremental changes through collisions (Coulomb collisions for a collisional plasma, pitch-angle scattering of fast particles in low frequency turbulence for a collisionless plasma, or Brownian motion in a fluid or gas), the variable x is replaced by the velocity v and Δx by Δv. Accordingly, the diffusion coefficients become

$$D^{(1)}(v,t) = \lim_{\Delta t \to 0} \frac{\langle \Delta v \rangle}{\Delta t};$$

$$D^{(2)}(v,t) = \lim_{\Delta t \to 0} \frac{\langle (\Delta v)^2 \rangle}{2\Delta t},$$

and the 1D Fokker-Planck equation in velocity space becomes

$$\frac{\partial f}{\partial t} = -\frac{\partial}{\partial v}\left(\frac{\langle \Delta v \rangle}{\Delta t}(v,t)f(v,t)\right) + \frac{\partial^2}{\partial v^2}\left(\frac{\langle (\Delta v)^2 \rangle}{2\Delta t}(v,t)f(v,t)\right) \equiv C(f). \quad (4.16)$$

The right-hand-side can be regarded as the collision operator. The first term on the right-hand-side describes the average change in v, and has the character of a drag force slowing the particle. The second term describes diffusion in velocity space due to particle scattering. In equilibrium, the collisional drag and diffusion will balance and result will be a Maxwellian distribution function describing the plasma. The extension of the Fokker-Planck equation to three dimensions is straightforward and is given by

$$C(f) = -\nabla_v \cdot \mathbf{J},$$

where ∇_v is the divergence in velocity space and $\mathbf{J}$ is the flux in velocity space

$$J_k \equiv \frac{\langle \Delta v_k \rangle}{\Delta t} f(\mathbf{v},\mathbf{x},t) - \frac{\partial}{\partial v_j}\left(\frac{\langle \Delta v_k \Delta v_j \rangle}{2\Delta t} f(\mathbf{v},\mathbf{x},t)\right).$$

The velocity diffusion coefficient is a tensor indicating that diffusion is not necessarily isotropic, and indeed may not even be governed by the same physical processes in different directions (see later the discussion about the parallel and perpendicular diffusion coefficients for energetic particles in a collisionless plasma).

The neglect of terms higher than second order is justified in that these terms are smaller by a factor of $(\ln \Lambda)^{-1}$, where Λ is proportional to the Debye length divided by the Coulomb impact parameter. Higher order terms describe the effects of close collisions that yield large deflections in the velocity vector. Most astrophysical and almost all laboratory plasmas satisfy $\ln \Lambda \gg 1$ but this is not necessarily true in a solar context.

Finally, note again that a plasma is comprised of multiple species typically, and each distribution experiences collisions with the others. Since collisions are additive, we have the velocity space flux of particles of species a expressed as the sum

$$\mathbf{J}^a = \sum_b \mathbf{J}^{ab},$$

due to contributions from collisions with each species b, including a.

Exercises

1. Consider a coin tossing event. Suppose

$$X(t_n) \equiv X_n = \sum_{i=1}^{n} x_i,$$

where x_i are given by $x_n = +1$ for a head and -1 for a tail. Calculate the pdf for the experiment.

4.3 Collision Dynamics, the Rosenbluth Potentials, and the Landau Collision Operator

To determine explicit forms of the diffusion coefficients,[2] we consider first the dynamics of single particle collisions and then evaluate the statistics of cumulative collisions, i.e., we compute the velocity-space flux of particle species a experiencing collisions with particles of species b. Consider the masses of two colliding particles a and b to be arbitrary. To evaluate the changes in velocity, we need to compute the particle deflection α – see Fig. 4.3. Introduce the Lagrangian

$$L = \frac{m_a \dot{\mathbf{x}}_a^2}{2} + \frac{m_b \dot{\mathbf{x}}_b^2}{2} - \frac{q_a q_b}{4\pi\varepsilon_0 |\mathbf{x}_a - \mathbf{x}_b|},$$

and express it in the coordinates of the center of mass and the relative position

$$\mathbf{R} \equiv \frac{m_a \mathbf{x}_a + m_b \mathbf{x}_b}{m_a + m_b}, \quad \mathbf{r} = \mathbf{x}_a - \mathbf{x}_b,$$

---

[2] The following sections deriving the collisional transport coefficients are based on the extensive monograph by Helander and Sigmar (2002).

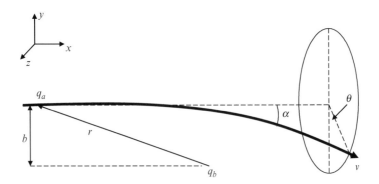

Fig. 4.3 Collision dynamics and angles in the rest frame of particle b. See text for details

so that

$$L = \frac{(m_a + m_b)\dot{\mathbf{R}}^2}{2} + \frac{1}{2}\frac{m_a m_b}{m_a + m_b}\dot{r}^2 - \frac{q_a q_b}{4\pi\varepsilon_0 r}.$$

L is independent of $\mathbf{R}$, so from the Euler equations,

$$\frac{d}{dt}\left(\frac{\partial L}{\partial \dot{\mathbf{R}}}\right) = \frac{\partial L}{\partial \mathbf{R}} \quad \text{and} \quad \frac{\partial L}{\partial \mathbf{R}} = 0.$$

Consequently, $(m_a + m_b)\ddot{\mathbf{R}} = 0$ implies that the center of mass moves at constant speed, $\dot{\mathbf{R}} = $ const. The first term in the center of mass form of the Lagrangian is therefore an additive constant, and the two remaining terms describe the motion around the center of gravity. Thus, this is simply the Lagrangian of a particle of reduced mass $m_* = m_a m_b/(m_a + m_b)$ moving in a fixed Coulomb field. Hence, the deflection angle of the relative velocity vector $\mathbf{V}_{\text{rel}} = \dot{\mathbf{r}}$ is equal to[3]

---

[3]Recall that if we have a particle of charge q_a and velocity $\mathbf{v}$ traveling past a particle of charge q_b, we can compute the deflection quite easily if the impact parameter b is assumed to be so large that the deviation is through only a small angle α, as in Fig. 4.3. The Coulomb force $q_a q_b/4\pi\varepsilon_0 r^2(t)$, where $r(t) \simeq (b^2 + v^2 t^2)^{1/2}$ is the separation between the particles at time t, implies that the momentum change in the y-direction to the moving charged particle is given by

$$m_a \Delta v_y = \int_{-\infty}^{\infty} \frac{q_a q_b}{4\pi\varepsilon_0 r^2(t)} \frac{b}{r(t)} dt = \frac{q_a q_b}{4\pi\varepsilon_0} \int_{-\infty}^{\infty} \frac{b}{(b^2 + v^2 t^2)^{3/2}} dt = \frac{q_a q_b}{2\pi\varepsilon_0 b v}.$$

The angle α is then given by

$$\alpha = \frac{\Delta v_y}{v} = \frac{q_a q_b}{2\pi\varepsilon_0 b m_a v^2} = \frac{b_{min}}{b},$$

where $b_{min} \equiv q_a q_b/2\pi\varepsilon_0 m_a v^2$. The deflection angle $\alpha \ll 1$ if the impact parameter $b \gg b_{min}$ i.e., large.

4.3 Collision Dynamics, the Rosenbluth Potentials...

$$\alpha = \frac{q_a q_b}{2\pi\varepsilon_0 r m_* V_{rel}^2},$$

in the collision. Introduce the orthogonal coordinate system (x, y, z) – Fig. 4.3 – with x in the direction of $\mathbf{v}_a$. The relative velocity $\mathbf{V}_{rel}$ then varies according to

$$\Delta V_{rel,x} = V_{rel}(\cos\alpha - 1);$$
$$\Delta V_{rel,y} = V_{rel}\sin\alpha\cos\theta;$$
$$\Delta V_{rel,z} = V_{rel}\sin\alpha\sin\theta,$$

as a result of the collision. Here θ denotes the angle of the perpendicular component of the deflection vector to the $x - y$-plane (Fig. 4.3). On using

$$\mathbf{x}_a = \mathbf{R} + \frac{m_b}{m_a + m_b}\mathbf{r},$$

we can compute the change $\Delta\mathbf{v}_a$ in the velocity vector of particle a as

$$\Delta\mathbf{v}_a = \frac{m_b}{m_a + m_b}\Delta\mathbf{V}_{rel}.$$

In a collision with impact parameter b, $\mathbf{v}_a$ changes as

$$\Delta v_x = \frac{m_b(\cos\alpha - 1)}{m_a + m_b}V_{rel} \simeq -\left(1 + \frac{m_a}{m_b}\right)\left(\frac{q_a q_b}{2\pi\varepsilon_0 m_a}\right)^2 \frac{1}{2r^2 V_{rel}^3},$$

in the x-direction, and

$$\Delta v_y = \frac{m_b \sin\alpha\cos\theta}{m_a + m_b}V_{rel} = \frac{q_a q_b}{2\pi\varepsilon_0 m_a}\frac{\cos\theta}{V_{rel}r};$$
$$\Delta v_z = \frac{m_b \sin\alpha\sin\theta}{m_a + m_b}V_{rel} = \frac{q_a q_b}{2\pi\varepsilon_0 m_a}\frac{\sin\theta}{V_{rel}r},$$

in the y- and z-directions, after approximating $\cos\alpha - 1 \sim -\alpha^2/2$ and $\sin\alpha \sim \alpha$ for assumed small angle deflections.

On the basis of the velocity deviation for a single collision, we now consider the cumulative effect of many collisions. To determine the number of collisions that can occur between a given particle a and particles of species b in time Δt with impact parameters in the interval $[r, r + dr]$ and angles in the interval $[\theta, \theta + d\theta]$, consider Fig. 4.4. The area spanned by dr and $d\theta$ is the cross-section $d\sigma = rdrd\theta$, and $\mathbf{V}_{rel}dt$ is the distance particle a travels relative to b in time dt. The volume corresponding to the cross section $d\sigma$ along the relative velocity vector $\mathbf{V}_{rel}$ is given by $dV = V_{rel}dt d\sigma$. On multiplying the volume by the distribution function f_b and integrating over all possible velocities that species b can have then yields the number

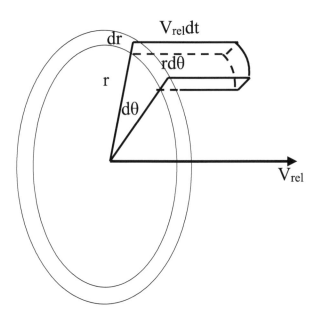

Fig. 4.4 Geometry to compute the collision frequency of a particle traveling through a gas of particles. See text for details

of particles in the volume. Consequently, the frequency of collisions of particle a with species b is given by

$$\Delta t \, d\theta \, r \, dr \int f_b(\mathbf{v}') V_{rel} d^3 v',$$

where $\mathbf{V}_{rel} = \mathbf{v} - \mathbf{v}'$ is the relative velocity of colliding particles.

The velocity change $\Delta \mathbf{v}$ resulting from a single collision of a particle with another can be used to estimate the average change in the velocity vector of particle a due to collisions with particles b. Thus, multiplying Δv_x, Δv_y and Δv_z by the number of collisions and integrating over r and θ yields the average change in velocity as

$$\frac{\langle \Delta v_x \rangle^{ab}}{\Delta t} = -\left(1 + \frac{m_a}{m_b}\right)\left(\frac{q_a q_b}{2\pi\varepsilon_0 m_a}\right)^2 \int_{r_{min}}^{\lambda_D} \int_0^{2\pi} \int \frac{1}{2r^2 V_{rel}^3} d\theta \, r \, dr \, f_b(\mathbf{v}') V_{rel} d^3 v'$$

$$= -\frac{1}{4\pi}\left(1 + \frac{m_a}{m_b}\right)\left(\frac{q_a q_b}{\varepsilon_0 m_a}\right)^2 \int_{r_{min}}^{\lambda_D} \frac{dr}{r} \int \frac{1}{V_{rel}^2} f_b(\mathbf{v}') d^3 v'$$

$$= -\frac{L^{ab}}{4\pi}\left(1 + \frac{m_a}{m_b}\right) \int \frac{1}{V_{rel}^2} f_b(\mathbf{v}') d^3 v',$$

4.3 Collision Dynamics, the Rosenbluth Potentials...

where the logarithmic integral has been "cut-off" at some minimum to avoid a divergent integral. Of course, particles somewhat further than the Debye radius or length

$$\lambda_D \equiv \sqrt{\frac{\varepsilon_0 k T}{n q^2}},$$

from the other particle do not experience an effective collision (i.e., the scattering particles remain on straight-line trajectories) so the integration regime is $[r_{min}, \lambda]$. Hence,

$$L^{ab} = \left(\frac{q_a q_b}{\varepsilon_0 m_a}\right)^2 \int_{r_{min}}^{\lambda_D} \frac{dr}{r} = \left(\frac{q_a q_b}{\varepsilon_0 m_a}\right)^2 \ln \Lambda,$$

where $\ln \Lambda \equiv \ln(\lambda_D/r_{min})$ is known as the Coulomb logarithm. Similarly, because of the $\cos\theta$ and $\sin\theta$ factors, we have

$$\frac{\langle \Delta v_y \rangle^{ab}}{\Delta t} = \frac{\langle \Delta v_z \rangle^{ab}}{\Delta t} = 0.$$

Since (Exercise),

$$\frac{\langle (\Delta v_y)^2 \rangle^{ab}}{2\Delta t} = \frac{\langle (\Delta v_z)^2 \rangle^{ab}}{2\Delta t} = \frac{L^{ab}}{8\pi} \int \frac{1}{V_{rel}} f_b(\mathbf{v}') d^3 v',$$

and

$$\frac{\langle (\Delta v_x)^2 \rangle^{ab}}{2\Delta t} = \frac{\pi}{8}\left(1 + \frac{m_a}{m_b}\right)^2 \left(\frac{q_a q_b}{2\pi\varepsilon_0 m_a}\right)^4 \left[\frac{1}{r_{min}^2} - \frac{1}{\lambda_D^2}\right] \int \frac{f_b(\mathbf{v}')}{V_{rel}^5} d^3 v',$$

we have crudely

$$\langle (\Delta v_y)^2 \rangle \sim r_{min}^2 \ln \Lambda \langle (\Delta v_x)^2 \rangle.$$

This result suggests that we may neglect $\langle (\Delta v_x)^2 \rangle / 2\Delta t$ under most circumstances. For the same reason, the higher-order terms in the Fokker-Planck expansion are similarly ordered and are thus typically neglected.

These expectations can be inserted into the Fokker-Planck collision operator. However, we have expressed $\langle \Delta v_i \rangle$ and $\langle \Delta v_i \Delta v_j \rangle$ in a coordinate system aligned with the velocity vector of one of the colliding particles, so we need to introduce an arbitrary orthogonal coordinate system with unit vectors $\mathbf{e}_k$. We therefore have

$$\frac{\langle \Delta v_i \rangle^{ab}}{\Delta t} = \frac{\langle \mathbf{e}_i \cdot \hat{\mathbf{x}} \Delta v_x \rangle^{ab}}{\Delta t} = -\frac{L^{ab}}{4\pi}\left(1 + \frac{m_a}{m_b}\right) \int \frac{V_{rel,i}}{V_{rel}^3} f_b(\mathbf{v}') d^3 v',$$

and, noting that $\langle \Delta v_x \Delta v_y \rangle / \Delta t = 0$ and $\langle (\Delta v_y)^2 \rangle / \Delta t = \langle (\Delta v_z)^2 \rangle / \Delta t$,

$$\frac{\langle \Delta v_i \Delta v_j \rangle^{ab}}{2\Delta t} = \frac{\langle \mathbf{e}_i \cdot (\hat{\mathbf{y}} \Delta v_y + \hat{\mathbf{z}} \Delta v_z) \mathbf{e}_j \cdot (\hat{\mathbf{y}} \Delta v_y + \hat{\mathbf{z}} \Delta v_z) \rangle^{ab}}{2\Delta t}$$

$$= \frac{\langle (\mathbf{e}_i \cdot \hat{\mathbf{y}})(\mathbf{e}_j \cdot \hat{\mathbf{y}})(\Delta v_y)^2 + (\mathbf{e}_i \cdot \hat{\mathbf{z}})(\mathbf{e}_j \cdot \hat{\mathbf{z}})(\Delta v_z)^2 \rangle^{ab}}{2\Delta t}$$

$$= \frac{\langle [\mathbf{e}_i \cdot \mathbf{e}_j - (\mathbf{e}_i \cdot \hat{\mathbf{x}})(\mathbf{e}_j \cdot \hat{\mathbf{x}})](\Delta v_y)^2 \rangle^{ab}}{2\Delta t}$$

$$= \frac{\langle (\delta_{ij} - V_{rel,i} V_{rel,j}/V_{rel}^2)(\Delta v_y)^2 \rangle^{ab}}{2\Delta t} = \frac{L^{ab}}{8\pi} \int V_{ij} f_b(\mathbf{v}') d^3 v',$$

where the tensor related to the relative velocities of the particles has been defined as

$$V_{ij} = \frac{V_{rel}^2 \delta_{ij} - V_{rel,i} V_{rel,j}}{V_{rel}^3}.$$

To complete the derivation of the Fokker-Planck collision operator, introduce the "Rosenbluth potentials[4]"

$$\phi_b(\mathbf{v}) \equiv -\frac{1}{4\pi} \int \frac{1}{V_{rel}} f_b(\mathbf{v}') d^3 v';$$

$$\psi_b(\mathbf{v}) \equiv -\frac{1}{8\pi} \int V_{rel} f_b(\mathbf{v}') d^3 v'.$$

Since

$$\frac{\partial V_{rel}}{\partial v_i} = \frac{\partial}{\partial v_i} \sqrt{\sum_k (v_k - v_k')^2} = \frac{V_{rel,i}}{V_{rel}},$$

and

$$\frac{\partial^2 V_{rel}}{\partial v_i \partial v_j} = V_{ij},$$

the expectation values can be expressed in terms of the potentials

$$A_i^{ab} \equiv -\frac{\langle \Delta v_i \rangle^{ab}}{\Delta t} = \left(1 + \frac{m_a}{m_b}\right) L^{ab} \frac{\partial \phi_b}{\partial v_i}; \qquad (4.17)$$

---

[4]Rosenbluth et al. (1965).

4.3 Collision Dynamics, the Rosenbluth Potentials...

$$D_{ij}^{ab} \equiv \frac{\langle \Delta v_i \Delta v_j \rangle^{ab}}{2\Delta t} = -L^{ab} \frac{\partial^2 \psi_b}{\partial v_i \partial v_j}. \tag{4.18}$$

Hence, the average force experienced by particle species a colliding with particle species b is simply $-m_a A_i^{ab}$ and D_{ij}^{ab} is a diffusion tensor in velocity space. The Fokker-Planck collision operator can therefore be expressed as

$$C_{ab}(f_a, f_b) = \frac{\partial}{\partial v_i} \left[A_i^{ab} f_a + \frac{\partial}{\partial v_j} \left(D_{ij}^{ab} f_a \right) \right]. \tag{4.19}$$

Since the Laplacian in velocity space is

$$\nabla_v^2 = \sum_i \frac{\partial^2}{\partial v_i^2},$$

then $\nabla_v^2 V_{rel} = V_{ii} = 2/V_{rel}$, so that

$$\nabla_v^2 \psi_b = \phi_b.$$

Hence, A_i^{ab} and D_{ij}^{ab} are related by the "Einstein relation"

$$A_i^{ab} = -\left(1 + \frac{m_a}{m_b}\right) \frac{\partial D_{ij}^{ab}}{\partial v_j}.$$

The collision operator (4.19) can then be written in terms of the Rosenbluth potentials as

$$C_{ab}(f_a, f_b) = \ln \Lambda \left(\frac{q_a q_b}{m_a \varepsilon_0}\right)^2 \frac{\partial}{\partial v_i} \left(\frac{m_a}{m_b} \frac{\partial \phi_b}{\partial v_i} f_a - \frac{\partial^2 \psi_b}{\partial v_i \partial v_j} \frac{\partial f_a}{\partial v_j} \right). \tag{4.20}$$

If we express this equation directly in the integral form of the Rosenbluth potentials, we obtain the Landau form of the collision operator,[5]

$$C_{ab}(f_a, f_b) = \frac{\ln \Lambda}{8\pi m_a} \left(\frac{q_a q_b}{\varepsilon_0}\right)^2 \frac{\partial}{\partial v_i} \int V_{ij} \left[\frac{f_a(\mathbf{v})}{m_b} \frac{\partial f_b(\mathbf{v}')}{\partial v'_j} - \frac{f_b(\mathbf{v}')}{m_a} \frac{\partial f_a(\mathbf{v})}{\partial v_j} \right] d^3v', \tag{4.21}$$

after using $\partial V_{ij}/\partial v_j = 2 V_{rel,i}/V_{rel}$.

---

[5] Landau (1936).

Exercises

1. Show that

$$\frac{\langle(\Delta v_y)^2\rangle^{ab}}{2\Delta t} = \frac{\langle(\Delta v_z)^2\rangle^{ab}}{2\Delta t} = \frac{L^{ab}}{8\pi}\int \frac{1}{V_{rel}}f_b(\mathbf{v}')d^3v',$$

and

$$\frac{\langle(\Delta v_x)^2\rangle^{ab}}{2\Delta t} = \frac{\pi}{8}\left(1+\frac{m_a}{m_b}\right)^2 \left(\frac{q_a q_b}{2\pi\varepsilon_0 m_a}\right)^4 \left[\frac{1}{r_{min}^2}-\frac{1}{\lambda_D^2}\right]\int \frac{f_b(\mathbf{v}')}{V_{rel}^5}d^3v'.$$

2. By direct substitution, show that the Landau collision operator (and hence the other forms) satisfy the conservation laws

$$\int C_{ab}(f_a)d^3v = 0;$$

$$\int m_a \mathbf{v} C_{ab}(f_a, f_b)d^3v = -\int m_b \mathbf{v} C_{ba}(f_b, f_a)d^3v;$$

$$\int \frac{1}{2}m_a v^2 C_{ab}(f_a, f_b)d^3v = -\int \frac{1}{2}m_b v^2 C_{ba}(f_b, f_a)d^3v.$$

4.4 Electron-Proton Collisions

The Coulomb collision operator can be simplified if the colliding particles move at very different speeds, such as electrons colliding with protons moving at some average speed $\mathbf{u}_p$. The velocity spread around $\mathbf{u}_p$ in the proton distribution function is the order of the proton thermal speed,

$$v_{Tp} = \sqrt{2kT_p/m_p} \ll \sqrt{2kT_e/m_e} = v_{Te}$$

unless $T_e \ll T_p$. To the electrons, the proton distribution is therefore a very narrowly peaked distribution function, which we may approximate as a delta function

$$f_p(\mathbf{v}) \simeq n_p \delta(\mathbf{v}-\mathbf{u}_p),$$

with $u_p \ll v_{Te}$. The Rosenbluth potentials then become

$$\phi_p \simeq -\frac{n_p}{4\pi}\frac{1}{|\mathbf{v}-\mathbf{u}_p|} \sim -\frac{n_p}{4\pi v}\left(1+\frac{\mathbf{v}\cdot\mathbf{u}_p}{v^2}\right);$$

$$\psi_p \simeq -\frac{n_p}{8\pi}|\mathbf{v}-\mathbf{u}_p| \sim -\frac{n_p}{8\pi}v\left(1-\frac{\mathbf{v}\cdot\mathbf{u}_p}{v^2}\right).$$

4.4 Electron-Proton Collisions

In the Rosenbluth potential form of the collision operator, the ϕ_p term is multiplied by $m_e/m_p \ll 1$, implying that only the ψ_p term is important to electron-proton collisions. Furthermore, in the expression approximating ψ_p, note that the second term is small since $u_p \ll v$. Hence,

$$C_{ep}(f_e) \simeq -L^{ep} \frac{\partial}{\partial v_i} \left(\frac{\partial^2 \psi_p}{\partial v_i \partial v_j} \frac{\partial f_e}{\partial v_j} \right)$$

$$= \frac{n_p}{8\pi} L^{ep} \frac{\partial}{\partial v_i} \left[\frac{\partial^2 v}{\partial v_i \partial v_j} \frac{\partial f_e}{\partial v_j} - \frac{\partial^2}{\partial v_i \partial v_j} \left(\frac{v_k u_{pk}}{v} \right) \frac{\partial f_{Me}}{\partial v_j} \right]$$

$$= C_{ep}^0 + C_{ep}^1, \qquad (4.22)$$

where the second term in the expansion was approximated using a Maxwellian electron distribution function f_{Me} since it is relatively small (and thus suitable to approximate the distribution function by the lowest-order distribution). Note that the tensor

$$\frac{\partial^2 v}{\partial v_i \partial v_j} = \frac{\partial}{\partial v_i} \left(\frac{\partial v}{\partial v_j} \right) = \frac{\partial}{\partial v_i} \left(\frac{v_j}{v} \right) = \frac{v^2 \delta_{ij} - v_i v_j}{v^3} \equiv W_{ij}$$

can be expressed as

$$\begin{pmatrix} v^2 - v_1^2 & -v_v v_2 & -v_1 v_3 \\ -v_2 v_1 & v^2 - v_2^2 & -v_2 v_3 \\ -v_3 v_1 & -v_3 v_2 & v^2 - v_3^2 \end{pmatrix}$$

which is easily seen to be orthogonal to the vector $\mathbf{v} = (v_1, v_2, v_3)$, and hence to

$$\frac{\partial f_{Me}}{\partial \mathbf{v}} = -\frac{m_e \mathbf{v}}{kT_e} f_{Me}.$$

This implies that both collision terms need to be retained in case the collisional term operates on a Maxwellian distribution since the first term vanishes, and so both terms may be of the same order. We can use this result to simplify the C_{ep}^0 term, expressing it as

$$C_{ep}^0 = \frac{n_p}{8\pi} L^{ep} \frac{\partial}{\partial v_i} \left(\frac{\partial^2 v}{\partial v_i \partial v_j} \frac{\partial f_e}{\partial v_j} \right) = \frac{n_p}{8\pi} L^{ep} \frac{\partial}{\partial \mathbf{v}} \cdot \left[\frac{1}{v} \frac{\partial f_e}{\partial \mathbf{v}} - \frac{\mathbf{v}}{v^3} \left(\mathbf{v} \cdot \frac{\partial f_e}{\partial \mathbf{v}} \right) \right].$$

We need only include the components of the square bracket that are perpendicular to $\mathbf{v}$. This implies that the second term in C_{ep}^0 will vanish identically, and only the non-radial terms in the first term will contribute. Physically, this is because collisions of electrons with ions do not change the magnitude of v but only the direction. Consequently, electrons will scatter on a sphere of fixed velocity radius, so it is sensible to introduce spherical coordinates in velocity space (v, θ, ϕ). In view of the comments above, on using spherical coordinates, we obtain

$$C_{ep}^0 = \frac{n_p L^{ep}}{8\pi v^3} \left[\frac{1}{\sin \theta} \frac{\partial}{\partial \theta} \left(\sin \theta \frac{\partial f_e}{\partial \theta} \right) + \frac{1}{\sin^2 \theta} \frac{\partial^2 f_e}{\partial \phi^2} \right] \equiv \frac{n_p L^{ep}}{4\pi v^3} \mathcal{L}(f_e),$$

where we have introduced the Lorentz scattering operator $\mathcal{L}(f_e)$, i.e.,

$$\begin{aligned} \mathcal{L}(f_e) &\equiv \frac{1}{2} \left[\frac{1}{\sin \theta} \frac{\partial}{\partial \theta} \left(\sin \theta \frac{\partial f_e}{\partial \theta} \right) + \frac{1}{\sin^2 \theta} \frac{\partial^2 f_e}{\partial \phi^2} \right] \\ &= \frac{1}{2} \left[\frac{\partial}{\partial \mu} \left[(1 - \mu^2) \frac{\partial f_e}{\partial \mu} \right] + \frac{1}{1 - \mu^2} \frac{\partial^2 f_e}{\partial \phi^2} \right] \\ &= \frac{1}{2} \frac{\partial}{\partial \mu} \left[(1 - \mu^2) \frac{\partial f_e}{\partial \mu} \right], \end{aligned}$$

where we have introduced the pitch-angle $\mu = \cos \theta$, and the last line follows if we assume that the electron distribution f_e is gyrotropic (i.e., independent of ϕ).

To evaluate C_{ep}^1, we need to evaluate the derivatives. We use the Einstein summation convention to obtain

$$\begin{aligned} \frac{\partial^2}{\partial v_i \partial v_j} \left(\frac{v_k}{v} \right) \frac{\partial f_{Me}}{\partial v_j} &= \frac{\partial}{\partial v_i} \left(\frac{\delta_{ij}}{v} - \frac{v_k v_j}{v^3} \right) \frac{\partial f_{Me}}{\partial v_j} \\ &= \left(\frac{v_i v_j}{v^3} \delta_{jk} + \frac{v_j^2}{v^3} \delta_{ik} + \frac{v_j v_k}{v^3} \delta_{ij} - 3 \frac{v_i v_k v_j^2}{v^5} \right) \frac{m_e}{kT_e} f_{Me} \\ &= \frac{v^2 \delta_{ik} - v_i v_k}{v^3} \frac{m_e}{kT_e} f_{Me} = W_{km} \frac{m_e}{kT_e} f_{Me}, \end{aligned}$$

and

$$\begin{aligned} \frac{\partial W_{ik}}{\partial v_i} &= \frac{\partial}{\partial v_i} \frac{v^2 \delta_{ik} - v_i v_k}{v^3} \\ &= -\frac{v_i}{v^3} \delta_{ik} - \frac{v_k}{v^3} - \frac{v_i}{v^3} \delta_{ik} + 3 \frac{v_i^2 v_k}{v^5} = \frac{2 v_k}{v^3}. \end{aligned}$$

Use of these two expressions allows us to find

$$C_{ep}^1 = -\frac{n_p L^{ep}}{8\pi} \frac{m_e u_{pk} f_{Me}}{kT_e} \frac{\partial W_{ik}}{\partial v_i} = -\frac{n_p L^{ep}}{4\pi} \frac{m_e}{kT_e} \frac{\mathbf{v} \cdot \mathbf{u}_p}{v^3} f_{Me}.$$

The total electron-proton scattering operator can therefore be expressed as the sum of the Lorentz scattering operator and C_{ep}^1,

$$C_{ep}(f_e) = \nu_{ep}(v) \left(\mathcal{L}(f_e) - \frac{m_e \mathbf{v} \cdot \mathbf{u}_p}{kT_e} f_{Me} \right), \qquad (4.23)$$

4.5 Collisions with a Maxwellian Background

where

$$\nu_{ep}(v) \equiv \frac{n_p L^{ep}}{4\pi v^3} = \frac{n_p e^4}{4\pi m_e^2 \varepsilon_0^2 v^3} \ln \Lambda = \frac{3\sqrt{\pi}}{4\tau_{ep}} \left(\frac{v_{Te}}{v}\right)^3,$$

is a velocity-dependent electron-proton collision frequency, and τ_{ep} is a measure of the electron-proton collision time,

$$\tau_{ep} = \frac{12(\pi k T_e)^{3/2} m_e^{1/2} \varepsilon_0^2}{\sqrt{2} n_p e^4 \ln \Lambda}.$$

The first part of the collision operator (4.23) describes the collision of electrons with "infinitely heavy" stationary protons, implying that only the electron direction and not the velocity changes in a collision. Consequently, there is only diffusion in velocity space on a sphere of constant radius $v = \text{constant}$, and the collision operator is spherically symmetric. Finally, note that the proton mass is completely absent from the collision operator, depending as it does only on charge e. This makes it straightforward to model electron collisions in a plasma comprising several different ion species.

Exercises

1. How would the result (4.23) change (i) if the electrons scattered off a background of α particles (He nuclei), and (ii) a mixture of protons and α particles, as found in the solar wind emitted by the Sun?

4.5 Collisions with a Maxwellian Background

Assume a stationary Maxwellian background population of charged particles b,

$$f_b(\mathbf{v}) = f_{b0}(v) = \frac{n_b}{\pi^{3/2} v_{Tb}^3} e^{-(v/v_{Tb})^2},$$

where $v_{Tb} = \sqrt{2 T_b / m_b}$ is the thermal speed. The background distribution is isotropic in v, making the Rosenbluth potentials dependent only on v, thus

$$\phi_b(\mathbf{v}) = \phi_b(v), \quad \psi_b(\mathbf{v}) = \psi_b(v).$$

Hence,

$$\frac{\partial \phi_b}{\partial v_i} = \frac{\partial \phi_b}{\partial v} \frac{\partial v}{\partial v_i} = \frac{v_i}{v} \frac{\partial \phi_b}{\partial v} = \frac{v_i}{v} \phi_b';$$

$$\frac{\partial^2 \psi_b}{\partial v_i \partial v_j} = \frac{\partial^2 v}{\partial v_i \partial v_j} \psi_b' + \frac{v_i v_j}{v^2} \psi_b'' = W_{ij} \psi_b' + \frac{v_i v_j}{v^2} \psi_b''.$$

where $W_{ij} = (v^2 \delta_{ij} - v_i v_j)/v^3$. Using these expressions, the Rosenbluth form of the collision operator for an isotropic background particle distribution reduces to

$$C_{ab}(f_a, f_{b0}) = \ln \Lambda \left(\frac{q_a q_b}{m_a \varepsilon_0}\right)^2 \frac{\partial}{\partial v_i} \left(\frac{m_a}{m_b} \frac{\partial \phi_b}{\partial v_i} f_a - \frac{\partial^2 \psi_b}{\partial v_i \partial v_j} \frac{\partial f_a}{\partial v_j}\right)$$

$$= L^{ab} \frac{\partial}{\partial v_i} \left(\frac{m_a}{m_b} \frac{v_i}{v} \phi_b' f_a - \left(W_{ij} \psi' + \frac{v_i v_j}{v^2} \psi_b''\right) \frac{\partial f_a}{\partial v_j}\right).$$

On using

$$\frac{\partial}{\partial v_i} \left(W_{ij} \frac{\partial f_a}{\partial v_j}\right) = \frac{2}{v^3} \mathcal{L}(f_a),$$

and

$$\nabla_v \cdot (A(v)\mathbf{v}) = \frac{1}{v^2} \frac{\partial (v^3 A)}{\partial v},$$

we obtain

$$C_{ab}(f_a, f_{b0}) = -\frac{2L^{ab}}{v^3} \psi_b' \mathcal{L}(f_a) + \frac{L^{ab}}{v^2} \frac{\partial}{\partial v} \left[v^3 \left(\frac{m_a}{m_b} \frac{\phi_b'}{v} f_a - \frac{\psi_b''}{v} \frac{\partial f}{\partial v}\right)\right],$$

where the velocity derivative is taken at fixed direction of $\mathbf{v}$, and we have used $v_j \partial f_a / \partial v_j = v \partial f_a / \partial v$. We will assume an orthogonal coordinate system in which the x-coordinate is parallel to the velocity vector.

To identify the role of each of the terms in the collision operator in this coordinate system, recall that Fokker-Planck collisions are described by the expectations

$$A_i^{ab} = -\frac{1}{\Delta t} \left\langle \begin{array}{c} \Delta v_\| \\ 0 \\ 0 \end{array} \right\rangle^{ab};$$

$$D_{ij}^{ab} = \frac{1}{2\Delta t} \left\langle \begin{array}{ccc} \Delta v_\|^2 & 0 & 0 \\ 0 & \Delta v_\perp^2/2 & 0 \\ 0 & 0 & \Delta v_\perp^2/2 \end{array} \right\rangle^{ab}.$$

The other two directions are orthogonal to x and denoted by the $\perp$ symbol. Hence, $\Delta v_\| = \Delta v_x$ and $\Delta v_\perp^2 = \Delta v_y^2 + \Delta v_z^2$. With the isotropic forms of the Rosenbluth potentials, we can evaluate the elements of the tensors A_i^{ab} and D_{ij}^{ab}. From (4.17), we have

$$\frac{1}{v} A_x^{ab} = -\frac{\langle \Delta v_\| / v \rangle^{ab}}{\Delta t} = \frac{1}{v} \left(1 + \frac{m_a}{m_b}\right) L^{ab} \frac{\partial \phi_b}{\partial v_i}$$

$$= \left(1 + \frac{m_a}{m_b}\right) L^{ab} \frac{\phi_b'}{v} \equiv v_s^{ab}(v) \quad (4.24)$$

4.5 Collisions with a Maxwellian Background

so that

$$A_i^{ab} = v \begin{pmatrix} v_s^{ab} \\ 0 \\ 0 \end{pmatrix}.$$

The frequency v_s^{ab} describes the rate at which a particle of species a is decelerated by collisions with particles of species b. From (4.18), the elements of the collisional diffusion tensor can be expressed through (using $\mathbf{v} = v\hat{\mathbf{x}}$)

$$\frac{\langle (\Delta v_\parallel / v)^2 \rangle^{ab}}{2\Delta t} = -\frac{L^{ab}}{v^2} \frac{\partial^2 \psi_b}{\partial v_x^2} = -\frac{L^{ab}}{v^2} \psi_b'' \equiv \frac{1}{2} v_\parallel^{ab}(v); \qquad (4.25)$$

$$\frac{\langle (\Delta v_\perp / v)^2 \rangle^{ab}}{2\Delta t} = -\frac{L^{ab}}{v^2} \frac{\partial^2 \psi_b}{\partial v_\perp^2} = -\frac{L^{ab}}{v^3} \psi_b'(v) \equiv v_D^{ab}(v), \qquad (4.26)$$

from which it follows that the diffusion tensor can be expressed as

$$D_{ij}^{ab} = \frac{v^2}{2} \begin{pmatrix} v_\parallel^{ab} & 0 & 0 \\ 0 & v_D^{ab} & 0 \\ 0 & 0 & v_D^{ab} \end{pmatrix}^{ab}.$$

The tensor $\mathbf{D}^{ab}$ gives the velocity diffusion frequency $v_\parallel^{ab}$ in the parallel direction and v_D^{ab} is the frequency with which particles are scattered from the parallel direction. Consequently, if, for example, $v_\parallel^{ab}$ can be neglected, v_D^{ab} describes how quickly particles are scattered onto a spherical shell of constant radius in velocity space i.e., onto a shell of constant energy. Parallel scattering acts to broaden the shell.

On replacing ϕ_b'/v and ψ_b''/v by the streaming and diffusion scattering frequencies allows the collision operator to be expressed as

$$C_{ab}(f_a, f_{b0}) = v_D^{ab} \mathcal{L}(f_a) + \frac{1}{v^2} \frac{\partial}{\partial v} \left[v^3 \left(\frac{m_a}{m_a + m_b} v_s^{ab} f_a + \frac{1}{2} v_\parallel^{ab} v \frac{\partial f_a}{\partial v} \right) \right]. \qquad (4.27)$$

Recall that the Lorentz operator $\mathcal{L}$ simply describes particle diffusion on the surface of a sphere $v = $ constant, so this term gives the rate at which particles scatter on a sphere – i.e., changing their direction while preserving v.

To complete the analysis, we need to evaluate the Rosenbluth potentials $\phi_b(v)$ and $\psi_b(v)$ for a Maxwellian distribution function, so that we can compute the collision frequencies v_s^{ab}, $v_\parallel^{ab}$, and v_D^{ab}. A useful comparison can be drawn between the Rosenbluth potentials and the electrostatic potential Φ, defined as usual by

$$\nabla^2 \Phi = -\frac{\rho}{\varepsilon_0} \Rightarrow \Phi(\mathbf{r}) = \int \frac{\rho(\mathbf{r}')}{4\pi\varepsilon_0 |\mathbf{r} - \mathbf{r}'|} d^3 r',$$

where ρ denotes the charge density. Since

$$\phi_b(\mathbf{v}) = -\frac{1}{4\pi} \int \frac{1}{V_{rel}} f_b(\mathbf{v}') d^3 v',$$

($\mathbf{V}_{rel} = \mathbf{v} - \mathbf{v}'$) we have that

$$\nabla_v^2 \phi_b = \frac{1}{v^2} \frac{\partial}{\partial v} \left(v^2 \frac{\partial \phi_b}{\partial v} \right) = f_b(v), \tag{4.28}$$

after expressing the Laplacian in spherical velocity-space coordinates and assuming isotropy. Thus, ϕ_b is the "potential" associated with the "charge distribution" f_b in velocity space. This is the origin of the term Rosenbluth *potential*. On assuming that the distribution f_b is a Maxwellian, we can integrate the Possion equation (4.20) to obtain (after integrating by parts)

$$\phi_b'(v) = \frac{m_b n_b}{4\pi k T_b} G(v/v_{Tb}),$$

where $G(x)$ is the Chandrasekhar function, $x_b \equiv v/v_{Tb}$, $x_a \equiv v/v_{Ta}$,

$$G(x) \equiv \frac{f(x) - xf'(x)}{2x^2}, \quad f(x) = \frac{2}{\sqrt{\pi}} \int_0^x e^{-z^2} dz = \mathrm{erf}(x),$$

$$\rightarrow \begin{cases} \frac{2x}{3\sqrt{\pi}} & \text{as } x \to 0 \\ \frac{1}{2x^2} & \text{as } x \to \infty \end{cases},$$

and $\mathrm{erf}(x)$ is the familiar error function.

Similarly, the second Rosenbluth potential ψ_b can be obtained by using the relationship $\nabla_v^2 \psi_b = \phi_b$, or in spherical velocity-space coordinates

$$\frac{1}{v^2} \frac{d}{dv} \left(v^2 \frac{d\psi_b}{dv} \right) = \phi_b(v).$$

It can then be established that

$$\frac{d\psi_b}{dv} = -\frac{n_b}{8\pi} [f(x_b) - G(x_b)],$$

(Exercise). On using these results in (4.24)–(4.26), we obtain the collision frequencies in terms of the error function and the Chandrasekhar function. Consider first the streaming scattering frequency, obtaining

$$v_s^{ab} = \frac{q_a^2 q_b^2 \ln \Lambda}{\varepsilon_0^2 m_a^2} \left(1 + \frac{m_a}{m_b} \right) \frac{m_b n_b}{4\pi k T_b} \frac{1}{v_{Ta}} \frac{G(v/v_{Tb})}{v/v_{Ta}}$$

4.5 Collisions with a Maxwellian Background

$$= 2\frac{n_b q_a^2 q_b^2 \ln \Lambda}{4\pi \varepsilon_0^2 m_a^2 v_{Ta}^3} \frac{m_b}{m_a} \left(1 + \frac{m_a}{m_b}\right) \frac{T_a}{T_b} \frac{G(x_b)}{x_a}$$

$$= 2\bar{v}_{ab} \frac{T_a}{T_b} \left(1 + \frac{m_b}{m_a}\right) \frac{G(x_b)}{x_a}, \qquad (4.29)$$

while the corresponding diffusion scattering frequencies are given by

$$v_D^{ab}(v) = \bar{v}_{ab} \frac{f(x_b) - G(x_b)}{x_a^3}; \qquad (4.30)$$

$$v_{\parallel}^{ab}(v) = 2\bar{v}_{ab} \frac{G(x_b)}{x_a^3}; \qquad (4.31)$$

$$\text{where} \quad \bar{v}_{ab} \equiv \frac{n_b q_a^2 q_b^2 \ln \Lambda}{4\pi \varepsilon_0^2 m_a^2 v_{Ta}^3}, \quad x_b = \frac{v}{v_{Tb}}, \quad x_a = \frac{v}{v_{Ta}}. \qquad (4.32)$$

The last equation defines the fundamental collision frequency $\bar{v}_{ab}$ of the system.

From the asymptotic form of the Chandrasekhar function (see the Exercises), we see that $G(x)$ decreases with increasing velocity ($x = v/v_T$) if x is sufficiently large. The average frictional force on a particle

$$m_a \frac{\langle \Delta v_{\parallel} \rangle^{ab}}{\Delta t} = -m_a v_{\parallel} v_s^{ab} \propto G(x_b),$$

therefore decreases with increasing velocity for x_b sufficiently large. This vanishes in the limit of infinite velocity, although relativistic effects eventually prevent this. This is a consequence of momentum exchange decreasing as an incident particle's speed increases if the impact parameter is held constant. This curious effect has an interesting implication. Consider an applied electric field in a plasma with a population of fast electrons. As a result of the decreasing frictional force on the electrons, the electrons can be accelerated to arbitrarily high energies, forming a population of *runaway electrons*. If the electric field is sufficiently large, then even thermal electrons can experience run away, and the bulk electron distribution will depart from a Maxwellian distribution. This will occur when, approximately,

$$eE > 2\bar{v}_{ee} m_e v_{Te},$$

or equivalently when the electric field exceeds the "Dreicer field"

$$E_D = \frac{n_e e^3 \ln \Lambda}{4\pi \varepsilon^2 T_e}.$$

Exercises

1. Show that the assumption of a Maxwellian distribution function $f_b(v)$ yields the solution to the partial differential equation

$$\frac{1}{v^2}\frac{\partial}{\partial v}\left(v^2\frac{\partial \phi_b}{\partial v}\right) = f_b(v)$$

as

$$\phi'_b(v) = \frac{m_b n_b}{4\pi T_b} G(v/v_{Tb}),$$

where $G(x)$ is the Chandrasekhar function defined above.

2. By using the definitions of $v_D^{ab}(v)$ and $v_\parallel^{ab}(v)$ and the relation $\nabla_v^2 \psi_b = \phi_b$ in spherical velocity-space coordinates, derive the collision frequencies (4.30) and (4.31).
3. Plot the Chandrasekhar function $G(x)$ from $[0, 5]$.

4.6 Collision Operator for Fast Ions

Energetic ions are ubiquitous in plasmas and often of considerable interest, frequently acting as a probe of high energy processes in astrophysics, space physics, or laboratory plasmas. Examples include solar energetic particles (accelerated in either solar flares or at interplanetary shock waves) or galactic cosmic rays. Cosmic rays especially propagate vast distances through a variety of interstellar environments experiencing numerous particle collisions that can modify their distribution. Note that this is distinct from the collisionless pitch-angle scattering that cosmic rays experience due to magnetic turbulence – this is discussed in the following chapter. For the present, we will assume a background comprised of thermal protons and electrons, and consider only moderately energetic particles that satisfy the ordering

$$v_{Tp} \ll v_\alpha \ll v_{Te},$$

where the subscript α identifies the energetic particle population.

The largest collision frequency in this case, assuming a background Maxwellian population, is the energetic ion-electron frictional drag, given by (4.29),

4.6 Collision Operator for Fast Ions

$$\nu_s^{\alpha e} = 2\bar{\nu}_{\alpha e} \frac{T_\alpha}{T_e} \left(1 + \frac{m_e}{m_\alpha}\right) \frac{G(x_e)}{x_\alpha}$$

$$= \bar{\nu}_{\alpha e} \frac{T_\alpha}{T_e} \frac{2v/v_{Te}}{3\sqrt{\pi}} \frac{v_{T\alpha}}{v}$$

$$= \frac{n_e q_\alpha^2 e^2 m_e^{1/2} \ln \Lambda}{3(2\pi)^{3/2} \varepsilon_0^2 m_\alpha (kT_e)^{3/2}} \equiv \tau_s^{-1},$$

where the small argument approximation to the Chandrasekhar function was used. The opposite limit, $x \to \infty$, is appropriate for frictional drag of energetic ions with thermal protons. The critical speed above which electron drag dominates proton drag is given by $\nu_s^{\alpha e} \geq \nu_s^{\alpha p}$, i.e.,

$$\tau_s^{-1} \geq 2\bar{\nu}_{\alpha p} \frac{T_\alpha}{T_p} \left(1 + \frac{m_p}{m_\alpha}\right) \frac{G(x_p)}{x_\alpha}$$

$$\simeq \bar{\nu}_{\alpha p} \left(1 + \frac{m_p}{m_\alpha}\right) \frac{T_\alpha}{T_p} \frac{v_{Tp}^2}{v^2} \frac{v_{T\alpha}}{v}$$

$$= \bar{\nu}_{\alpha p} \left(1 + \frac{m_p}{m_\alpha}\right) \left(\frac{v_{T\alpha}}{v}\right)^3.$$

Hence, we find a critical particle velocity v_c such that

$$\left(\frac{v_c}{v_{T\alpha}}\right)^3 = \bar{\nu}_{\alpha p} \left(1 + \frac{m_p}{m_\alpha}\right) \tau_s$$

$$= \frac{3\sqrt{\pi}}{4} \frac{n_p}{n_e} \frac{1}{m_\alpha m_e^{1/2}} \left(1 + \frac{m_p}{m_\alpha}\right) \frac{(2kT_e)^{3/2}}{v_{T\alpha}^3},$$

or

$$v_c = \left(\frac{3\sqrt{\pi}}{4} \frac{n_p}{n_e} \frac{m_e}{m_\alpha} \left(1 + \frac{m_p}{m_\alpha}\right)\right)^{1/3} v_{Te}.$$

This corresponds to an energetic particle energy of approximately (on setting $v_{T\alpha} \simeq v_{Te} = \sqrt{2kT_e/m_e}$)

$$\frac{1}{2} m_\alpha v_c^2 \sim 50 k T_e.$$

A massive particle moving through a cloud of electrons experiences a drag force but little deflection (think of a cyclist riding through air) as it propagates. This can be verified by estimating the diffusion frequencies. On the other hand, collisions between energetic ions and thermal protons can lead to large deflections because their masses are comparable. In this case,

$$\nu_D^{\alpha p} = \left(\frac{v_b}{v}\right)^3 \frac{1}{\tau_s},$$

and the critical speed for pitch angle scattering

$$v_b \equiv \left(\frac{m_p}{m_\alpha}\right)^{1/3} v_c,$$

is of the same order of magnitude as v_c. Thus pitch-angle scattering of energetic particles is comparable to proton drag since both are determined by energetic ion and thermal proton collisions.

The collision operator for energetic ions is therefore given by the electron drag, proton drag, and ion-proton scattering expressions above, thus substituting these results into the collision operator (4.27) yields

$$C_\alpha(f_\alpha) = \frac{1}{v^2 \tau_s} \frac{\partial}{\partial v} \left[\left(v^3 + v_c^3\right) f_\alpha(v)\right] + \left(\frac{v_b}{v}\right)^3 \frac{1}{\tau_s} \mathcal{L}(f_\alpha),$$

after using

$$\frac{m_\alpha}{m_\alpha + m_p} v_s^{\alpha p} = \left(\frac{v_c}{v}\right)^3 \tau_s^{-1}.$$

If the energetic particles are isotropically distributed, the scattering operator will vanish by symmetry.

Exercises

1. Suppose that energetic particles are introduced as an isotropic distribution with speed U at a rate η per unit volume. Since the energetic particles are isotropically distributed in velocity space, the kinetic equation may be expressed as

$$\frac{\partial f_\alpha}{\partial t} = \frac{1}{v^2 \tau_s} \frac{\partial}{\partial v} \left[\left(v^3 + v_c^3\right) f_\alpha(v)\right] + \eta \frac{\delta(v - U)}{4\pi U^2}.$$

Subject to the boundary condition $f_\alpha(v > U) = 0$, show that the steady-state energetic particle distribution function is given by

$$f_\alpha(v) = \eta \frac{\tau_s}{4\pi (v^3 + v_c^3)} \quad \text{for} \quad v < U.$$

4.7 Proton-Electron Collisions

Unless the proton temperature is sufficiently high that $T_e \ll T_p$, electrons, being much lighter, move far faster than protons. Protons (and ions in general) therefore experience multiple collisions by light fast particles, and this problem is thus similar to the mathematical description of Brownian motion. In view of the different particle speeds, we introduce the electron distribution function as a sum of a Maxwellian distribution with mean velocity equal to that of the proton mean velocity f_{Me} and a remainder f_{e1},

4.7 Proton-Electron Collisions

$$f_e(\mathbf{x}, t, \mathbf{v}) = f_{Me}(\mathbf{v} - \mathbf{v}_p) + f_{e1}(\mathbf{v}).$$

Similarly, we separate the collision operator into two parts,

$$C_{pe}(f_p, f_e) = C_{pe}[f_p, f_{Me}(\mathbf{v} - \mathbf{v}_p)] + C_{pe}(f_p, f_{e1}).$$

Consider the latter contribution to the collision integral first. From the form of the collision operator,

$$C_{pe}(f_p, f_{e1}) = L^{pe} \frac{\partial}{\partial v_i} \left(\frac{m_p}{m_e} \frac{\partial \phi_e}{\partial v_i} f_p - \frac{\partial^2 \psi_e}{\partial v_i \partial v_j} \frac{\partial f_p}{\partial v_j} \right)$$

$$\simeq L^{pe} \frac{\partial}{\partial v_i} \left(\frac{m_p}{m_e} \frac{\partial \phi_e}{\partial v_i} f_p \right),$$

since the first term is large. Recall that the first term is the streaming/frictional drag term

$$L^{pe} \frac{m_p}{m_e} \frac{\partial \phi_e}{\partial v_i} \simeq -\frac{\langle \Delta v_i \rangle^{pe}}{\Delta t} = -\frac{F_i^{pe}(\mathbf{v})}{m_p},$$

where $\mathbf{F}^{pe}(\mathbf{v})$ is the averaged force acting on a proton as it propagates through an electron distribution f_{e1} at velocity $\mathbf{v}$ while experiencing collisions. The force acting on the proton through electron collisions must be essentially independent of the proton velocity since the electrons in this case are fast, $v_{Te} \gg v_{Tp}$, meaning that f_{e1} varies on this time scale only. Hence, $\mathbf{F}^{pe}(\mathbf{v})$ is approximately unchanged for all ion velocities, and so

$$\mathbf{F}^{pe} = \frac{\mathbf{R}_{pe}}{n_p} = -\frac{\mathbf{R}_{ep}}{n_p},$$

with the result that

$$C_{pe}(f_p, f_{e1}) = -\frac{\partial}{\partial v_i} \left(\frac{F_i^{pe}}{m_p} f_p \right) = \frac{\mathbf{R}_{ep}}{m_p n_p} \cdot \frac{\partial f_p}{\partial \mathbf{v}}.$$

Recall that $\mathbf{R} \equiv \int m_p \mathbf{v} C(f) d^3v$ and that $C_{ep}(f_e)$ was computed in Sect. 3.4.

Consider now the first term in the proton-electron collision operator. In this case, the electron distribution is given by the Maxwellian $f_{Me}(\mathbf{v} - \mathbf{v}_p)$. Nonetheless, protons move more slowly than the electrons, so we may use the Rosenbluth potential solution for ϕ_e' in terms of the Chandrasekhar function $G(x)$ in the limit that $x \to 0$,

$$\frac{\partial \phi_e}{\partial v_j} = \frac{v_j - v_{pj}}{v} \phi_e' = \frac{v_j - v_{pj}}{v} \frac{m_e n_e}{4\pi k T_e} \frac{2}{3\sqrt{\pi}} \frac{v}{v_{Te}} = (v_j - v_{pj}) \frac{n_e}{3} \left(\frac{m_e}{2\pi k T_e} \right)^{3/2}.$$

To compute ψ_e, we use the relation

$$\nabla_v^2 \psi_e = \phi_e(\mathbf{v}) \simeq -\frac{1}{4\pi} \int \frac{f_{Me}(v')}{v'} d^3v'$$

$$= -\frac{1}{4\pi} \int_0^{2\pi} \int_0^{\pi} \int_0^{\infty} \frac{v'^2 e^{-v'^2/v_{Te}'^2}}{v'} dv' d\theta d\phi$$

$$= -\frac{n_e}{(2\pi)^{3/2}} \sqrt{\frac{m_e}{kT_e}},$$

which yields

$$\frac{\partial^2 \psi_e}{\partial v_i \partial v_j} \simeq -\frac{n_e}{3(2\pi)^{3/2}} \sqrt{\frac{m_e}{kT_e}} \delta_{ij}.$$

We therefore obtain

$$C_{pe}(f_p, f_e) = \frac{\mathbf{R}_{ep}}{m_p n_p} \cdot \frac{\partial f_p}{\partial \mathbf{v}} + \ln \Lambda \left(\frac{e^2}{m_p \varepsilon_0}\right)^2 \frac{\partial}{\partial \mathbf{v}} \cdot \left[\frac{m_p}{m_e}(\mathbf{v}-\mathbf{v}_p)\frac{n_e}{3}\left(\frac{m_e}{2\pi kT_e}\right)^{3/2} f_p\right.$$

$$\left. + \frac{n_e}{3(2\pi)^{3/2}} \sqrt{\frac{m_e}{kT_e}} \frac{\partial f_p}{\partial \mathbf{v}}\right]$$

$$= \frac{\mathbf{R}_{ep}}{m_p n_p} \cdot \frac{\partial f_p}{\partial \mathbf{v}} + \frac{m_e n_e}{m_p n_p \tau_{ep}} \frac{\partial}{\partial \mathbf{v}} \cdot \left[(\mathbf{v} - \mathbf{v}_p) f_p + \frac{kT_e}{m_p} \frac{\partial f_p}{\partial \mathbf{v}}\right],$$

where the electron-proton collision time

$$\tau_{ep} \equiv \frac{12\pi^{3/2}}{\sqrt{2}} \frac{\sqrt{m_e}(kT_e)^{3/2}\varepsilon_0^2}{n_p e^4 \ln \Lambda},$$

has been introduced. The first term in the square brackets is the proton-electron frictional drag term, and is inversely proportional to τ_{ep}. The second term in the square brackets describes energy exchange between protons and electrons, mediated by collisions. This term is has a time scale

$$\frac{m_p \tau_{ep}}{m_e} \sim \tau_{pe}$$

that is significantly slower than that associated with frictional drag.

4.8 Transport Equations for a Collisional Electron-Proton Plasma

As with the ideal gas problem, the Chapman-Enskog expansion is one approach to deriving plasma fluid transport equations in the limit of strong collisionality. This approach was developed in considerable detail in a classic article by Braginskii (1965). As with the case of gas dynamics, the collision frequency ν is assumed to be sufficiently high that local thermodynamical equilibrium is attained more rapidly than the relaxation of the plasma due to the presence of macroscopic gradients that drive plasma transport. This requires that the orderings

$$\frac{\partial}{\partial t} \ll \nu, \quad \lambda \ll \nabla^{-1} \equiv L,$$

be satisfied, where $\lambda \equiv v_T/\nu$ is the mean free path and L defines a macroscopic length scale over which the mean plasma parameters (density, velocity, temperature, magnetic field) vary. Unlike a gas of neutral particles, gradients can vary significantly along and perpendicular to magnetic field lines, so one needs to distinguish between parallel ($L_\parallel$) and perpendicular ($L_\perp$) length scales. Furthermore, the particle gyrofrequency ($\Omega \equiv qB/m$) introduces a further time scale into the system, often larger than the collision frequency, thus

$$\Delta \equiv \frac{\nu}{\Omega} \ll 1.$$

Obviously, this implies that the particle gyro- or Larmor radius $r_g = m_a v/(q_a B) = v/|\Omega_a| \simeq v_{Ta}/|\Omega_a| \ll \lambda$. For this ordering, the ordering perpendicular to the mean magnetic field can be relaxed with

$$\lambda \ll L_\parallel, \quad \delta \equiv \frac{r_g}{L_\perp} \ll 1.$$

Thus, for a magnetized plasma, the mean free path need only be short compared to the parallel length scale, and this is the ordering that we shall assume henceforth in this subsection. The distance that a particle travels nominally before experiencing a collision is at least a mean free path along the field and a Larmor radius across the field. Macroscopic gradients must be essentially absent over these scales i.e., the plasma is essentially homogeneous on these scales.

We consider a plasma comprising electrons and protons only and develop a transport theory in the presence of proton-proton, electron-proton, and electron-electron collisions. Since the electrons do not collide with a stationary background, we need to transform the kinetic equation for each species a to a coordinate frame moving with the mean or bulk flow velocity $\mathbf{u}_a(\mathbf{r}, t)$ of each species. This requires a transformation $(\mathbf{x}, \mathbf{v}, t) \to (\mathbf{x}, \mathbf{c}_a, t)$ where $\mathbf{c}_a = \mathbf{v} - \mathbf{u}_a$, so the derivatives become

$$\frac{\partial}{\partial t} \to \frac{\partial}{\partial t} - \frac{\partial \mathbf{u}_a}{\partial t} \cdot \frac{\partial}{\partial \mathbf{c}_a};$$

$$\nabla \to \nabla - \frac{\partial u_{aj}}{\partial \mathbf{x}} \frac{\partial}{\partial c_{aj}};$$

$$\nabla_\mathbf{v} \to \nabla_{\mathbf{c}_a}.$$

The kinetic equation therefore becomes

$$\frac{\partial f_a}{\partial t} + (\mathbf{u}_a + \mathbf{c}_a) \cdot \nabla f_a + \left[\frac{q_a}{m_a} (\mathbf{E}' + \mathbf{c}_a \times \mathbf{B}) - \left(\frac{\partial}{\partial t} + \mathbf{u}_a \cdot \nabla \right) \mathbf{u}_a \right] \cdot \frac{\partial f_a}{\partial \mathbf{c}_a}$$
$$- c_{aj} \frac{\partial u_{ak}}{\partial x_j} \frac{\partial f_a}{\partial c_{ak}} = C_a(f_a), \quad (4.33)$$

where $\mathbf{E}' = \mathbf{E} + \mathbf{u}_a \times \mathbf{B}$ is the motional electric field.

The dominant time scales are those associated with the collision frequency and the gyrofrequency. For electrons, we can order the kinetic equation as

$$C_{ee}(f_e) + C_{ep}^0(f_e) + \left(\frac{e}{m_e} \mathbf{c}_e \times \mathbf{B} \right) \cdot \frac{\partial f_e}{\partial \mathbf{c}_e} = \frac{\partial f_e}{\partial t} + \mathbf{u}_e \cdot \nabla f_e$$
$$+ \mathbf{c}_e \cdot \nabla f_e - \left(\frac{e}{m_e} \mathbf{E}' + \left(\frac{\partial \mathbf{u}_e}{\partial t} + \mathbf{u}_e \cdot \nabla \mathbf{u}_e \right) \right) \cdot \frac{\partial f_e}{\partial \mathbf{c}_e} - c'_{aj} \frac{\partial u_{ek}}{\partial x_j} \frac{\partial f_a}{\partial c_{ek}} - C_{ep}^1(f_e),$$

where the higher order correction to the collision operator has been included on the right because it acts more slowly than the leading order term.

Following the Chapman-Enskog expansion procedure, we solve the above equation by expanding the distribution function as $f_e = f_{e0} + f_{e1} + \ldots$. To the lowest order, the left hand side must vanish, which requires the distribution function to be a Maxwellian at rest in the moving frame,

$$f_{e0} = n_e \left(\frac{m_e}{2\pi k T_e} \right)^{3/2} e^{-\beta^2},$$

where $\beta^2 = m_e c_e^2 / 2k T_e$. On using the zeroth order solution on the right hand side, we obtain an equation for the next order solution f_{e1},

$$C_{ee}(f_{e1}) + C_{ep}^0(f_{e1}) + \left(\frac{e}{m_e} \mathbf{v} \times \mathbf{B} \right) \cdot \frac{\partial f_{e1}}{\partial \mathbf{v}} =$$
$$\left(\frac{\partial}{\partial t} + \mathbf{u}_e \cdot \nabla \right) \ln n_e f_{e0} + \left[\left(\beta^2 - \frac{3}{2} \right) \left(\frac{\partial}{\partial t} + \mathbf{u}_e \cdot \nabla \right) \ln k T_e \right.$$

4.8 Transport Equations for a Collisional Electron-Proton Plasma

$$+ \mathbf{v} \cdot \nabla \ln n_e + \left(\beta^2 - \frac{3}{2}\right) \mathbf{v} \cdot \nabla \ln kT_e + \frac{m_e \mathbf{v}}{kT_e} \cdot \left(\frac{e}{m_e} \mathbf{E}'\right)$$

$$+ \left(\frac{\partial}{\partial t} + \mathbf{u}_e \cdot \nabla \mathbf{u}_e\right) + \frac{m_e v_j v_k}{T_e} \frac{\partial u_{ek}}{\partial x_j} + v_{ep} \frac{m_e \mathbf{v} \cdot (\mathbf{u}_e - \mathbf{u}_p)}{kT_e}\right] f_{e0}, \quad (4.34)$$

where, for notational convenience, we have written $\mathbf{v}$ for $\mathbf{c}_e$. Integrating Eq. (4.34) over velocity space yields the continuity equation (see Exercise below), showing that $d \ln n_e / dt$ (where d/dt is the convective derivative) can be replaced by $-\nabla \cdot \mathbf{u}_e$. Similarly, taking the first moment of (4.34) yields the momentum equation without the viscous term, showing that (Exercise)

$$\left(\frac{\partial}{\partial t} + \mathbf{u}_e \cdot \nabla\right) \mathbf{u}_e + \frac{e\mathbf{E}'}{m_e} = \frac{\mathbf{R}_e - \nabla(n_e kT_e)}{m_e n_e}.$$

The second moment of (4.34) yields the energy equation without the heat conduction, viscous heating, and energy exchange terms. Thus

$$\frac{3}{2}\left(\frac{\partial}{\partial t} + \mathbf{u}_e \cdot \nabla\right) \ln kT_e + \nabla \cdot \mathbf{u}_e = 0,$$

which allows us to eliminate the $d(kT_e)/dt$ in (4.34).

Using the above results allows us to eliminate the time derivatives from (4.34), giving the kinetic equation

$$C_{ee}(f_{e1}) + C_{ep}^0(f_{e1}) + \left(\frac{e}{m_e}\mathbf{v} \times \mathbf{B}\right) \cdot \frac{\partial f_{e1}}{\partial \mathbf{v}}$$

$$= \left[\left(\beta^2 - \frac{5}{2}\right)\mathbf{v} \cdot \nabla \ln kT_e + \mathbf{v} \cdot \left(\frac{\mathbf{R}_e}{p_e} + \frac{m_e v_{ep}}{kT_e}(\mathbf{u_e} - \mathbf{u}_p)\right)\right.$$

$$\left. + \frac{m_e}{2kT_e}\left(v_j v_k - \frac{v^2}{3}\delta_{jk}\right) W_{jk}^e\right] f_{e0}, \quad (4.35)$$

where

$$W_{jk}^a \equiv \frac{\partial v_{aj}}{\partial x_k} + \frac{\partial v_{ak}}{\partial x_j} - \frac{2}{3}(\nabla \cdot \mathbf{u}_a)\delta_{jk},$$

is the rate-of-strain tensor, as before. As with a neutral gas, this term introduces the plasma viscous terms.

A closely related analysis yields an equation corresponding to the electron correction kinetic equation (4.35) for the proton kinetic problem,

$$C_{pp}(f_{p1}) - \left(\frac{e}{m_p} \mathbf{v} \times \mathbf{B}\right) \cdot \frac{\partial f_{p1}}{\partial \mathbf{v}}$$

$$= \left[\left(\beta^2 - \frac{5}{2}\right) \mathbf{v} \cdot \nabla \ln kT_p + \frac{m_p}{2kT_p}\left(v_j v_k - \frac{v^2}{3}\delta_{jk}\right) W^p_{jk}\right] f_{p0}, \quad (4.36)$$

where the proton-electron collision operator has been assumed small, allowing the frictional term $\mathbf{R}_{pe}$ to be neglected.

Exercises

1. Show that integrating (4.34) over velocity space yields the continuity equation

$$\frac{\partial n_e}{\partial t} + \nabla \cdot (n_e \mathbf{u}_e) = 0.$$

2. Show that the first moment of (4.34) yields the momentum equation without the viscous term, and that

$$\left(\frac{\partial}{\partial t} + \mathbf{u}_e \cdot \nabla\right) \mathbf{U}_e + \frac{e\mathbf{E}'}{m_e} = \frac{\mathbf{R}_e - \nabla(n_e kT_e)}{m_e n_e}.$$

3. Show that the second moment of (4.34) yields the energy equation without the heat conduction, viscous heating, and energy exchange terms, and hence that

$$\frac{3}{2}\left(\frac{\partial}{\partial t} + \mathbf{u}_e \cdot \nabla\right) \ln kT_e + \nabla \cdot \mathbf{u}_e = 0.$$

4. Eliminate the time derivatives in (4.34) using the results from the Exercises above to derive (4.35).

The kinetic equations (4.35) and (4.36) must be solved for the $f_{e/i1}$ correction so that a two-fluid description can be determined. The approach used by Braginskii is to expand the distribution function as a series of orthogonal polynomials, which is, as we have seen earlier, an attractive and systematic approach that yields accurate results to any order, in principle. Typically, generalized Laguerre polynomials, also known as Sonine polynomials, are used. The Sonine polynomials tend to converge rather rapidly, so although the reduced system of equations is of infinite order, a truncation at about order 3 is generally sufficient. Nonetheless, the full procedure is tedious, so we simply provide the main results without derivation – the details can be found in Braginskii's review.

4.8 Transport Equations for a Collisional Electron-Proton Plasma

We assume from the outset that the gyrofrequency is much greater than the collision frequency i.e., $\Delta \ll 1$, for both electrons and protons. Following Braginskii, we introduce the collision times

$$\tau_e = \tau_p, \quad \tau_p = \sqrt{2}\tau_{pp},$$

and assume the strongly magnetized limit $\Omega_a \tau_a \to \infty$.

The force $\mathbf{R}_e$ for electrons comprises both the drag force $\mathbf{R}_u$ and the thermal force $\mathbf{R}_T$,

$$\mathbf{R}_e = \mathbf{R}_u + \mathbf{R}_T;$$

$$\mathbf{R}_u = -\frac{m_e n_e}{\tau_e}\left(0.51\left(\mathbf{u}_\parallel^e - \mathbf{u}_\parallel^p\right) + \left(\mathbf{u}_\perp^e - \mathbf{u}_\perp^p\right)\right);$$

$$\mathbf{R}_T = -0.71 n_e \nabla_\parallel k T_e + \frac{3 n_e}{2 \Omega_e \tau_e} \mathbf{b} \times \nabla k T_e,$$

where $\mathbf{b} \equiv \mathbf{B}/B$, $\mathbf{u}^{e/p}$ are the electron and proton bulk velocities respectively, and the electron gyrofrequency $\Omega = -eB/m_e$ is negative. Observe that the term $(\Omega_e \tau_e)^{-1}$ is of higher order than the preceding term for $\mathbf{R}_T$. Since momentum is conserved in Coulomb collisions, $\mathbf{R}_e = -\mathbf{R}_p$.

The electron heat flux also comprises a drag term and a thermal term,

$$\mathbf{q}_e = \mathbf{q}_u^e + \mathbf{q}_T^e;$$

$$\mathbf{q}_u^e = 0.71 n_e k T_e \left(\mathbf{u}_\parallel^e - \mathbf{u}_\parallel^p\right) - \frac{3 n_e k T_e}{2\Omega_e \tau_e}\mathbf{b} \times \left(\mathbf{u}_\perp^e - \mathbf{u}_\perp^p\right);$$

$$\mathbf{q}_T^e = -\kappa_\parallel^e \nabla_\parallel k T_e - \kappa_A^e \mathbf{b} \times \nabla k T_e - \kappa_\perp^e \nabla_\perp k T_e,$$

and the heat conductivities are defined by

$$\kappa_\parallel^e = 3.16 \frac{n_e k T_e \tau_e}{m_e}, \quad \kappa_A^e = -\frac{5}{2}\frac{n_e k T_e}{m_e \Omega_e}, \quad \kappa_\perp^e = 4.66 \frac{n_e k T_e}{m_e \Omega_e^2 \tau_e}.$$

Since electrons are strongly tied to the magnetic field, fast electrons streaming along the field can significantly distort a Maxwellian distribution because the collision frequency decreases with velocity according to $\tau_e \sim v^3$. This is reflected in both the electron force term $\mathbf{R}_u$ (because fast electrons contribute to the bulk velocity more than they do to friction) and the electron flux along the parallel direction (first term in $\mathbf{q}_T$ since fast electrons will stream more in one direction and slow electrons more in the opposite direction).

The proton heat flux has terms proportional to the ion temperature gradient only,

$$\mathbf{q}_p = -\kappa_\parallel^p \nabla_\parallel k T_p + \kappa_A^p \mathbf{b} \times \nabla k T_p - \kappa_\perp^p \nabla_\perp k T_p,$$

with the heat conductivities defined by

$$\kappa_\parallel^p = 3.9 \frac{n_p k T_p \tau_p}{m_p}, \quad \kappa_A^p = -\frac{5}{2} \frac{n_p k T_p}{m_p \Omega_p}, \quad \kappa_\perp^p = 2 \frac{n_p k T_p}{m_p \Omega_p^2 \tau_p}.$$

The magnetization factor $\Omega\tau$ separates the various heat conduction coefficients κ. The time scales are therefore significantly different for protons and electrons, and the proton contribution is dominant perpendicular to the field whereas along the magnetic field, the electron heat flux dominates. Note that the *diamgnetic heat flux* term, carrying heat across the field perpendicular to the gradient, is not the consequence of collisions ($\tau_{e/p}$ is absent), and we shall discuss this further below.

Heat exchange between protons and electrons results from temperature equilibration on the slow time scale τ_{pe} and the frictional generation of heat, and is expressed through

$$Q_p = -Q_e - \mathbf{R}_e \cdot (\mathbf{u}_e - \mathbf{u}_p) = \frac{3 n_e m_e}{m_p \tau_e} (k T_e - k T_p).$$

Like the Navier-Stokes equations for a non-magnetized gas, the viscosity tensor is given by

$$\pi_{ij} = \mu W_{ij} + \lambda \nabla \cdot \mathbf{u} \delta_{ij},$$

where μ and λ are the viscosity coefficients and W_{ij} is the rate-of-strain tensor, as before. Because the transport of momentum occurs at different rates in different directions for a magnetized plasma, the viscosity tensor is more complicated, and has the form

$$\pi_{xx} = -\frac{1}{2}\mu_0 (W_{xx} + W_{yy}) - \frac{1}{2}\mu_1 (W_{xx} - W_{yy}) - \mu_3 W_{xy};$$

$$\pi_{yy} = -\frac{1}{2}\mu_0 (W_{xx} + W_{yy}) - \frac{1}{2}\mu_1 (W_{yy} - W_{xx}) + \mu_3 W_{xy};$$

$$\pi_{xy} = \pi_{yx} = -\mu_1 W_{xy} + \frac{1}{2}\mu_3 (W_{xx} - W_{yy});$$

$$\pi_{xz} = \pi_{zx} = -\mu_2 W_{xz} - \mu_4 W_{yz};$$

$$\pi_{yz} = \pi_{zy} = -\mu_2 W_{yz} + \mu_4 W_{xz};$$

$$\pi_{zz} = -\mu_0 W_{zz},$$

for a coordinate system (x, y, z) defined by the magnetic field orientation $\mathbf{b} = (0, 0, 1)$. Consistent with the Chapman-Enskog ordering, the viscous terms enter the magnetized fluid equations at an order lower (either in terms of the ratio of the gyroradius to the macroscopic perpendicular length scale, $r_g/L_\perp$, or the ratio of the mean free path to the parallel macroscopic scale, $\lambda/L_\parallel$), i.e., $\nabla p \sim \mathbf{R} \gg \nabla \cdot \pi$. The viscosity coefficients for protons are

$$\mu_0^p = 0.96 n_p k T_p \tau_p; \quad \mu_1^p = \frac{1}{4}\mu_2^p = \frac{3 n_p k T_p}{10 \Omega_p^2 \tau_p}; \quad \mu_3^p = \frac{1}{2}\mu_4^p = \frac{n_p k T_p}{2 \Omega_p},$$

and for electrons

$$\mu_0^e = 0.73 n_3 T_e \tau_e; \quad \mu_1^e = \frac{1}{4}\mu_2^e = 0.51 \frac{n_e T_e}{\Omega_e^2 \tau_e}; \quad \mu_3^e = \frac{1}{2}\mu_4^e = \frac{n_e T_e}{2\Omega_e}.$$

As with the heat flux, different scalings are present depending on the direction of momentum transport. Along the magnetic field, the μ_0 are entirely due to particle collisions and therefore resemble transport in a nonmagnetized gas. Transport perpendicular to the magnetic field ($\mu_{1,2}$) involves a scaling with the ratio of the scattering time to the particle gyroradius, introduced through the factor $(\Omega\tau)^2$. The remaining viscous terms, $\mu_{3,4}$ depend only on the gyrofrequency i.e., are independent of the scattering time τ, and correspond to a diamagnetic momentum flux across the field.

4.9 Application 1: Transport Perpendicular to a Mean Magnetic Field

Here we consider an alternative approach to the transport of particles, momentum, and energy across a mean magnetic field in a plasma. Such flows are called diamagnetic flows. In the section above, we saw that the perpendicular heat flux and momentum flux were independent of particle collisions. The momentum equation is given by

$$m_a n_a \left(\frac{\partial \mathbf{u}_a}{\partial t} + \mathbf{u}_a \cdot \nabla \mathbf{u}_a \right) = -\nabla p_a - \nabla \cdot \pi_a + q_a n_a (\mathbf{E} + \mathbf{u}_a \times \mathbf{B}) + \mathbf{R}_a.$$

By taking the cross product of the momentum equation with $\mathbf{B}$ (and dropping the subscript a) yields ($\mathbf{b} \equiv \mathbf{B}/B$, and $\mathbf{b} \times (\mathbf{u} \times \mathbf{b}) = \mathbf{u} - (\mathbf{u} \cdot \mathbf{b})\mathbf{b} \equiv \mathbf{u} - \mathbf{u}_\| = \mathbf{u}_\perp$)

$$\mathbf{u}_\perp = \frac{\mathbf{E} \times \mathbf{B}}{B^2} + \frac{\mathbf{b} \times (\nabla p + \nabla \cdot \pi - \mathbf{R} + mn d\mathbf{u}/dt)}{mn\Omega},$$

where d/dt denotes the convective derivative. This expression shows that the bulk flow velocity perpendicular to the mean magnetic field is the sum of the $\mathbf{E} \times \mathbf{B}$ drift and a diamagnetic drift velocity. Present in the diamagnetic term is the small parameter $\delta = r_g/L_\perp$. The viscous terms and resistive force are generally small, and the inertial term typically varies slowly, so that to leading order we have approximately

$$\mathbf{u}_\perp = \frac{\mathbf{E} \times \mathbf{B}}{B^2} + \frac{\mathbf{b} \times \nabla p}{mn\Omega}.$$

This ordering is certainly true whenever the particle distribution is close to a Maxwellian. The perpendicular force $\mathbf{R} \sim \nu m n \mathbf{u}_\perp$ is small if $\nu/\Omega \ll 1$. Neglecting the inertial term is valid if the thermal speed and time variability is small. If a plasma flow does not satisfy this criterion, then the convective term $mn\mathbf{u} \cdot \nabla \mathbf{u}$ must be included.

If the magnetic field $\mathbf{B}$ is constant and $\mathbf{E} = -\nabla \Phi$, then the approximate flow velocity $\mathbf{u}_\perp$ is incompressible, i.e., $\nabla \cdot \mathbf{u}_\perp = 0$. If the flow is in general incompressible, then it follows that

$$\nabla \cdot \left(\mathbf{u}_\| + \mathbf{u}_\perp\right) = 0,$$

indicating that the divergence of the parallel flow balances that of the perpendicular flow.

A related approach can be used to derive the viscosity tensor if we approximate the collision operator by the BGK operator,

$$C(f) = \nu(f_0 - f),$$

as we did for the gas kinetic problem. The kinetic equation can be expressed in components as

$$\frac{\partial f}{\partial t} + \frac{\partial}{\partial x_i}\left[(u_i + c_i)f\right] + \frac{\partial}{\partial v_i}\left[\left(\frac{eE_i}{m} + \epsilon_{ijk}(u_j + c_j)\Omega_k\right)f\right] = \nu(f_0 - f),$$

where $\mathbf{c} \equiv \mathbf{v} - \mathbf{u}$ is the particle velocity relative to the bulk velocity, as before. The gyrofrequency is eB_k/m and ϵ_{ijk} is the Levi-Civita tensor, a completely antisymmetric unit tensor in 3D such that $\epsilon_{ijk} = 1$ if ijk is an even permutation of 123, $\epsilon_{ijk} = -1$ for odd permutations, and $\epsilon_{ijk} = 0$ for repeated indices. Taking the $c_l c_m$-moment of the kinetic equation yields an equation for the pressure tensor P_{lm} and conductive heat flux term, where

$$P_{lm} \equiv m \int c_l c_m f d^3v, \quad q_{ilm} \equiv m \int c_i c_l c_m f d^3v,$$

in the form

$$\frac{\partial P_{lm}}{\partial t} + \frac{\partial}{\partial x_i}(u_i P_{lm} + q_{ilm}) + \frac{\partial u_l}{\partial x_i} P_{im} + \frac{\partial u_m}{\partial x_i} P_{il}$$
$$- \Omega_k \left(\epsilon_{ljk} P_{jm} + \epsilon_{mjk} P_{jl}\right) = \nu(p\delta_{lm} - P_{lm}). \tag{4.37}$$

The non-diagonal elements of the pressure tensor are given simply by

$$P_{lm} = p\delta_{lm} + \pi_{lm}.$$

4.9 Application 1: Transport Perpendicular to a Mean Magnetic Field

Let us impose the Braginskii short-mean-free-path orderings discussed above to the evolution equation for P_{lm}. Thus, terms proportional to ν and Ω will dominate, and so to lowest order the distribution function is a Maxwellian $f = f_0$. At this order, the pressure tensor is symmetric and

$$P_{lm} = p\delta_{lm}, \quad q_{ilm} = 0.$$

We can therefore use these expressions in the small terms of (4.37) to obtain the corrections at the next order for P_{lm} – these are of course the terms proportional to ν and Ω. This may be represented formally as[6]

$$\mathbf{K}(P) \equiv \frac{\mathbf{S}}{\Omega},$$

where S is the tensor defined by

$$S_{lm} = \left[\frac{\partial p}{\partial t} + \frac{\partial}{\partial x_i}(u_i p)\right]\delta_{lm} + p\left(\frac{\partial u_l}{\partial x_m} + \frac{\partial u_m}{\partial x_l}\right) + \nu(P_{lm} - p\delta_{lm}),$$

and K is an operator defined by

$$K(P) = P \times \mathbf{b} + (P \times \mathbf{b})^t = \frac{\Omega_k}{\Omega}\left(\epsilon_{ljk}P_{jm} + \epsilon_{mjk}P_{lj}\right)$$

$$= \begin{pmatrix} 2P_{xy} & P_{yy} - P_{xx} & P_{yz} \\ P_{yy} - P_{xx} & -2P_{xy} & -P_{xz} \\ P_{yz} & -P_{xz} & 0 \end{pmatrix}.$$

As usual, the z-axis is defined by the magnetic field direction, so that $\Omega_k/\Omega = \delta_{k3}$. The K-operator of Kaufmann has two useful properties. Since $\pi_{lm} = P_{lm} - p\delta_{lm}$, $K(P) = K(\pi)$. Secondly, at lowest order, we have seen that the viscosity, heat flux, and collisional terms vanish, so that to lowest order, entropy is conserved and

$$\frac{d\ln p}{dt} = \frac{5}{3}\frac{d\ln n}{dt}.$$

Consequently, the first term in S becomes

$$\frac{\partial p}{\partial t} + \nabla \cdot (\mathbf{u}p) = p\left(\frac{d\ln p}{dt} + \nabla \cdot \mathbf{u}\right) = -\frac{2}{3}p\nabla \cdot \mathbf{u},$$

from which we obtain

$$\mathbf{S} = p\mathbf{W} + \nu\pi; \quad W_{lm} = \frac{\partial u_l}{\partial x_m} + \frac{\partial u_m}{\partial x_l} - \frac{2}{3}\nabla \cdot \mathbf{u}\delta_{lm}.$$

---

[6] Kaufmann (1960).

We can therefore rewrite $\mathbf{K} = \mathbf{S}/\Omega$ as a system of algebraic equations

$$\begin{pmatrix} 2\pi_{xy} & \pi_{yy} - \pi_{xx} & \pi_{yz} \\ \pi_{yy} - \pi_{xx} & -2\pi_{xy} & -\pi_{xz} \\ \pi_{yz} & -\pi_{xz} & 0 \end{pmatrix} = \frac{1}{\Omega}(pW + \nu\pi),$$

from which we can solve for π. Solving this system of six equations yields precisely the same results as derived by Braginskii and quoted above, except that the viscosity coefficients are a little different,

$$\mu_0 = p\tau; \quad \mu_1 = \frac{\mu_0}{1 + 4\Omega^2\tau^2}; \quad \mu_2 = \frac{\mu_0}{1 + \Omega^2\tau^2}; \quad \mu_3 = 2\mu_4 = \frac{2\Omega\tau}{1 + 4\Omega^2\tau^2}\mu_0,$$

where $\tau = \nu^{-1}$. The form of the viscosity tensor is therefore recovered and the qualitative form of the dependence of the viscous coefficients on gyrofrequency and collision frequency is correct.

In the collisionless limit, $\Omega\tau \gg 1$, the non-diagonal elements of the viscosity tensor become

$$\pi_{xy} = \pi_{yx} = \frac{p}{4\Omega}(W_{xx} - W_{yy});$$

$$\pi_{xz} = \pi_{zx} = -\frac{p}{\Omega}W_{yz};$$

$$\pi_{yz} = \pi_{zy} = \frac{p}{\Omega}W_{xz},$$

which agrees exactly with Braginskii since $p/2\Omega$ coincides with the gyroviscosity coefficient μ_3 when $\Omega\tau \gg 1$. Gyroviscosity is a consequence of the Larmor gyration of particles and not collisions, and in the limit of $\Omega\tau \gg 1$ is unaffected by collisions.

Consider now the diamagnetic flow of heat. Determining heat transport is entirely analogous to determining the transport of momentum across the magnetic field in a collisionally dominated plasma. In this case, we take the energy moment $(mv^2/2)\mathbf{v}$ of the kinetic equation, and assume that the flow velocity is small $v \sim \delta v_T \ll v_T$. This assumption yields the simplification that there is little to distinguish between $\mathbf{v}$ and $\mathbf{c} = \mathbf{v} - \mathbf{u}$ at the lowest order in the pressure tensor. We therefore obtain

$$\frac{\partial \mathbf{Q}}{\partial t} + \nabla \cdot H - \frac{e}{m}\left(\mathbf{E} \cdot \mathbf{P} + \frac{3}{2}p\mathbf{E} + \mathbf{Q} \times \mathbf{B}\right) = \mathbf{G},$$

where

$$Q_i = q_i + \frac{5}{2}pu_i + \pi_{ij}u_j + \frac{1}{2}mnu^2u_j \simeq q_i + \frac{5}{2}pu_i,$$

4.9 Application 1: Transport Perpendicular to a Mean Magnetic Field

since $\Pi_{ij} = p\delta_{ij}$ and $\pi_{ij} = 0$ in a Maxwellian plasma. A "heat stress" tensor

$$H_{ij} \equiv \int \frac{1}{2} mv^2 v_i v_j \, f d^3v$$

has been introduced, together with the collisional rate of change of the total heat flux

$$\mathbf{G} \equiv \int \frac{1}{2} mv^2 \mathbf{v} C(f) d^3v.$$

If the particle distribution function is close to a Maxwellian, then $H_{ij} = (5pT/2m)\delta_{ij}$ and $P_{ij} = p\delta_{ij}$ (Exercises). The perpendicular heat flux $\mathbf{Q}_\perp$ is obtained from the cross product of the transport equation with $\mathbf{B}$,

$$\mathbf{Q}_\perp = \frac{5p}{2}\left(\frac{\mathbf{E} \times \mathbf{B}}{B^2} + \frac{\mathbf{b} \times \nabla p}{mn\Omega}\right) + \frac{5p}{2m}\frac{\mathbf{b} \times \nabla T}{\Omega},$$

where the time derivative $\partial \mathbf{Q}/\partial t$ has been neglected, and we have assumed that

$$\mathbf{G} \times \frac{\mathbf{b}}{\Omega} \sim \frac{\nu}{\Omega}\mathbf{q}_\perp,$$

is small. On using the expression for the diamagnetic drift velocity and the slow flow approximation $\mathbf{Q} \simeq \mathbf{q} + 5p/2\mathbf{u}$, the lowest order conductive part of the diamagnetic heat flux is given by

$$\mathbf{q}_\perp = \frac{5p}{2m}\frac{\mathbf{b} \times \nabla T}{\Omega}.$$

1. By taking the $c_l c_k$-moment of the kinetic equation, derive the evolution equation for the pressure tensor P_{lm}, Eq. (4.37).
2. Complete the steps in deriving the Kaufmann representation for S_{lm} and the K-operator.
3. Show that for a Maxwellian distribution,

$$H_{ij} \equiv \int \frac{1}{2} mv^2 v_i v_j \, f d^3v = \frac{5}{2m} pT\delta_{ij}.$$

4.10 Application 2: The Equations of Magnetohydrodynamics

One of the most important simplifications of the system of transport equations is the single magnetized fluid system of equations – essentially the ideal gas dynamic equations modified by the inclusion of the Maxwell equations. Let us assume that the distribution function for both electrons and protons is a Maxwellian distribution, and for simplicity we restrict ourselves to a simple two-component plasma. It is straightforward to include more particle species. The equations for the number density and momentum transport are given by

$$\frac{\partial n_a}{\partial t} + \nabla \cdot (n_a \mathbf{u}_a) = 0;$$

$$m_a n_a \left(\frac{\partial \mathbf{u}_a}{\partial t} + \mathbf{u}_a \cdot \nabla \mathbf{u}_a \right) = -\nabla p_a + q_a n_a (\mathbf{E} + \mathbf{u}_a \times \mathbf{B}) + \mathbf{R}_a,$$

where $a \equiv e, p$ identifies the species (electron, proton). As before, $\mathbf{R}_p = -\mathbf{R}_e$. This set of equations corresponds to a *two-fluid* model. We can combine the electron and ion fluids to obtain a one-fluid model, which will be the *magnetohydrodynamic (MHD)* model.

The single fluid is characterized by a mass density

$$\rho(\mathbf{x}, t) \equiv m_e n_e + m_p n_p \simeq m_p n_p(\mathbf{x}, t),$$

where the last approximation results from assuming $m_e/m_p \ll 1$, a charge density

$$\rho_q \equiv q_e n_e + q_p n_p = e(n_p - n_e),$$

and a center-of-flow velocity

$$\mathbf{u} \equiv \frac{m_p n_p \mathbf{u}_p + m_e n_e \mathbf{u}_e}{\rho} \simeq \mathbf{u}_p,$$

(again assuming $m_e/m_p \ll 1$), a current density

$$\mathbf{J} \equiv q_p n_p \mathbf{u}_p + q_e n_e \mathbf{u}_e = e(n_p \mathbf{u}_p - n_e \mathbf{u}_e),$$

and a total pressure

$$P \equiv p_p + p_e.$$

By combining the individual two-fluid equations appropriately, we can derive a set of equations that relate and describe the evolution of these quantities. Of particular note will be an equation, *Ohm's law*, relating the magnetic field, electric field, and the single fluid flow velocity.

4.10 Application 2: The Equations of Magnetohydrodynamics

By multiplying the proton and electron continuity equations by their respective masses and adding yields immediately

$$\frac{\partial \rho}{\partial t} + \nabla \cdot (\rho \mathbf{u}) = 0,$$

which is the conservation of mass equation. Similarly, by multiplying the proton and electron continuity equations by their respective charges and adding yields

$$\frac{\partial \rho_q}{\partial t} + \nabla \cdot (\rho_q \mathbf{J}) = 0,$$

which is the conservation of charge equation.

The construction of the single fluid momentum equation results from summing the proton and electron force equations, using $\mathbf{R}_p = -\mathbf{R}_e$, regarding u_a and $\partial n_a/\partial t$ as small and neglecting the products of small quantities. This then yields the total momentum equation as

$$\rho \left(\frac{\partial \mathbf{u}}{\partial t} + \mathbf{u} \cdot \nabla \mathbf{u} \right) = -\nabla P + \rho_q \mathbf{E} + \mathbf{J} \times \mathbf{B}.$$

The final equation that is needed is of course for the current. This equation, known as the *generalized Ohm's law*, follows from multiplying the proton and electron momentum equations by q_a/m_a respectively, summing, and neglecting quadratic terms in the small quantities u_a and $\partial n_a/\partial t$, and using $q_p = -q_e = e$. This yields

$$\frac{\partial \mathbf{J}}{\partial t} = -\frac{e}{m_p} \nabla p_p + \frac{e}{m_e} \nabla p_e + \left(\frac{e^2 n_e}{m_e} + \frac{e^2 n_p}{m_p} \right) \mathbf{E} + \frac{e^2 n_e}{m_e} \mathbf{u}_e \times \mathbf{B}$$

$$+ \frac{e^2 n_p}{m_p} \mathbf{u}_p \times \mathbf{B} + \left(\frac{e}{m_p} + \frac{e}{m_e} \right) \mathbf{R}_p.$$

The term multiplying the electron velocity can be expressed as

$$\frac{e^2 n_e}{m_e} \mathbf{u}_e = \frac{e}{m_e} \left(n_e e \mathbf{u}_e - n_p e \mathbf{u}_p \right) + \frac{e^2}{m_p m_e} m_p n_p \mathbf{u}_p$$

$$\simeq -\frac{e}{m_e} \mathbf{J} + \frac{e^2}{m_p m_e} \left(m_p n_p \mathbf{u}_p + m_e n_e \mathbf{u}_e \right)$$

$$= -\frac{e}{m_e} \mathbf{J} + \frac{e^2}{m_p m_e} (\rho \mathbf{u}).$$

We can use this expression to simplify the generalized Ohm's law, together with the following approximations, $m_e/m_p \ll 1$, $n_p \simeq n_e$, and $p_p \simeq p_e \simeq P/2$, to obtain

$$\frac{\partial \mathbf{J}}{\partial t} = -\frac{e}{2m_e} \nabla P + \frac{e^2 \rho}{m_e m_p} (\mathbf{E} + \mathbf{u} \times \mathbf{B}) - \frac{e}{m_e} \mathbf{J} \times \mathbf{B} + \frac{e}{m_e} \mathbf{R}_p.$$

We suppose for simplicity that we may approximate

$$\mathbf{R}_p \propto \mathbf{u}_p - \mathbf{u}_e \propto \mathbf{J} \quad \text{so that} \quad \mathbf{R}_p = -\frac{\rho e}{m_p \sigma}\mathbf{J},$$

where we have introduced the conductivity σ. A negative sign was chosen to reflect the expected decrease in current $\mathbf{J}$ due to proton-electron collisions. Thus, the generalized Ohm's law can be expressed in the commonly used form

$$\frac{m_p m_e}{\rho e^2}\frac{\partial \mathbf{J}}{\partial t} = \frac{m_p}{2\rho e}\nabla P + \mathbf{E} + \mathbf{u} \times \mathbf{B} - \frac{m_p}{\rho e}\mathbf{J} \times \mathbf{B} - \frac{\mathbf{J}}{\sigma}. \tag{4.38}$$

The term "Ohm's law" derives from the neglect of all but the two terms in (4.38) that give

$$\mathbf{J} = \sigma \mathbf{E},$$

which is Ohm's law with conductivity σ. For very low frequencies, one can typically ignore the $\partial \mathbf{J}/\partial t$ term in the generalized Ohm's law. For low temperature plasmas, the ∇P term can also be neglected, and if the current is small, we can neglect the *Hall term* $\mathbf{J} \times \mathbf{B}$ compared to the $\mathbf{u} \times \mathbf{B}$ term. Subject to these assumptions, Ohm's law becomes

$$\mathbf{J} = \sigma \left(\mathbf{E} + \mathbf{u} \times \mathbf{B}\right).$$

Finally, in the limit of vanishing collisions, the conductivity becomes infinite, so that we must have

$$\mathbf{E} = -\mathbf{u} \times \mathbf{B}.$$

The one-fluid equations are coupled to Maxwell's equations,

$$\nabla \times \mathbf{E} = -\frac{\partial \mathbf{B}}{\partial t};$$

$$\nabla \times \mathbf{B} = \mu_0 \mathbf{J} + \varepsilon_0 \mu_0 \frac{\partial \mathbf{E}}{\partial t};$$

$$\nabla \cdot \mathbf{B} = 0.$$

Maxwell's equations and the one-fluid or MHD equations correspond to 14 equations in 14 unknowns, ρ, ρ_q, $\mathbf{u}$, $\mathbf{J}$, $\mathbf{E}$, and $\mathbf{B}$, provided we assume that the pressure P can be expressed in terms of ρ. ε_0 is the vacuum permittivity and μ_0 the permeability.

In the low frequency limit and infinite conductivity, no charge imbalance is allowed, i.e., $\rho_q = 0$. These conditions then yield the equations of *ideal MHD*,

4.10 Application 2: The Equations of Magnetohydrodynamics

$$\frac{\partial \rho}{\partial t} + \nabla \cdot (\rho \mathbf{u}) = 0; \tag{4.39}$$

$$\rho \left(\frac{\partial \mathbf{u}}{\partial t} + \mathbf{u} \cdot \nabla \mathbf{u} \right) = -\nabla P + \mathbf{J} \times \mathbf{B}; \tag{4.40}$$

$$\frac{\partial \mathbf{B}}{\partial t} = \nabla \times (\mathbf{u} \times \mathbf{B}); \tag{4.41}$$

$$\nabla \times \mathbf{B} = \mu_0 \mathbf{J}; \tag{4.42}$$

$$\nabla \cdot \mathbf{B} = 0, \tag{4.43}$$

where the low-frequency assumption allowed us to neglect the $\partial \mathbf{E}/\partial t$ term. Equations (4.39)–(4.42) are typically augmented by an equation for the pressure, the simplest possibility being the adiabatic equation of state,

$$\frac{\partial P}{\partial t} + \mathbf{u} \cdot \nabla P + \gamma P \nabla \cdot \mathbf{u} = 0, \tag{4.44}$$

where γ is the adiabatic index of the system.

Exercises

1. Linearize the ideal MHD equations (4.39)–(4.44) with $\rho = \rho_0 = $ const., $\mathbf{B} = B_0 \hat{\mathbf{z}} + \delta B \hat{\mathbf{y}}$, $\mathbf{u} = \delta u \hat{\mathbf{y}}$, $\mathbf{J} = \delta J \hat{\mathbf{x}}$. Seek solutions of the linearized 1D MHD equations in the form $\exp(i\omega t - kz)$, where ω is the wave frequency and k the corresponding wave number and derive the Alfvén wave dispersion relation.
2. For $\mathbf{B} = B\hat{\mathbf{z}}$, linearize the ideal 1D (say x) MHD equations about a stationary constant state. Seek solutions of the linearized 1D MHD equations in the form $\exp(i\omega t - kx)$, where ω is the wave frequency and k the corresponding wave number and derive the magnetosonic wave dispersion relation.

Before concluding this section, we discuss briefly the conservation form of the Eqs. (4.39)–(4.44). Evidently, Eq. (4.39) is already in conservation form (mass conservation), as is (4.43). By using (4.39), we have

$$\rho \left(\frac{\partial \mathbf{u}}{\partial t} + \mathbf{u} \cdot \nabla \mathbf{u} \right) = \frac{\partial \rho \mathbf{u}}{\partial t} + \nabla \cdot (\rho \mathbf{u} \mathbf{u}),$$

and, using the vector identity $\mathbf{a} \times (\nabla \times \mathbf{b}) = (\nabla \mathbf{b}) \cdot \mathbf{a} - \nabla \cdot (\mathbf{a}\mathbf{b}) - \mathbf{b} \nabla \cdot \mathbf{a}$, we have

$$-\mathbf{J} \times \mathbf{B} = \frac{1}{\mu_0} \mathbf{B} \times (\nabla \times \mathbf{B}) = \frac{1}{\mu_0} \nabla \left(\frac{1}{2} B^2 \right) - \frac{1}{\mu_0} \nabla \cdot (\mathbf{B}\mathbf{B}),$$

which yields the conservation form of the momentum equation,

$$\frac{\partial (\rho \mathbf{u})}{\partial t} + \nabla \cdot \left(\rho \mathbf{u} \mathbf{u} + \left(P + \frac{1}{2\mu_0} B^2 \right) \mathbf{I} - \frac{1}{\mu_0} \mathbf{B}\mathbf{B} \right) = 0.$$

By introducing the internal energy e as before with $P = (\gamma - 1)\rho e$, where γ denotes the adiabatic index for the magnetized gas, we can rewrite the gas pressure equation (4.44) as

$$\frac{\partial(\rho e)}{\partial t} + \mathbf{u} \cdot \nabla(\rho e) + \gamma \rho e \nabla \cdot \mathbf{u} = 0,$$

or

$$\frac{\partial(\rho e)}{\partial t} + \nabla \cdot (\rho e \mathbf{u}) + (\gamma - 1)\rho e \nabla \cdot \mathbf{u} = 0,$$

giving

$$\frac{\partial(\rho e)}{\partial t} + \nabla \cdot (\rho e \mathbf{u}) + P \nabla \cdot \mathbf{u} = 0; \qquad (4.45)$$

$$\frac{\partial e}{\partial t} + \mathbf{u} \cdot \nabla e + (\gamma - 1) e \nabla \cdot \mathbf{u} = 0. \qquad (4.46)$$

The conservation of energy equation corresponds to the temporal evolution of the kinetic, internal, and magnetic energy. Considering each of these separately and in turn yields first for the kinetic energy (taking $\mathbf{u}\cdot$ (4.40))

$$\frac{\partial}{\partial t}\left(\frac{1}{2}\rho u^2\right) + \nabla \cdot \left(\frac{1}{2}\rho u^2 \mathbf{u}\right) + \mathbf{u} \cdot \nabla P - \mathbf{u} \cdot \mathbf{J} \times \mathbf{B} = 0.$$

The internal energy expression is given by (4.45). Finally, taking $\mathbf{B}\cdot$ (4.41) yields

$$\frac{\partial}{\partial t}\left(\frac{1}{2}B^2\right) + \nabla \cdot [\mathbf{B} \times (\mathbf{u} \times \mathbf{B})] - (\mathbf{u} \times \mathbf{B}) \cdot \nabla \times \mathbf{B} = 0,$$

which implies[7]

$$\frac{\partial}{\partial t}\left(\frac{1}{2}B^2\right) + \nabla \cdot (\mathbf{B} \cdot \mathbf{B}\mathbf{u} - \mathbf{u} \cdot \mathbf{B}\mathbf{B}) + \mu_0 \mathbf{u} \cdot \mathbf{J} \times \mathbf{B} = 0.$$

Adding these results together for the kinetic, internal, and magnetic energy yields the conservation of energy equation (4.49) below.

---

[7] The following identities are used,

$$\nabla \times (\mathbf{a} \times \mathbf{b}) = \mathbf{b} \cdot \nabla \times \mathbf{a} - \mathbf{a} \cdot \nabla \times \mathbf{b},$$

and

$$\mathbf{a} \times (\mathbf{b} \times \mathbf{c}) = (\mathbf{a} \cdot \mathbf{c})\mathbf{b} - (\mathbf{a} \cdot \mathbf{b})\mathbf{c}.$$

4.10 Application 2: The Equations of Magnetohydrodynamics

Finally, since

$$\nabla \times \mathbf{E} = -\nabla \times (\mathbf{u} \times \mathbf{B}) = \nabla \cdot (\mathbf{uB} - \mathbf{Bu}),$$

Equation (4.41) can be expressed in the conservation of magnetic flux form (4.50) below.

We collect the results above to express the conservation form of the ideal MHD equations through

$$\frac{\partial \rho}{\partial t} + \nabla \cdot (\rho \mathbf{u}) = 0; \tag{4.47}$$

$$\frac{\partial (\rho \mathbf{u})}{\partial t} + \nabla \cdot \left(\rho \mathbf{uu} + \left(P + \frac{1}{2\mu_0} B^2 \right) \mathbf{I} - \frac{1}{\mu_0} \mathbf{BB} \right) = 0; \tag{4.48}$$

$$\frac{\partial}{\partial t} \left(\frac{1}{2} \rho u^2 + \rho e + \frac{1}{2\mu_0} B^2 \right)$$

$$+ \nabla \cdot \left[\left(\frac{1}{2} \rho u^2 + \rho e + P + \frac{1}{\mu_0} B^2 \right) \mathbf{u} - \frac{1}{\mu_0} \mathbf{u} \cdot \mathbf{BB} \right] = 0; \tag{4.49}$$

$$\frac{\partial \mathbf{B}}{\partial t} + \nabla \cdot (\mathbf{uB} - \mathbf{Bu}) = 0; \tag{4.50}$$

$$\nabla \cdot \mathbf{B} = 0. \tag{4.51}$$

Finally, note that we may express the entropy (Sect. 3.1) generally as $S = C_v \ln(P\rho^{-\gamma}) + $ const. (C_v the specific heat at constant volume) and so, from (4.39) and (4.44), the evolution equation for S is given by

$$\frac{\partial S}{\partial t} + \mathbf{u} \cdot \nabla S = 0.$$

We can then derive an additional conservation law, the conservation of entropy per unit volume,

$$\frac{\partial}{\partial t}(\rho S) + \nabla \cdot (\rho S \mathbf{u}) = 0. \tag{4.52}$$

These conservation relations are important in that it allows us to introduce the notion of weak solutions to the system of MHD equations, these being critical to treating discontinuities in a magnetized gas flow.

Before concluding this section, we derive two further conservation results that are fundamental elements of the theory of MHD. The first is the *frozen-in flux theorem* derived originally by H. Alfvén. Consider a surface S bounded by a closed contour C moving with the local magnetized fluid velocity $\mathbf{u}^8$ (Fig. 4.5). We may consider

---

[8]Consider a line element $d\mathbf{l}$ in a fluid flow $\mathbf{u}(\mathbf{x}, t)$ with ends at $\mathbf{x}$ and $\mathbf{x} + d\mathbf{l}$. The motion of $d\mathbf{l}$ is given by the Lagrangian derivative of $d\mathbf{l}$,

$$\frac{d}{dt}(d\mathbf{l}) = \frac{d(\mathbf{x} + d\mathbf{l})}{dt} - \frac{d\mathbf{x}}{dt} = \mathbf{u}(\mathbf{x} + d\mathbf{l}) - \mathbf{u}(\mathbf{x}, t) = d\mathbf{l} \cdot \nabla \mathbf{u}.$$

Fig. 4.5 Magnetic flux passing through a closed surface S bounded by a curve C

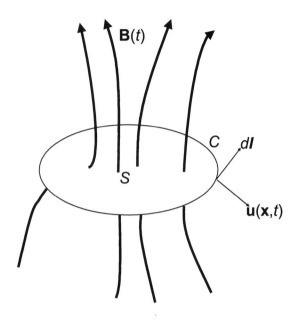

We may define a surface element $dS \equiv d\mathbf{l}_1 \times d\mathbf{l}_2$ and determine its equation of motion from the Lagrangian derivative

$$\frac{d}{dt}(dS) = \frac{d}{dt}(d\mathbf{l}_1) \times d\mathbf{l}_2 + d\mathbf{l}_1 \times \frac{d}{dt}(d\mathbf{l}_2) = d\mathbf{l}_1 \cdot \nabla\mathbf{u} \times d\mathbf{l}_2 - d\mathbf{l}_2 \cdot \nabla\mathbf{u} \times d\mathbf{l}_1,$$

after using the result for the motion of a line. The vector identity $(\mathbf{a} \times \mathbf{b}) \times \nabla = \mathbf{ba} \cdot \nabla - \mathbf{ab} \cdot \nabla$ applied to the fluid velocity $\mathbf{u}$ gives

$$[(\mathbf{a} \times \mathbf{b}) \times \nabla] \times \mathbf{u} = (\mathbf{ba} \cdot \nabla) \times \mathbf{u} - (\mathbf{ab} \cdot \nabla) \times \mathbf{u} = -\mathbf{a} \cdot (\nabla\mathbf{u}) \times \mathbf{b} + \mathbf{b} \cdot (\nabla\mathbf{u}) \times \mathbf{a}.$$

With $\mathbf{a} = d\mathbf{l}_1$ and $\mathbf{b} = d\mathbf{l}_2$ we have $d\mathbf{l}_1 \cdot \nabla\mathbf{u} \times d\mathbf{l}_2 - d\mathbf{l}_2 \cdot \nabla\mathbf{u} \times d\mathbf{l}_1 = -(dS \times \nabla) \times \mathbf{u} = -(\nabla\mathbf{u}) \cdot dS + \nabla \cdot \mathbf{u} dS$ (using $(\mathbf{a} \times \nabla) \times \mathbf{b} = (\nabla\mathbf{b}) \cdot \mathbf{a} - \mathbf{a}\nabla \cdot \mathbf{b}$), we can obtain the equation for the kinematic motion of a surface element,

$$\frac{d}{dt}(dS) = -(dS \times \nabla) \times \mathbf{u} = -(\nabla\mathbf{u}) \cdot dS + \nabla \cdot \mathbf{u} dS.$$

Finally, we note that the equation for the kinematic motion of a volume element $dV \equiv dS \cdot d\mathbf{l}_3 = (d\mathbf{l}_1 \times d\mathbf{l}_2) \cdot d\mathbf{l}_3$ can be derived using the above result as

$$\frac{d}{dt}(dV) = \nabla \cdot \mathbf{u} dV.$$

This last result can be used to show that the mass of a fluid element $dM \equiv \rho dV$ is constant, that momentum of a fluid element $d(\rho\mathbf{u})dV$ is not constant, and that the total energy density of a fluid element is not constant.

4.10 Application 2: The Equations of Magnetohydrodynamics

the magnetic flux through a surface element $d\Psi \equiv \mathbf{B} \cdot d\mathbf{S}$, using the Lagrangian form for the magnetic flux, or induction equation,

$$\frac{d\mathbf{B}}{dt} = \frac{\partial \mathbf{B}}{\partial t} + \mathbf{u} \cdot \nabla \mathbf{B} = \mathbf{B} \cdot \nabla \mathbf{u} - \mathbf{B}\nabla \cdot \mathbf{u},$$

and the kinematic equation of motion for the convected surface element $d\mathbf{S}$, to obtain

$$\frac{d}{dt}(d\Psi) = \frac{d}{dt}(\mathbf{B} \cdot d\mathbf{S}) = \frac{d\mathbf{B}}{dt} \cdot d\mathbf{S} + \mathbf{B} \cdot \frac{d}{dt}(d\mathbf{S})$$
$$= (\mathbf{B} \cdot \nabla \mathbf{u} - \mathbf{B}\nabla \cdot \mathbf{u}) \cdot d\mathbf{S} + \mathbf{B} \cdot (-(\nabla \mathbf{u}) \cdot d\mathbf{S} + \nabla \cdot \mathbf{u} d\mathbf{S}) = 0. \quad (4.53)$$

Consequently, the magnetic flux through a co-moving surface element is constant, and since this holds for any surface element, the flux through any surface bounded by a contour C moving with the fluid is conserved, i.e.,

$$\Psi = \int_C \mathbf{B} \cdot \mathbf{n} dS = \text{const.}$$

Note that we can combine the Lagrangian form of the conservation of mass and the induction equation to obtain

$$\frac{d}{dt}\left(\frac{\mathbf{B}}{\rho}\right) = \frac{1}{\rho}(\mathbf{B} \cdot \nabla \mathbf{u} - \mathbf{B}\nabla \cdot \mathbf{u}) + \frac{\mathbf{B}}{\rho}\nabla \cdot \mathbf{u} = \left(\frac{\mathbf{B}}{\rho}\right) \cdot \nabla \mathbf{u}. \quad (4.54)$$

This equation is structurally identical to the kinematic equation of motion for a line element $d\mathbf{l}$, so a line element $d\mathbf{l} \parallel \mathbf{B}$ moves exactly as $\mathbf{B}/\rho$. Thus, plasma on this line element and magnetic field line move together, often described as the magnetic field lines are frozen into the plasma. The concept of field lines and flux tubes, although a mathematical construction, offers considerable insight into our understanding of magnetic fields. At every point, the field lines follow the direction of the magnetic field and are therefore defined by the characteristic equations

$$\frac{dx}{B_x} = \frac{dy}{B_y} = \frac{dz}{B_z}.$$

A magnetic or flux surface is one that is everywhere tangential to the field i.e., the normal to the surface is everywhere perpendicular to $\mathbf{B}$. An open ended cylindrical magnetic surface defines a flux tube, and the density of field lines through a flux tube can represent the strength of the field.

The divergence-free or solenoidal condition for the magnetic field, i.e., $\nabla \cdot \mathbf{B} = 0$ indicates that the a vector potential $\mathbf{A}$ can be used to represent the magnetic field,

$$\mathbf{B} = \nabla \times \mathbf{A}.$$

The electric field (Ohm's law) then becomes

$$\mathbf{E} = -\mathbf{u} \times (\nabla \times \mathbf{A}),$$

and the induction equation can be expressed in terms of $\mathbf{A}$,

$$\frac{\partial \mathbf{A}}{\partial t} = \mathbf{u} \times (\nabla \times \mathbf{A}) - \nabla \Phi,$$

where Φ is a scalar potential. The above equations are invariant under a gauge transformation so we can choose $\Phi = 0$ without loss of generality. We introduce the *magnetic helicity*

$$K = \int_V \mathbf{A} \cdot \mathbf{B} \, dV, \tag{4.55}$$

where the integration is taken over the volume of some flux tube. Consider now the Lagrangian rate of change of the magnetic helicity,

$$\begin{aligned}
\frac{dK}{dt} &= \int \left(\frac{d\mathbf{A}}{dt} \cdot \mathbf{B} + \mathbf{A} \cdot \frac{d\mathbf{B}}{dt} \right) dV + \int \mathbf{A} \cdot \mathbf{B} \frac{d}{dt}(dV) \\
&= \int \left(\frac{\partial \mathbf{A}}{\partial t} \cdot \mathbf{B} + \mathbf{u} \cdot (\nabla \mathbf{A}) \cdot \mathbf{B} + \mathbf{A} \cdot \frac{\partial \mathbf{B}}{\partial t} + \mathbf{u} \cdot (\nabla \mathbf{B}) \cdot \mathbf{A} + \mathbf{A} \cdot \mathbf{B} \nabla \cdot \mathbf{u} \right) dV \\
&= \int \left[(\mathbf{u} \times \mathbf{B}) \cdot \mathbf{B} + \mathbf{A} \cdot \nabla \times (\mathbf{u} \times \mathbf{B}) + \nabla \cdot (\mathbf{A} \cdot \mathbf{B} \mathbf{u}) \right] dV \\
&= \int \nabla \cdot \left[(\mathbf{u} \times \mathbf{B}) \times \mathbf{A} + \mathbf{A} \cdot \mathbf{B} \mathbf{u} \right] dV \\
&= \int (\mathbf{A} \cdot \mathbf{u} \mathbf{B} - \mathbf{A} \cdot \mathbf{B} \mathbf{u} + \mathbf{A} \cdot \mathbf{B} \mathbf{u}) \cdot \mathbf{n} \, dS = 0,
\end{aligned}$$

because $\mathbf{B} \cdot \mathbf{n} = 0$ on the boundary of a flux tube and certain vector identities were used to obtain the second last line.[9] Thus, the magnetic helicity of any flux tube is conserved in ideal MHD. The concept of magnetic helicity is of particular importance to MHD studies of plasma fusion devices and coronal magnetic fields.

---

[9]The following vector identity was used,

$$\nabla \cdot (\mathbf{a} \times \mathbf{b}) = \mathbf{b} \cdot (\nabla \times \mathbf{a}) - \mathbf{a} \cdot \nabla \times \mathbf{b},$$

and identifying $\mathbf{b} \equiv \mathbf{u} \times \mathbf{B}$ and $\mathbf{a} \equiv \mathbf{A}$. Thus,

$$(\mathbf{u} \times \mathbf{B}) \cdot \nabla \mathbf{A} - \mathbf{A} \cdot \nabla \times (\mathbf{u} \times \mathbf{B}) = \nabla \cdot (\mathbf{A} \times \mathbf{u} \times \mathbf{B}).$$

Exercises

1. Show using the kinematic equation of motion for a volume element that mass in a fluid element $dM \equiv \rho dV$ is conserved. Show that the momentum of a fluid element $d(\rho \mathbf{u})dV$ is not constant, and that the total energy density $\rho u^2/2 + P/(\gamma - 1) + B^2/2\mu_0$ of a fluid element is not constant.
2. Derive the frozen-in field equation (4.54).

4.11 Application 3: MHD Shock Waves

In this section, we discuss the theory of magnetohydrodynamic (MHD) shocks based on the conservation form of the MHD equations (4.47)–(4.51). An extensive discussion of MHD shock theory can be found in the classic works of Cabannes (1970) and Anderson (1963). We confine our attention to steady planar shocks moving in a direction normal to the plane of the shock (see Fig. 4.6). As with gas dynamic shocks, we can derive the MHD form of the Rankine-Hugoniot conditions from the conservation form of the MHD equations. We can choose a coordinate system in which the velocity and magnetic field vectors on both sides of the shock are co-planar i.e., lying in the same plane. If we suppose the $(\mathbf{u}, \mathbf{B})$-plane to be in the (x, y)-plane and the shock wave to lie in the (x, z)-plane (Fig. 4.6), the generalized Rankine-Hugoniot conditions are given by

$$[\rho u_x] = 0; \tag{4.56}$$

$$\left[\rho u_x^2 + P + \frac{1}{2\mu_0} B_y^2\right] = 0; \tag{4.57}$$

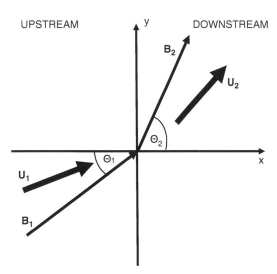

Fig. 4.6 Schematic of an oblique MHD shock located in the $x = 0$ plane of a rectangular Cartesian coordinate system OXYZ. The fluid velocity $\mathbf{u}$ and magnetic field $\mathbf{B}$ lie in the (x, y)-plane and the electric field $\mathbf{E}$ lies along the z-axis

$$\left[\rho u_x u_y - \frac{1}{\mu_0} B_x B_y\right] = 0; \tag{4.58}$$

$$\left[\rho u_x \left(\frac{\gamma}{\gamma-1} P + \frac{1}{2} u^2\right) + \frac{1}{\mu} (\mathbf{E} \times \mathbf{B})_x\right] = 0; \tag{4.59}$$

$$[\mathbf{E}_t] = 0; \tag{4.60}$$

$$[B_x] = 0, \tag{4.61}$$

where (u_x, u_y) is the fluid velocity, $u^2 = u_x^2 + u_y^2$. The square brackets $[\cdot]$ are defined in the usual way to mean the difference between the quantities evaluated upstream and downstream of the shock.

If a transformation is made to a new reference frame with constant transverse velocity $\mathbf{u}_t$, the corresponding transformation of the transverse electric field is

$$\mathbf{E}'_t = \mathbf{E}_t + \mathbf{u}_t \times \mathbf{B},$$

or

$$E'_y = E_y + u_z B_x, \qquad E'_z = E_z - u_y B_x.$$

Thus, for $B_x = \text{const.} \neq 0$, u_z and u_y can be chosen so that $E'_y = E'_z = 0$. The generalized Rankine-Hugoniot conditions are simplified considerably in such a reference frame, known as the de Hoffmann-Teller frame, in which $E'_y = E'_z = 0$, ensuring that the magnetic field and velocity vectors are parallel (i.e., co-planar), and allowing the transverse electric field to be replaced by

$$u_x B_y - u_y B_x = 0, \tag{4.62}$$

on both sides of the shock.

On introducing the specific volume of the fluid, $\tau = 1/\rho$ and the mass flux density $m = \rho u_x$, the conservation of energy condition (4.59) becomes

$$[w] + \frac{1}{2} m^2 \left[\tau^2\right] + \frac{1}{2} \left[\left(u_y - \frac{1}{m\mu_0} B_x B_y\right)^2\right] + \frac{1}{\mu_0} \left[\tau B_y^2\right] - \frac{B_x^2}{2m^2 \mu_0^2} \left[B_y^2\right] = 0,$$

where $w = \frac{\gamma}{\gamma-1} P/\rho$ denotes the enthalpy per unit mass. From the transverse momentum condition (4.58), the third term above is zero. On using the normal momentum relation for m, we obtain the MHD Hugoniot relation

$$[w] + \langle P \rangle [\tau] + \frac{1}{2\mu_0} [B_y]^2 [\tau] = 0, \tag{4.63}$$

4.11 Application 3: MHD Shock Waves

where $\langle Q \rangle \equiv \frac{1}{2}(Q_1 + Q_2)$. Equation (4.63) differs from the hydrodynamic form only in the third term. The MHD Hugoniot relation (4.63) can be used directly to show that entropy increases or decreases across a shock as the density ρ increases or decreases across a shock,[10] showing that only compressive shocks are physically admissible. We do not provide a direct proof here.

For a perfect gas with adiabatic index γ, the square of the sound speed and the enthalpy can be expressed as

$$a_s^2 = \frac{\gamma P}{\rho}; \qquad w = \frac{a_s^2}{\gamma - 1}.$$

We introduce the dimensionless parameters

$$r = \frac{\rho_2}{\rho_1};$$

$$M_A = \frac{u_x}{V_{An}} = \frac{1}{r^{1/2}}\sqrt{\frac{\mu_0 m^2}{B_x^2}},$$

where r is the shock compression ratio, M_A the Alfvénic Mach number of the flow and V_{An} the component of the Alfvén velocity normal to the shock. On rewriting (4.56)–(4.61), using the sound speed and enthalpy relations for an ideal gas, we have

$$\rho_1 u_{x1} = \rho u_{x2} = m; \tag{4.64}$$

$$\frac{\rho_1 a_{s1}^2}{\gamma} + \rho_1 u_{x1}^2 + \frac{1}{2\mu_0} B_{y1}^2 = \frac{\rho_2 a_{s2}^2}{\gamma} + \rho_2 u_{x2}^2 + \frac{1}{2\mu_0} B_{y2}^2; \tag{4.65}$$

$$\rho_1 u_{x1} u_{y1} - \frac{1}{\mu_0} B_x B_{y1} = \rho_2 u_{x2} u_{y2} - \frac{1}{\mu_0} B_x B_{y2}; \tag{4.66}$$

$$\rho_1 u_{x1} \left(\frac{a_{s1}^2}{\gamma - 1} + \frac{1}{2} u_1^2\right) + \frac{1}{\mu_0} B_{y1} \left(u_{x1} B_{y1} - u_{y1} B_x\right)$$

$$= \rho_2 u_{x2} \left(\frac{a_{s2}^2}{\gamma - 1} + \frac{1}{2} u_2^2\right) + \frac{1}{\mu_0} B_{y2} \left(u_{x2} B_{y2} - u_{y2} B_x\right); \tag{4.67}$$

$$u_{y1} B_x - u_{x1} B_{y1} = u_{y2} B_x - u_{x2} B_{y2}; \tag{4.68}$$

$$B_{x1} = B_{x2} = B_x. \tag{4.69}$$

The transverse momentum equation (4.66) and the transverse electric field equation (4.68) can be solved for u_{y2} and B_{y2},

$$u_{y2} = u_{y1} + \frac{(r-1)\tan\theta_1}{M_{A1}^2 - r} u_{x1}; \tag{4.70}$$

---

[10] See e.g., Boyd and Sanderson (2003).

$$B_{y2} = \frac{M_{A1}^2 - 1}{M_{A1}^2/r - 1} B_{y1}, \qquad (4.71)$$

where $\tan\theta \equiv B_y/B_x$, and, in this geometry, is the angle between the magnetic field and the shock normal upstream of the shock. On using (4.70) and (4.71), we can rewrite parts of the normal momentum (4.65) and total energy (4.67) equation as follows,

$$\frac{1}{2\mu_0}\left(B_{y2}^2 - B_{y1}^2\right) = (r-1)M_{A1}^2 \frac{(r+1)M_{A1}^2 - 2r}{(M_{A1}^2 - r)^2} \frac{B_{y1}^2}{2\mu_0};$$

$$\frac{B_{y2}^2}{\mu_0\rho_2} - \frac{B_{y1}^2}{\mu_0\rho_1} = (r-1)\frac{M_{A1}^4 - r}{(M_{A1}^2 - r)^2} \frac{B_{y1}^2}{\mu_0\rho_1};$$

$$\frac{1}{2}\left(u_{y2}^2 - u_{y1}^2\right) - \frac{B_x}{\mu_0 m}\left(B_{y2}u_{y2} - B_{y1}u_{y1}\right) = -\frac{1}{2}(r-1)\frac{(r+1)M_{A1}^2 - 2r}{(M_{A1}^2 - r)^2} \frac{B_{y1}^2}{\mu_0\rho_1}.$$

By using the above results in the normal momentum equation and dividing by B_x^2/μ_0, we obtain a dimensionless equation,

$$\frac{a_{s2}^2/V_{An2}^2 - a_{s1}^2/V_{An1}^2}{\gamma} + M_{A2}^2 - M_{A1}^2 + (r-1)M_{A1}^2 \frac{(r+1)M_{A1}^2 - 2r}{2(M_{A1}^2 - r)^2} \tan^2\theta_1 = 0. \qquad (4.72)$$

On dividing the energy equation (4.67) by $mB_x^2/\mu_0\rho_2$ and using the above relations, we find

$$\frac{1}{\gamma-1}\left(\frac{a_{s2}^2}{V_{An2}^2} - \frac{a_{s1}^2}{V_{An1}^2}\right) - \frac{r-1}{\gamma-1}\frac{a_{s1}^2}{V_{An1}^2} + \frac{1}{2}\left(M_{A2}^2 - rM_{A1}^2\right)$$

$$+ \frac{r(r-1)}{2}\frac{2M_{A1}^2 - r - 1}{(M_{A1}^2 - r)^2} M_{A1}^2 \tan^2\theta_1 = 0. \qquad (4.73)$$

Since

$$\frac{M_{A1}^2}{M_{A2}^2} = \frac{\rho_2}{\rho_1} = r,$$

we can eliminate M_{A2}^2 from (4.73) to obtain the equation for the shock adiabatic,

$$(M_{A1}^2 - r)^2 \left\{ M_{A1}^2 - \frac{2ra_{s1}^2/V_{An1}^2}{r+1-\gamma(r-1)} \right\}$$

$$- rM_{A1}^2 \left\{ \frac{2r - \gamma(r-1)}{r+1-\gamma(r-1)} M_{A1}^2 - r \right\} \tan^2\theta_1 = 0. \qquad (4.74)$$

4.11 Application 3: MHD Shock Waves

Alternatively, since $M_{A1}^2/M_{A2}^2 = r$, we can express the shock adiabatic as the so-called shock polar relation or shock cubic,

$$M_{A1}^2 \left\{(\gamma-1)\left(M_{A2}^2-1\right)^2 + \left[(2-\gamma)M_{A2}^2 + \gamma - 1\right]\tan^2\theta_1\right\}$$
$$= \left\{(\gamma+1)M_{A2}^2 - 2\frac{a_{s1}^2}{V_{A1}^2}\sec^2\theta_1\right\}\left(M_{A1}^2-1\right)^2 - M_{A2}^2\left[\gamma M_{A2}^2 - \gamma - 1\right]\tan^2\theta_1, \tag{4.75}$$

where V_A is the Alfvén velocity based on the magnetic field magnitude.

The importance of the shock polar relation is that, by specifying the upstream variables a_{s1}, V_{A1}, θ_1, and M_{A1}, the downstream Alfvén Mach number can be computed easily, and the remaining downstream variables determined from

$$B_{y2}\left(M_{A2}^2 - 1\right) = B_{y1}\left(M_{A1}^2 - 1\right);$$

$$u_{y2} = u_{y1} + \frac{B_x}{\mu_0 m}(B_{y2} - B_{y1});$$

$$r = \frac{\rho_2}{\rho_1} = \frac{u_{x1}}{u_{x2}} = \frac{M_{A1}^2}{M_{A2}^2};$$

$$P_2 = P_1 + \rho_1 u_{x1}^2\left(1 - \frac{1}{r}\right) + \frac{1}{2\mu_0}\left(B_{y1}^2 - B_{y2}^2\right).$$

The nature of the shock polar can be made more transparent by expressing the variables a_{s1}^2/V_{An1}^2 and $\tan^2\theta_1$ in terms of the fast and slow magnetosonic speeds in the upstream plasma. The shock adiabatic is a bicubic in M_{A1}^2, which implies that three finite amplitude shocks are possible, corresponding to slow, intermediate, and fast mode shocks. In the weak shock limit, $r \simeq 1$, Eq. (4.74) reduces to

$$(M_{A1}^2 - 1)\left[(M_{A1}^2 - 1)\left(M_{A1}^2 - \frac{a_{s1}^2}{V_{An1}^2}\right) - M_{A1}^2\tan^2\theta_1\right] = 0, \tag{4.76}$$

which, on introducing a phase velocity V_p, is reminiscent of the well-known expression for the speeds of propagation of perturbations in a perfectly conducting fluid embedded in a magnetic field, i.e.,

$$(V_p^2 - V_A^2)\left[V_p^4 - (V_A^2 + a_s^2)V_p^2 + V_{An}^2 a_s^2\right] = 0. \tag{4.77}$$

The first factor in both (4.76) and (4.77) represents Alfvén wave propagation and the second magnetosonic propagation. Thus, to be consistent with (4.77), fast and slow magnetosonic Mach numbers M_+ and M_- must satisfy

$$(M_A^2 - 1)\left(M_A^2 - \frac{a_s^2}{V_{An}^2}\right) - M_A^2\tan^2\theta_1 = (M_A^2 - M_+^2)(M_A^2 - M_-^2),$$

where

$$M_\pm^2 = \frac{1}{2}\left\{\frac{a_s^2}{V_{An}^2} + \tan^2\theta_1 + 1 \pm \sqrt{\left(\frac{a_s^2}{V_{An}^2} + \tan^2\theta_1 + 1\right)^2 + 4\tan^2\theta_1}\right\},$$

or equivalently

$$M_\pm = \frac{V_{f,s}}{V_{An}},$$

where V_f and V_s are the fast and slow magnetosonic speeds satisfying the dispersion relation (4.76). Thus, we have established that weak shocks correspond to magnetosonic waves and have phase velocities $V_p = \pm V_f$, $V_p = \pm V_s$.

On employing the relations

$$\frac{a_s^2}{V_{An}^2} = M_+^2 M_-^2; \qquad \tan^2\theta_1 = (M_+^2 - 1)(1 - M_-^2),$$

the shock polar relation (4.75) reduces to

$$M_{A1}^2 - M_{A2}^2 = \frac{2(M_{A2}^2 - 1)(M_{A2}^2 - M_{1+}^2)(M_{A2}^2 - M_{1-}^2)}{(\gamma + 1)(M_{A2}^2 - 1)^2 + (1 - M_{1-}^2)(M_{1+}^2 - 1)\left[(2 - \gamma)M_{A2}^2 + \gamma - 1\right]}. \quad (4.78)$$

Hence, specifying the fast and slow magnetosonic Mach numbers ahead of the shock yields a single curve along which all shocks must lie. Furthermore, in general, $0 < M_{1-} < 1 < M_{1+}$.

As discussed above, entropy increases across a shock if and only if the shock is compressive. Since $M_A^2 \propto 1/\rho$, the shock is compressive, provided $M_{A1}^2 > M_{A2}^2$ in Eq. (4.78); therefore M_{A2}^2 must satisfy one of the inequalities

$$M_{1-}^2 < M_{A2}^2 < 1, \qquad (4.79)$$

$$1 < M_{1+}^2 < M_{A2}^2. \qquad (4.80)$$

Shocks in which the downstream normal fluid speed is sub-Alfvénic, (4.79), are called *slow mode shocks*, whilst those that satisfy the super-Alfvénic inequality (4.80) are called *fast mode shocks*. The weak shock solutions for which $r = 1$ correspond to an *Alfvénic shock* ($M_{A2} = 1$) and a fast and slow magnetosonic wave ($M_{A2} = M_{1+}$ and $M_{A2} = M_{1-}$ respectively). A shock with $\theta_1 \neq 0°$ is generally referred to as an *oblique shock*. Shocks that have either $\theta_1 = 0°$ or $\theta_1 = 90°$ are called *parallel* or *perpendicular* shock waves respectively, and shocks for which $\theta_1 \leq 45°$ are described as *quasi-parallel* whereas shocks for which $\theta_1 > 45°$ are described as *quasi-perpendicular*.

4.11 Application 3: MHD Shock Waves

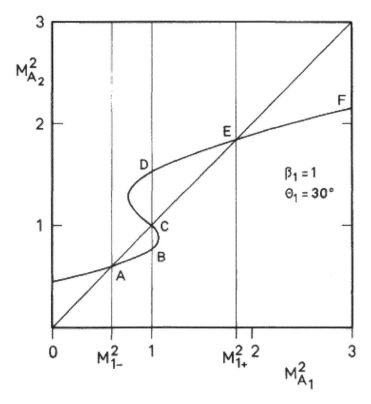

Fig. 4.7 An example of the Alfvén Mach number shock polar (4.75) in which the upstream plasma beta $\beta_{p1} = 1$ and the upstream magnetic field angle (or shock obliquity) is $\theta_1 = 30°$. For compressive shocks, $M_{A1}^2 > M_{A2}^2$ (Webb et al. 1987)

A schematic[11] of the shock polar (4.75) or (4.78) is illustrated in Fig. 4.7 for an upstream state $\theta_1 = 30°$ and plasma beta $\beta_{p1} = 1$, where

$$\beta_p \equiv \frac{P}{B^2/2\mu_0}; \qquad \frac{a_s^2}{V_A^2} = \frac{\gamma}{2}\beta_p.$$

The straight line $r = 1$ lies in the M_{A1}^2-M_{A2}^2 plane and intersects the polar profile at the square of the small disturbance speeds. The asymptote of the shock curve as M_{A1} tends to infinity is given by $r = (\gamma + 1)/(\gamma - 1) = 4$ if $\gamma = 5/3$. This is the maximum possible compression ratio for a strong MHD shock. Those sections of the shock curve above the line $r = 1$ correspond to expansion shocks, which are inadmissible physically since they violate the second law of thermodynamics. Hence

---

[11]Relativistic oblique magnetohydrodynamic shocks by Webb et al. (1987).

the only physically acceptable solutions are those below the line $r = 1$ i.e., the arcs ABC and EF that correspond to compressive solutions. The arc AB is the locus of slow mode shocks and EF of fast mode shocks. The double-valued regime BC (two possible downstream states corresponding to a single upstream state) represent unstable shocks which disintegrate spontaneously into several waves. The points A, B, and E correspond respectively to the slow magnetosonic wave, the Alfvén or intermediate shock, and the fast magnetosonic wave. The point B at which $M_{A1} = 1$ corresponds to the slow mode switch-off shock in which the downstream magnetic field angle $\theta_2 = 0°$ and $B_{t2} = 0$. On the arc AB, $B_{t2} > 0$ whereas on the arc BC, $B_{t2} < 0$.

Figure 4.8 illustrates the effect of varying the upstream field angle θ_1 with $\beta_{p1} = 1$ and $\gamma = 5/3$ on MHD shocks. When $\theta_1 = 0$ in the shock polar relation (4.75), one obtains either

$$M_{A2}^2 = 1,$$

or

$$M_{A2}^2 = \frac{\gamma + 1}{\gamma - 1} M_{A1}^2 + \frac{2}{\gamma + 1} \frac{a_{s1}^2}{V_{A1}^2}.$$

This relation gives the inverse compression ratio of a pure gas shock[12] i.e.,

$$r^{-1} = \frac{\rho_1}{\rho_2} = \frac{\gamma - 1}{\gamma + 1} + \frac{2}{\gamma + 1} \frac{1}{M_{s1}^2},$$

where $M_{s1} = u_{x1}/a_{s1}$ is the acoustic Mach number of the gas upstream of the shock.

As $\theta_1 \longrightarrow 0$, the kink in the shock polar curve near $M_{A1}^2 = 1$ (Fig. 4.8c) degenerates into the straight line segment AB of Fig. 4.8d. The line AB of Fig. 4.8d corresponds to switch-on shocks for which the upstream tangential magnetic field is zero yet downstream is non-zero and $M_{A2} = 1$. This implies

$$\tan \theta_2 = \frac{M_{A1}^2 - 1}{M_{A2}^2 - 1} \tan \theta_1. \tag{4.81}$$

To determine the behavior of θ_2 as $\theta_1 \longrightarrow 0$ and $M_{A2} \longrightarrow 1$, we substitute (4.81) into the shock polar relation (4.75) and let $\theta_1 \longrightarrow 0$ and $M_{A2} \longrightarrow 1$ simultaneously, to obtain

$$\tan^2 \theta_2 = 2(M_{A1}^2 - 1)\left[1 - \frac{a_{s1}^2}{V_{A1}^2} - \frac{\gamma - 1}{2}(M_{A1}^2 - 1)\right], \tag{4.82}$$

---

[12]Landau and Lifshitz (2000).

4.11 Application 3: MHD Shock Waves

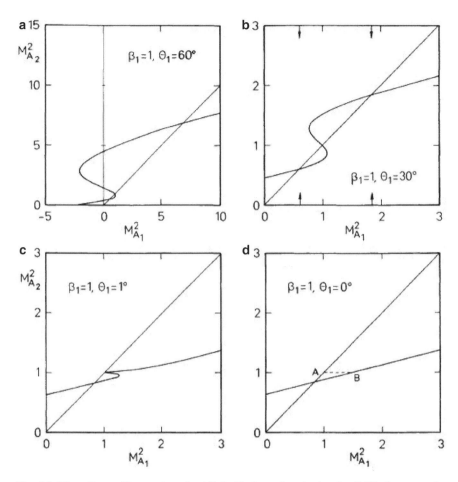

Fig. 4.8 These figures illustrate how the Alfvén Mach number shock polar (4.75) changes as the upstream magnetic field angle θ_1 varies ($\theta_1 = 60°, 30°, 1°$, and $0°$) for $\beta_{p1} = 1$. Case (**d**) shows both the pure gas dynamical shock (the *solid line* passing through B) and switch-on shocks (the *dashed line* AB) (Webb et al. 1987)

which implies that θ_2 is non-zero in this limit. Furthermore, Eq. (4.82) implies that the Alfvénic Mach number must satisfy the inequality

$$1 < M_{A1}^2 < 1 + \frac{2}{\gamma - 1}\left(1 - \frac{a_{s1}^2}{V_{A1}^2}\right), \qquad \frac{a_{s1}^2}{V_{A1}^2} < 1. \tag{4.83}$$

This implies that switch-on shocks occur only in low beta plasmas satisfying $\beta_p < 2/\gamma$. The lower and upper limits on M_{A1}^2 in (4.83) correspond to the points A and B of Fig. 4.8d. From the auxiliary relations listed above, it is easily seen that the pressure jump across a switch-on shock is given by

$$\frac{[P]}{P_1} = (\gamma - 1)(r - 1)\left\{1 + \frac{(\gamma - 1)(r - 1)}{2a_{s1}^2/V_{A1}^2}\right\},$$

where in this case, the compression ratio is $r = M_{A1}^2$.

Thus, besides providing a convenient method of calculating the downstream state of the fluid after specifying the upstream fluid parameters, the shock polar relation furnishes an elegant geometric classification of the possible shocked downstream states and also contains the dispersion relation for small amplitude waves.

Exercises

1. Derive the shock polar relation (4.75) from the shock adiabatic (4.74).
2. Derive the alternative form of the shock polar relation (4.78).
3. Solve the shock polar relation numerically and plot curves corresponding to those of Fig. 4.8 for $\beta_{p1} = 0.1$ and $\beta_{p1} = 4$.

References

J.E. Anderson, *Magnetohydrodynamic Shock Waves* (The MIT Press Classics, Cambridge, 1963)

T.J.M. Boyd, J.J. Sanderson, *The Physics of Plasmas* (Cambridge University Press, Cambridge/New York, 2003), p. 184

S.I. Braginskii, in *Reviews of Plasma Physics*, vol. 1, ed. by M.A. Leontovich (Consultants Bureau, New York, 1965), p. 205

H. Cabannes, *Theoretical Magnetofluiddynamics*. Applied Mathematics and Mechanics Series, vol. 13 (Academic, New York/London, 1970)

S. Chandrasekhar, Stochastic problems in physics and astronomy. Rev. Mod. Phys. **15**(1), 1–89 (1943)

P. Helander, D.J. Sigmar, *Collisional Transport in Magnetized Plasmas*. Cambridge Monographs on Plasma Physics (Cambridge University Press, Cambridge, 2002)

A.N. Kaufmann, Phys. Fluids **3**, 630 (1960)

L.D. Landau, Physikalische Zeitschrift der Sowjetunion **10**, 154 (1936)

L.D. Landau, E.M. Lifshitz, *Fluid Mechanics*, 2nd edn. (Butterworth-Heinemann, Oxford/Boston, 2000)

M.N. Rosenbluth et al., Phys. Rev. **107**, 1 (1957); B.A. Trubnikov, in *Reviews of Plasma Physics*, vol. 1, ed. by M. Leontovich (Consultants Bureau, New York, 1965), p. 105

G.M. Webb, G.P. Zank, J.F. McKenzie, J. Plasma Phys. **37**, 117–141 (1987)

Chapter 5
Charged Particle Transport in a Collisionless Magnetized Plasma

The solar wind, undoubtedly like many other astrophysical plasmas, is essentially collisionless and the description of a plasma based on particle collisions as described in Chap. 3 is inappropriate. Instead, interplanetary plasmas and astrophysical plasmas typically possess numerous waves and fluctuations that include a fluctuating magnetic field. Many studies, beginning with the landmark observational studies of Belcher and Davis (1971) and Coleman (1968) revealed the presence of both Alfvén waves and extended power-law spectra for the energy density of solar wind magnetic fluctuations. These low frequency magnetic field fluctuations can be interpreted in terms of an MHD turbulence description. The turbulence description of fluctuations in the solar wind has become increasingly refined, and today, solar wind turbulence is thought to comprise a superposition of propagating Alfvénic fluctuations (the minority component, sometimes called the *slab* component) and a dominant 2D component that is non-propagating. The two-dimensional component has velocity and magnetic field components and wave numbers nearly perpendicular to the background magnetic field with "zero frequency." The origin of the terminology is from the shear Alfvén wave dispersion relation $\omega = V_A |k_\parallel|$, where $k_\parallel$ is the wave number parallel to the background mean magnetic field $\mathbf{B}_0$ and V_A is the Alfvén speed. For perpendicular wave numbers, $\omega = 0$ so 2D magnetic turbulence has zero frequency modes and non-propagating fluctuations. A turbulent magnetic field in a moving medium, such as the solar wind, scatters charged particles in pitch-angle. Thus, unlike the collisional plasma considered in Chap. 3, the Fokker-Planck coefficients for scattering must now account for particle scattering by forward and backward propagating electro-magnetic waves (Alfvén waves, magnetosonic modes, etc.), randomly propagating magnetic fluctuations for which a wave description is inappropriate (MHD turbulence), waves that may be resonant with certain particles, and magnetic fields that experience complicated non-resonant temporal changes in the background plasma. This remains a very active area of current research, and we present only some of the most basic results here.

5.1 Transport Equations for Non-relativistic Particles Scattered by Plasma Fluctuations

5.1.1 *The Focussed Transport Equation*

Electromagnetic fluctuations in a flowing medium such as the solar wind act to scatter particles, in pitch angle, gyrophase, or in energy. Although we do not explicitly restrict our attention to any particular form of electromagnetic waves in this subsection, we will implicitly consider particles scattered in pitch angle by magnetic fluctuations – either Alfvén waves or convected magnetic fluctuations. In this subsection, we derive a general equation for a gyrotropic distribution function that describes non-relativistic particles scattering in a flowing medium. Such a model was developed by Isenberg (1997) based on an approach by Skilling (1971) to describe the propagation of pickup ions in the solar wind. Although particles may eventually scatter towards isotropy in the frame of the medium, we not assume an isotropic distribution in this subsection. Following Isenberg, we begin with the Boltzmann equation for the distribution function $f(\mathbf{x}, \mathbf{v}, t)$ of non-relativistic particles in the inertial frame,1

$$\left(\frac{\partial f}{\partial t} + \mathbf{v} \cdot \nabla f + \frac{\mathbf{F}}{m} \cdot \nabla_v f = \frac{\delta f}{\delta t} \right)_s + S. \tag{5.1}$$

The force term can be quite general, but we restrict our attention to $\mathbf{F} = q/c(\mathbf{E} + \mathbf{v} \times \mathbf{B})$ i.e., the inertial frame electromagnetic force acting on a particle of charge q, mass m, with c the speed of light. In the Boltzmann equation, S is a particle source term. Of note is that (5.1) has been implicitly separated into mean and fluctuating parts with the fluctuating components being treated as "scattering" terms and relegated to the right-hand-side. The scattering term $\delta f/\delta t)_s$ acts to stochastically scatter particles towards isotropy. In later subsections, we explicitly calculate various forms of the scattering operator. Here, we focus on the left-hand-side of (5.1).

Let us consider a frame of reference that propagates in the inertial "rest" frame at a velocity $\mathbf{U}$. Strictly speaking, this new frame comprises both the background convection velocity and the "average" velocity of the scattering "centers" (Alfvén waves, for example). Certainly in the supersonic solar wind, the convection velocity is much larger than the velocity of the background scattering fluctuations and so the additional velocity of the fluctuations is often neglected. Most importantly, a velocity transformation $\mathbf{U}$ can be identified with the velocity of the background conducting plasma, in which case the motional electric field $\mathbf{E} = -\mathbf{U} \times \mathbf{B}/c$ exactly cancels the electric field and leaves $\mathbf{F} = q\mathbf{v} \times \mathbf{B}/c$. It is important to recognize that the scattering term in this frame conserves energy since all macroscopic electric fields are transformed away. With no electric fields, particles can only scatter in pitch angle. However, energy is not conserved in the "rest" frame and this has

5.1 Transport Equations for Non-relativistic Particles Scattered by Plasma Fluctuations

important consequences, as we discuss later in considering particle acceleration at shock waves. Let us write

$$\mathbf{v} = \mathbf{c} + \mathbf{U} \iff \mathbf{c} = \mathbf{v} - \mathbf{U},$$

for which the following transformations hold,

$$\frac{\partial}{\partial t} \to \frac{\partial}{\partial t} + \frac{\partial c_i}{\partial t}\frac{\partial}{\partial c_i} = \frac{\partial}{\partial t} - \frac{\partial U_i}{\partial t}\frac{\partial}{\partial c_i};$$

$$\frac{\partial}{\partial x_j} \to \frac{\partial}{\partial x_j} - \frac{\partial U_i}{\partial x_j}\frac{\partial}{\partial c_i}; \qquad \frac{\partial}{\partial v_i} \to \frac{\partial}{\partial c_i}.$$

On applying these frame transformations to the inertial form of the Boltzmann equation (5.1), we obtain an equation in mixed coordinates for the distribution function $f(\mathbf{x},\mathbf{c},t)$,

$$\frac{\partial f}{\partial t} + (U_i + c_i)\frac{\partial f}{\partial x_i} + \left[\frac{q}{m}(\mathbf{c}\times\mathbf{B})_i - \frac{\partial U_i}{\partial t} - (U_j + c_j)\frac{\partial U_i}{\partial x_j}\right]\frac{\partial f}{\partial c_i} = \left(\frac{\delta f}{\delta t}\right)_s. \quad (5.2)$$

The subscripts refer to vector components and the summation convention holds.

Let us now suppose that the particle gyroradius is much smaller than any other spatial scales in the system and similarly that their gyroperiod is smaller than other time scales. Thus, the particle distribution function can be regarded as nearly gyrotropic, and so $f(\mathbf{x},\mathbf{c},t)$ is essentially independent of gyrophase i.e., $f(\mathbf{x},\mathbf{c},t) \simeq f(\mathbf{x},c,\mu,t)$, where the particle pitch angle $\mu \equiv \cos\theta = \mathbf{c}\cdot\mathbf{b}/c$ and the direction vector $\mathbf{b} \equiv \mathbf{B}/|B|$ is the unit vector along the large-scale magnetic field. Since we are assuming gyrotropy of the distribution function, we may average (5.2) over gyrophase. By gyrophase averaging, we neglect the action of perpendicular drifts on the distribution function. It is convenient to introduce spherical coordinates ($\theta \equiv$ pitch-angle, $\phi \equiv$ gyrophase, $c \equiv |\mathbf{c}|$),

$$c_x = c\sin\theta\cos\phi; \qquad c_y = c\sin\theta\sin\phi; \qquad c_z = c\cos\theta = c\mu;$$

$$c^2 = c_x^2 + c_y^2 + c_z^2; \qquad \cos\theta = \frac{c_z}{c}; \qquad \mu = \frac{c_i b_i}{c}; \qquad \mathbf{c} = c_x\hat{e}_x + c_y\hat{e}_y + c_z\hat{e}_z,$$

and $\mu = \mu(\mathbf{x})$ and $\phi = \phi(\mathbf{x})$. Consequently, we have the following transformations,

$$\nabla = \nabla + \nabla\mu\frac{\partial}{\partial\mu} + \nabla\phi\frac{\partial}{\partial\phi};$$

$$\frac{\partial}{\partial c_i} = \frac{\partial c}{\partial c_i}\frac{\partial}{\partial c} + \frac{\partial\mu}{\partial c_i}\frac{\partial}{\partial\mu} + \frac{\partial\phi}{\partial c_i}\frac{\partial}{\partial\phi}$$

$$= \frac{c_i}{c}\frac{\partial}{\partial c} + \left(\frac{b_i}{c} - \frac{\mu c_i}{c^2}\right)\frac{\partial}{\partial\mu} + \frac{\partial\phi}{\partial c_i}\frac{\partial}{\partial\phi},$$

which yields, on assuming that $f(\mathbf{x}, c, \mu, \phi, t) = f(\mathbf{x}, c, \mu, t)$,

$$\frac{\partial f}{\partial t} + (U_i + c_i)\left(\frac{\partial f}{\partial x_i} + \frac{\partial \mu}{\partial x_i}\frac{\partial f}{\partial \mu}\right) + \left[\Omega \varepsilon_{ijk} c_j b_k - \frac{\partial U_i}{\partial t} - (U_j + c_j)\frac{\partial U_i}{\partial x_j}\right]$$
$$\times \left(\frac{c_i}{c}\frac{\partial f}{\partial c} + \left(\frac{b_i}{c} - \frac{\mu c_i}{c^2}\right)\frac{\partial f}{\partial \mu}\right) = \left(\frac{\delta f}{\delta t}\right)_s, \qquad (5.3)$$

where the gyrofrequency $\Omega = q|\mathbf{B}|/m$ has been introduced and ε is the Levi-Civita tensor. We introduce an averaging operator for ϕ such that $\varepsilon_{ijk} = 1/2\pi \int_0^{2\pi} Q d\phi$ and average (5.3) term-by term. Thus, since

$$\left\langle \frac{\partial f}{\partial t} \right\rangle = \frac{\partial f}{\partial t}; \quad \left\langle U_i \frac{\partial f}{\partial x_i} \right\rangle = U_i \frac{\partial f}{\partial x_i}; \quad \left\langle c_i \frac{\partial f}{\partial x_i} \right\rangle = \langle c_i \rangle \frac{\partial f}{\partial x_i},$$

and $\hat{e}_z = \mathbf{b}$, we obtain

$$\langle c_i \rangle \frac{\partial f}{\partial x_i} = c\mu b_i \frac{\partial f}{\partial x_i} \Rightarrow \left\langle (U_i + c_i)\frac{\partial f}{\partial x_i} \right\rangle = (U_i + c\mu b_i)\frac{\partial f}{\partial x_i}.$$

Here we used

$$\langle \mathbf{c} \rangle = c\langle \sin\theta \cos\phi \hat{e}_x + \sin\theta \sin\phi \hat{e}_y + \cos\theta \hat{e}_z \rangle = c\mu \mathbf{b},$$

since $\langle \sin\phi \rangle = \langle \cos\phi \rangle = 0$. Use of

$$\left\langle \frac{\partial \mu}{\partial x_i} \right\rangle = \left\langle \frac{c_j}{c} \right\rangle \frac{\partial b_j}{\partial x_i} = \mu b_j \frac{\partial b_j}{\partial x_i},$$

and $b_j b_j = 1$, or $b_j \partial b_j/\partial x_i = 0$, shows that

$$\left\langle U_i \frac{\partial \mu}{\partial x_i}\frac{\partial f}{\partial \mu} \right\rangle = 0.$$

Now consider

$$\left\langle c_i \frac{\partial \mu}{\partial x_i} \right\rangle = c\left\langle \frac{c_i c_j}{c^2} \right\rangle \frac{\partial b_j}{\partial x_i},$$

and the gyrophase averaged term $\langle c_i c_j/c^2 \rangle$ term-by term. We have

$$\left\langle \frac{c_x^2}{c^2} \right\rangle = \frac{1-\mu^2}{2}\hat{e}_x \hat{e}_x; \quad \left\langle \frac{c_y^2}{c^2} \right\rangle = \frac{1-\mu^2}{2}\hat{e}_y \hat{e}_y; \quad \left\langle \frac{c_z^2}{c^2} \right\rangle = \mu^2 \hat{e}_z \hat{e}_z,$$

5.1 Transport Equations for Non-relativistic Particles Scattered by Plasma Fluctuations

and the cross terms $\langle c_i c_j / c^2 \rangle = 0$ for all $i \neq j$, $i, j = x, y, z$. Recalling that $\hat{e}_z \hat{e}_z = \mathbf{bb}$ we obtain

$$c\left\langle \frac{c_i c_j}{c^2} \right\rangle \frac{\partial b_j}{\partial x_i} = c\left(\frac{1-\mu^2}{2}\hat{e}_x\hat{e}_x + \frac{1-\mu^2}{2}\hat{e}_y\hat{e}_y + \mu^2 \mathbf{bb}\right)\frac{\partial b_j}{\partial x_i}$$

$$= c\left[\frac{1-\mu^2}{2}(\mathbf{I} - \mathbf{bb}) + \mu^2 \mathbf{bb}\right]\frac{\partial b_j}{\partial x_i}$$

$$= c\left[\frac{1-\mu^2}{2}(\delta_{ij} - b_i b_j) + \mu^2 \mathbf{bb}\right]\frac{\partial b_j}{\partial x_i}$$

$$= c\frac{1-\mu^2}{2}\delta_{ij}\frac{\partial b_j}{\partial x_i} = c\frac{1-\mu^2}{2}\frac{\partial b_i}{\partial x_i},$$

since $b_j \partial b_j / \partial x_i = 0$. Here we used $\hat{e}_x\hat{e}_x + \hat{e}_y\hat{e}_y + \hat{e}_z\hat{e}_z = \hat{e}_x\hat{e}_x + \hat{e}_y\hat{e}_y + \mathbf{bb} = \mathbf{I}$ or $\hat{e}_x\hat{e}_x + \hat{e}_y\hat{e}_y = \mathbf{I} - \mathbf{bb}$, where $\mathbf{I}$ is the identity matrix. Consequently, we have

$$\left\langle c_i \frac{\partial \mu}{\partial x_i}\frac{\partial f}{\partial \mu}\right\rangle = c\frac{1-\mu^2}{2}\frac{\partial b_i}{\partial x_i}\frac{\partial f}{\partial \mu}.$$

On using the results $\langle \hat{e}_j \rangle = \langle c_j/c \rangle = \mu b_j$ and $\langle \hat{e}_j \hat{e}_i \rangle = (1-\mu^2)/2(\delta_{ij} - b_i b_j) + \mu^2 b_i b_j$ of before, we find

$$\left\langle \left[-\frac{\partial U_i}{\partial t} - U_j\frac{\partial U_i}{\partial x_j}\right]\left(\frac{c_i}{c}\frac{\partial f}{\partial c} + \left(\frac{b_i}{c} - \frac{\mu c_i}{c^2}\right)\frac{\partial f}{\partial \mu}\right)\right\rangle$$

$$= \left(-\frac{\partial U_i}{\partial t} - U_j\frac{\partial U_i}{\partial x_j}\right)\left(\mu b_i\frac{\partial f}{\partial c} + \frac{1-\mu^2}{c}b_i\frac{\partial f}{\partial \mu}\right),$$

and

$$\left\langle c_j\frac{\partial U_i}{\partial x_j}\left(\frac{c_i}{c}\frac{\partial f}{\partial c} + \left(\frac{b_i}{c} - \frac{\mu c_i}{c^2}\right)\frac{\partial f}{\partial \mu}\right)\right\rangle$$

$$= \frac{\partial U_i}{\partial x_j}\left[c\left(\frac{1-\mu^2}{2}(\delta_{ij} - b_i b_j) + \mu^2 b_i b_j\right)\frac{\partial f}{\partial c}\right.$$

$$\left. + \left[\mu b_i b_j - \mu\left(\frac{1-\mu^2}{2}(\delta_{ij} - b_i b_j) + \mu^2 b_i b_j\right)\right]\frac{\partial f}{\partial \mu}\right].$$

Finally, the Lorentz force terms yield

$$\Omega \varepsilon_{ijk}\left\langle \frac{c_i c_j}{c^2}\right\rangle b_k c\frac{\partial f}{\partial c} = 0;$$

$$-\Omega \varepsilon_{ijk}\left\langle \frac{c_i c_j}{c^2}\right\rangle b_k \mu\frac{\partial f}{\partial \mu} = 0,$$

because $\langle c_i c_j / c^2 \rangle = 0$ for all $i \neq j$ and $\varepsilon_{ijk} = 0$ if and only if $i \neq j \neq k$. The final term,

$$\Omega \varepsilon_{ijk} \left\langle \frac{c_j}{c} \right\rangle b_i b_k \frac{\partial f}{\partial \mu} = \Omega \varepsilon_{ijk} b_i b_j b_k \mu \frac{\partial f}{\partial \mu} = 0,$$

because $\varepsilon_{ijk} = 0$ if and only if $i \neq j \neq k$ and $\sum_i \sum_j \sum_j \varepsilon_{ijk} = 0$.

On using the above gyrophase-averaged results and collecting terms, we obtain the reduced gyrophase-averaged transport equation

$$\frac{\partial f}{\partial t} + (U_i + c\mu b_i) \frac{\partial f}{\partial x_i} + \left[\frac{1 - 3\mu^2}{2} b_i b_j \frac{\partial U_i}{\partial x_j} - \frac{1 - \mu^2}{2} \nabla \cdot \mathbf{U} \right.$$

$$- \frac{\mu b_i}{c} \left(\frac{\partial U_i}{\partial t} + U_j \frac{\partial U_i}{\partial x_j} \right) \right] c \frac{\partial f}{\partial c} + \frac{1 - \mu^2}{2} \left[c \nabla \cdot \mathbf{b} + \mu \nabla \cdot \mathbf{U} \right.$$

$$\left. - 3\mu b_i b_j \frac{\partial U_i}{\partial x_j} - \frac{2b_i}{c} \left(\frac{\partial U_i}{\partial t} + U_j \frac{\partial U_i}{\partial x_j} \right) \right] \frac{\partial f}{\partial \mu} = \left\langle \frac{\delta f}{\delta t} \bigg|_s \right\rangle. \quad (5.4)$$

The transport equation (5.4) is also known as the "focussed transport equation" and this non-relativistic form, derived by Isenberg (1997), differs from the earlier relativistically correct form derived by Skilling (1971) in that it contains the convective derivative of $\mathbf{U}$ since Skilling assumed that $U \ll c$.

le Roux and Webb (2012) present a particularly nice discussion of the meaning of the terms in the focussed transport equation (5.4). As discussed above, Eq. (5.4) is in the solar wind flow frame, which is noninertial. Since the plasma flow is non-uniform and non-stationary, scattered particles undergo velocity or momentum changes as measured in the flow frame due to pseudoforces associated with the non-uniform non-stationary nature of the flow. Recall from Chap. 2 that the gradient of the flow velocity can be expressed as the sum of the flow divergence, the flow shear, and the flow rotation, i.e.,

$$\frac{\partial U_i}{\partial x_j} = \frac{1}{3} \frac{\partial U_i}{\partial x_i} \delta_{ij} + \frac{1}{2} \left(\frac{\partial U_i}{\partial x_j} + \frac{\partial U_j}{\partial x_i} - \frac{2}{3} \frac{\partial U_i}{\partial x_i} \delta_{ij} \right) + \frac{1}{2} \left(\frac{\partial U_i}{\partial x_j} - \frac{\partial U_j}{\partial x_i} \right)$$

$$= \frac{1}{3} \frac{\partial U_i}{\partial x_i} \delta_{ij} + \sigma_{ij} + \omega_{ij},$$

where σ_{ij} and ω_{ij} denote the shear and rotation tensors of the flow respectively. On expressing the flow gradient terms in the focussed transport equation (5.4) by the general representation above, the method of characteristics shows that

$$\frac{1}{c} \left\langle \frac{\partial c}{\partial t} \right\rangle = -\frac{1}{3} \frac{\partial U_i}{\partial x_i} + \frac{1 - 3\mu^2}{2} b_i b_j (\sigma_{ij} + \omega_{ij}) - \frac{\mu b_i}{c} \frac{dU_i}{dt}$$

$$= -\frac{1}{3} \frac{\partial U_i}{\partial x_i} + \frac{1 - 3\mu^2}{2} b_i b_j \sigma_{ij} - \frac{\mu b_i}{c} \frac{dU_i}{dt} \quad (5.5)$$

5.1 Transport Equations for Non-relativistic Particles Scattered by Plasma Fluctuations

where dU_i/dt is the convective derivative, and we recognize that the rotation tensor is antisymmetric ($\omega_{ij} = -\omega_{ji}$), so that the sum $b_i b_j \omega_{ij} = 0$. Thus, flow rotation does not contribute to changes in particle speed. Similarly, we find that

$$\left\langle \frac{\partial \mu}{\partial t} \right\rangle = \frac{1-\mu^2}{2}\left[c\frac{\partial b_i}{\partial x_i} - 3\mu b_i b_j \sigma_{ij} - \frac{2b_i}{c}\frac{dU_i}{dt} \right], \tag{5.6}$$

and flow rotation does not contribute to changes in particle pitch-angle either. Expressions (5.5) and (5.6) describe the gyrophase averaged rate of change of the particle velocity c and pitch-angle μ. If particle velocity or momentum is measured in a nonuniform nonstationary plasma (noninertial) flow frame, the magnitude of the particle velocity or momentum will be modified if the flow diverges ($\nabla \cdot \mathbf{U}$), experiences shear (σ_{ij}), or rotation (ω_{ij}), or accelerates (dU_i/dt), while the particle pitch angle varies in response to flow shear, rotation, or acceleration. It is interesting to note (recall the telegrapher equation discussion, Chap. 2) that the shear and rotation tensor terms in Eq. (5.5) are multiplied by the term $-(3\mu^2 - 1)/2$, which is the second-order Legendre polynomial $P_2(\mu)$, whereas the divergence of the flow, $\nabla \cdot \mathbf{U}$ is multiplied by the zeroth-order Legendre polynomial $P_0 = 1$, and the acceleration term dU_i/dt by the first-order Legendre polynomial $P_1(\mu) = \mu$. For distributions f that are close to isotropic, this ordering of the terms associated with the Legendre polynomials gives the order of the importance in terms of energy change with respect to a physical process.

The flow divergence term $\nabla \cdot \mathbf{U}$ in Eq. (5.5) is nothing more than the well known adiabatic momentum change term in the standard cosmic ray transport equation that will be discussed below. Evidently, the divergence of the flow has no effect on the particle pitch angle. Physically, the effect of the divergence of a collisionless flow on energetic particles is consistent with the notion that particles are coupled to the flow through their interaction (scattering) with electromagnetic fields embedded in a highly conductive flow, but when the electromagnetic fields are neglected the divergence of the flow still affects the particle momentum simply because momentum is measured in the frame of a nonuniform plasma flow. As we discuss in more detail below, the rapid (negative) divergence of a flow across a shock wave leads to a convergence of the flow and the compression of electromagnetic fields embedded in the flow. As shown explicitly in the formulation of the focused transport equation, particles respond to the compression of electromagnetic fields embedded in the flow, and experience adiabatic compression. Notice that all of the effects due to a nonuniform nonstationary flow frame vanish if particle momentum is measured in an inertial frame, but if one is interested in what happens to the random component of the particle velocity at a shock, for example, noninertial effects must be taken into account.

Most investigations are currently restricted to the 1D version of the focussed transport equation. If one assumes for example a constant radial flow, such as the solar wind, with $\mathbf{U} = U\hat{\mathbf{r}}$ and a large-scale radial magnetic field pointing away from the Sun, $\mathbf{b} = \hat{\mathbf{r}}$, then Eq. (5.4) simplifies to

$$\frac{\partial f}{\partial t} + (U + \mu c)\frac{\partial f}{\partial r} - \frac{1-\mu^2}{r}Uc\frac{\partial f}{\partial c} + \frac{1-\mu^2}{r}(c + \mu U)\frac{\partial f}{\partial \mu} = \left\langle \frac{\delta f}{\delta t}\bigg|_s \right\rangle. \quad (5.7)$$

Exercises

1. By collecting all the terms associated with the gyrophase-averaging of (5.2), derive the general form of the gyrophase-averaged transport equation (5.4).
2. By assuming a constant radial flow velocity for the solar wind and a radial interplanetary magnetic field, derive the 1D focussed transport equation (5.7).
3. Assume that the one spatial dimensional gyrotropic distribution function can be expressed as

$$f(r, c, \mu) = f_-(r, c)H(-\mu) + f_+(r, c)H(\mu),$$

where $H(x)$ denotes the Heaviside step function and $f_\pm$ refer to anti-sunward (f_+)/sunward (f_-) hemispherical distributions. By substituting $f = f_- H(-\mu) + f_+ H(\mu)$ in the 1D focussed transport equation

$$\frac{\partial f}{\partial t} + (U+\mu c)\frac{\partial f}{\partial r} - \frac{1-\mu^2}{r}Uc\frac{\partial f}{\partial c} + \frac{1-\mu^2}{r}(c+\mu U)\frac{\partial f}{\partial \mu} = \frac{\partial}{\partial \mu}\left(\nu(1-\mu^2)\frac{\partial f}{\partial \mu}\right),$$

and integrating over μ separately from -1 to 0 and then from 0 to 1, show that

$$\frac{\partial f_\pm}{\partial t} + \left(U \pm \frac{c}{2}\right)\frac{\partial f_\pm}{\partial r} - \frac{2U}{r}\frac{c}{3}\frac{\partial f_\pm}{\partial c} + \frac{c}{r}(f_+ - f_-) = \mp \Gamma(f_+ - f_-)$$

where $\Gamma \equiv \nu(\mu = 0)$ gives the rate of scattering across $\mu = 0$. Note that the form of the scattering term is of diffusion in pitch-angle, and this is discussed below. The term ν is the scattering frequency.

5.1.2 The Diffusive Transport Equation

The solution of the general gyrophase-averaged transport equation is a formidable task for almost any physically interesting system so considerable effort has been invested in trying to simplify (5.4) by means of several additional assumptions. Let us assume that the scattering operator can be represented by a diffusion operator in pitch-angle,

$$\left\langle \frac{\delta f}{\delta t}\bigg|_s \right\rangle = \frac{\partial}{\partial \mu}\left(\nu\left(1 - \mu^2\right)\frac{\partial f}{\partial \mu}\right), \quad (5.8)$$

where ν is a characteristic scattering frequency. The scattering term is discussed further in more general terms in the following subsections.

5.1 Transport Equations for Non-relativistic Particles Scattered by Plasma Fluctuations

The dependence of the gyrophase-averaged particle distribution function f on the pitch-angle $\mu = \cos\theta$ with $\mu \in [-1, 1]$ suggests a natural expansion in terms of Legendre polynomials. The orthogonality properties of the complete set of Legendre polynomials allow us to rewrite the focussed transport equation (5.4) as an infinite set of partial differential equations in terms of the polynomial coefficients of the expansion. To ensure tractability, one typically truncates the infinite set at a low order, which is a form of closure. Accordingly, we expand the gyrophase-averaged particle distribution function f as

$$f(\mathbf{x}, t, c, \mu) = \sum_{n=0}^{\infty} \frac{1}{2}(2n+1)P_n(\mu)f_n(\mathbf{x}, t, c), \text{ where } f_n(\mathbf{x}, t, c) = \int_{-1}^{1} f P_n(\mu) d\mu.$$

The orthogonality condition is given by

$$\int_{-1}^{1} P_m(\mu) P_n(\mu) d\mu = \begin{cases} 0 & m \neq n \\ \frac{2}{2n+1} & m = n \end{cases},$$

and some useful recurrence relations that will be used below are

$$(n+1)P_{n+1}(\mu) = (2n+1)\mu P_n(\mu) - n P_{n-1}(\mu);$$

$$(1-\mu^2)\frac{d}{d\mu} P_n(\mu) = n P_{n-1}(\mu) - n\mu P_n(\mu) = \frac{n(n+1)}{2n+1}[P_{n-1}(\mu) - P_{n+1}(\mu)];$$

$$\frac{d}{d\mu} P_{n+1}(\mu) - \mu \frac{d}{d\mu} P_n(\mu) = (n+1)P_n(\mu);$$

$$\mu \frac{d}{d\mu} P_n(\mu) - \frac{d}{d\mu} P_{n-1}(\mu) = n P_n(\mu);$$

$$\frac{d}{d\mu}[P_{n+1}(\mu) - P_{n-1}(\mu)] = (2n+1)P_n(\mu).$$

We systematically project and expand each of the terms in (5.4) from left to right using the Legendre polynomial $P_m(\mu)$ and the expansion for f.

The first (time-derivative) term becomes

$$\frac{\partial f}{\partial t} : \int_{-1}^{1} P_m \frac{\partial f}{\partial t} d\mu = \sum_{n=0}^{\infty} \frac{1}{2}(2n+1) \int_{-1}^{1} P_m P_n d\mu \frac{\partial f_n}{\partial t} = \frac{\partial f_m}{\partial t},$$

after using the orthogonality relation. Similarly, the second (convective) term becomes

$$U_i \frac{\partial f}{\partial x_i} : U_i \sum_{n=0}^{\infty} \frac{1}{2}(2n+1) \int_{-1}^{1} P_m P_n d\mu \frac{\partial f_n}{\partial x_i} = U_i \frac{\partial f_m}{\partial x_i}.$$

The third term in (5.4) is a little more interesting in that we need to use the first of the recurrence relations. Thus,

$$c\mu b_i \frac{\partial f}{\partial x_i} : cb_i \int_{-1}^{1} \mu P_m \frac{\partial f}{\partial x_i} d\mu = cb_i \int_{-1}^{1} \left[\frac{m+1}{2m+1} P_{m+1} + \frac{m}{2m+1} P_{m-1} \right] \frac{\partial f}{\partial x_i} d\mu.$$

On expanding f, we find

$$cb_i \int_{-1}^{1} \mu P_m \frac{\partial f}{\partial x_i} d\mu = cb_i \int_{-1}^{1} \frac{m+1}{2m+1} P_{m+1} \frac{\partial f}{\partial x_i} d\mu + cb_i \int_{-1}^{1} \frac{m}{2m+1} P_{m-1} \frac{\partial f}{\partial x_i} d\mu$$

$$= cb_i \sum_{n=0}^{\infty} \frac{2n+1}{2} \int_{-1}^{1} \frac{m+1}{2m+1} P_{m+1} P_n \frac{\partial f_n}{\partial x_i} d\mu$$

$$+ cb_i \sum_{n=0}^{\infty} \frac{2n+1}{2} \int_{-1}^{1} \frac{m}{2m+1} P_{m-1} P_n \frac{\partial f_n}{\partial x_i} d\mu.$$

The first term on the right-hand side contributes only when $n = m+1$ and the second term only when $n = m-1$, so yielding

$$cb_i \int_{-1}^{1} \mu P_m \frac{\partial f}{\partial x_i} d\mu = cb_i \frac{2m+3}{2} \frac{m+1}{2m+1} \frac{2}{2m+3} \frac{\partial f_{m+1}}{\partial x_i}$$

$$+ cb_i \frac{2m-1}{2} \frac{m}{2m+1} \frac{2}{2m-1} \frac{\partial f_{m-1}}{\partial x_i}$$

$$= cb_i \frac{m+1}{2m+1} \frac{\partial f_{m+1}}{\partial x_i} + cb_i \frac{m}{2m+1} \frac{\partial f_{m-1}}{\partial x_i}.$$

The third term can therefore be expressed as

$$c\mu b_i \frac{\partial f}{\partial x_i} : \frac{cb_i}{2m+1} \left[(m+1) \frac{\partial f_{m+1}}{\partial x_i} + m \frac{\partial f_{m-1}}{\partial x_i} \right].$$

The fourth term in the focused transport equation (5.4) is

$$\frac{1-3\mu^2}{2} b_i b_j \frac{\partial U_j}{\partial x_i} c \frac{\partial f}{\partial c} : cb_i b_j \frac{\partial U_j}{\partial x_i} \frac{1}{2} \frac{\partial f}{\partial c} - cb_i b_j \frac{\partial U_j}{\partial x_i} \frac{3\mu^2}{2} \frac{\partial f}{\partial c}.$$

The first term can be rewritten immediately as

$$cb_i b_j \frac{\partial U_j}{\partial x_i} \frac{1}{2} \frac{\partial f}{\partial c} : cb_i b_j \frac{\partial U_j}{\partial x_i} \frac{1}{2} \frac{\partial f_m}{\partial c}.$$

5.1 Transport Equations for Non-relativistic Particles Scattered by Plasma Fluctuations

We need to use the first of the recurrence relations to infer

$$\mu^2 P_n = \frac{(n+1)(n+2)}{(2n+1)(2n+3)} P_{n+2} + \left[\frac{(n+1)^2}{2n+3} + \frac{n^2}{2n-1} \right] \frac{P_n}{2n+1}$$
$$+ \frac{n(n-1)}{(2n-1)(2n+1)} P_{n-2}.$$

On using this identity for the second term above, we obtain

$$-cb_i b_j \frac{\partial U_j}{\partial x_i} \frac{3\mu^2}{2} \frac{\partial f}{\partial c} : -cb_i b_j \frac{\partial U_j}{\partial x_i} \frac{3}{2} \int_{-1}^{1} \mu^2 P_m \frac{\partial f}{\partial c} d\mu$$

$$= -cb_i b_j \frac{\partial U_j}{\partial x_i} \frac{3}{2} \sum_{n=0}^{\infty} \frac{2n+1}{2} \int_{-1}^{1} \mu^2 P_n P_m \frac{\partial f_n}{\partial c} d\mu$$

$$= -cb_i b_j \frac{\partial U_j}{\partial x_i} \frac{3}{2} \sum_{n=0}^{\infty} \frac{2n+1}{2} \frac{\partial f_n}{\partial c} \int_{-1}^{1} \left[\frac{(n+1)(n+2)}{(2n+1)(2n+3)} P_{n+2} P_m \right.$$

$$\left. + \left(\frac{(n+1)^2}{2n+3} + \frac{n^2}{2n-1} \right) \frac{P_n}{2n+1} P_m + \frac{n(n-1)}{(2n-1)(2n+1)} P_{n-2} P_m \right] d\mu.$$

The first integral contributes only when $n = m - 2$, the second when $n = m$ and the last when $n = m + 2$, so yielding

$$-cb_i b_j \frac{\partial U_j}{\partial x_i} \frac{3}{2} \int_{-1}^{1} \mu^2 P_m \frac{\partial f}{\partial c} d\mu$$

$$= -cb_i b_j \frac{\partial U_j}{\partial x_i} \frac{3}{2} \left[\frac{(m-1)m}{(2m+1)(2m-1)} \frac{\partial f_{m-2}}{\partial c} + \left(\frac{(m+1)^2}{2m+3} + \frac{m^2}{2m-1} \right) \right.$$

$$\left. \times \frac{1}{2m+1} \frac{\partial f_m}{\partial c} + \frac{(m+2)(m+1)}{(2m+3)(2m+1)} \frac{\partial f_{m+2}}{\partial c} \right].$$

On assembling the various terms, we obtain for the fourth term of the focused transport equation,

$$\frac{1-3\mu^2}{2} b_i b_j c \frac{\partial U_j}{\partial x_i} \frac{\partial f}{\partial c} : cb_i b_j \frac{\partial U_j}{\partial x_i} \frac{1}{2} \frac{\partial f_m}{\partial c} - vb_i b_j \frac{\partial U_j}{\partial x_i} \frac{3}{2} \frac{(m-1)m}{(2m+1)(2m-1)} \frac{\partial f_{m-2}}{\partial c}$$

$$- cb_i b_j \frac{\partial U_j}{\partial x_i} \frac{3}{2} \left(\frac{(m+1)^2}{2m+3} + \frac{m^2}{2m-1} \right) \frac{1}{2m+1} \frac{\partial f_m}{\partial c}$$

$$- vb_i b_j \frac{\partial U_j}{\partial x_i} \frac{3}{2} \frac{(m+2)(m+1)}{(2m+3)(2m+1)} \frac{\partial f_{m+2}}{\partial c}.$$

The fifth term of (5.4) can be expanded as

$$-c\frac{1-\mu^2}{2}\frac{\partial U_i}{\partial x_i}\frac{\partial f}{\partial c} : -\frac{c}{2}\frac{\partial U_i}{\partial x_i}\frac{\partial f_m}{\partial c} + \frac{c}{2}\frac{\partial U_i}{\partial x_i}\frac{(m-1)m}{(2m+1)(2m-1)}\frac{\partial f_{m-2}}{\partial c}$$

$$+ \frac{c}{2}\frac{\partial U_i}{\partial x_i}\left(\frac{(m+1)^2}{2m+3} + \frac{m^2}{2m-1}\right)\frac{1}{2m+1}\frac{\partial f_m}{\partial c}$$

$$+ \frac{c}{2}\frac{\partial U_i}{\partial x_i}\frac{(m+2)(m+1)}{(2m+3)(2m+1)}\frac{\partial f_{m+2}}{\partial c}.$$

On using the results of expressing the third term in terms of a Legendre polynomial expansion, we have for the sixth term in (5.4)

$$-b_i\frac{DU_i}{Dt}\mu\frac{\partial f}{\partial c} : -\frac{DU_i}{Dt}\frac{b_i}{2m+1}\left[(m+1)\frac{\partial f_{m+1}}{\partial c} + m\frac{\partial f_{m-1}}{\partial c}\right].$$

The computation of the seventh term in the focused transport equation is also straightforward. We can immediately express

$$\frac{c}{2}\frac{\partial b_i}{\partial x_i}(1-\mu^2)\frac{\partial f}{\partial \mu} : \frac{c}{2}\frac{\partial b_i}{\partial x_i}\sum_{n=0}^{\infty}\frac{2n+1}{2}f_n\int_{-1}^{1}d\mu\, P_m(1-\mu^2)\frac{\partial P_n}{\partial \mu}.$$

Use of the second recursion relation above yields

$$\frac{c}{2}\frac{\partial b_i}{\partial x_i}(1-\mu^2)\frac{\partial f}{\partial \mu} : \frac{c}{2}\frac{\partial b_i}{\partial x_i}\sum_{n=0}^{\infty}\frac{2n+1}{2}f_n\int_{-1}^{1}d\mu\,\frac{n(n+1)}{2n+1}\left[P_m P_{n-1} - P_m P_{n+1}\right],$$

so that, since the first summand contributes only for $n = m+1$ and the second for $n = m-1$,

$$\frac{c}{2}\frac{\partial b_i}{\partial x_i}(1-\mu^2)\frac{\partial f}{\partial \mu} : \frac{c}{2}\frac{\partial b_i}{\partial x_i}\left[\frac{(m+1)(m+2)}{2m+1}f_{m+1} - \frac{m(m-1)}{2m+1}f_{m-1}\right].$$

Consider now the eighth term in (5.4). We have

$$\frac{1}{2}\frac{\partial U_i}{\partial x_i}(1-\mu^2)\mu\frac{\partial f}{\partial \mu} : \frac{1}{2}\frac{\partial U_i}{\partial x_i}\sum_{n=0}^{\infty}\frac{2n+1}{2}f_n\int_{-1}^{1}d\mu\, P_m(1-\mu^2)\mu\frac{\partial P_n}{\partial \mu}.$$

5.1 Transport Equations for Non-relativistic Particles Scattered by Plasma Fluctuations

The second of the recursion relations yields

$$\mu(1-\mu^2)\frac{\partial P_n}{\partial \mu} = \frac{n(n+1)}{2n+1}[\mu P_{n-1} - \mu P_{n+1}]$$

$$= \frac{n(n+1)}{2n+1}\left[\frac{n-1}{2n-1}P_{n-2} + \frac{n}{2n-1}P_n\right]$$

$$- \frac{n(n+1)}{2n+1}\left[\frac{n+2}{2n+3}P_{n+2} + \frac{n+1}{2n+3}P_n\right]$$

$$= \frac{n(n+1)}{2n+1}\frac{n-1}{2n-1}P_{n-2} + \left[\frac{n^2(n+1)}{(2n+1)(2n-1)}\right.$$

$$\left. - \frac{n(n+1)^2}{(2n+1)(2n+3)}\right]P_n - \frac{n(n+1)}{2n+1}\frac{n+2}{2n+3}P_{n+2},$$

from which we obtain

$$\frac{1}{2}\frac{\partial U_i}{\partial x_i}\sum_{n=0}^{\infty}\frac{2n+1}{2}f_n\int_{-1}^{1}d\mu\, P_m(1-\mu^2)\mu\frac{\partial P_n}{\partial \mu}$$

$$= \frac{1}{2}\frac{\partial U_i}{\partial x_i}\sum_{n=0}^{\infty}\frac{2n+1}{2}f_n\int_{-1}^{1}d\mu\, P_m\frac{n(n+1)}{2n+1}\frac{n-1}{2n-1}P_{n-2}$$

$$+ \frac{1}{2}\frac{\partial U_i}{\partial x_i}\sum_{n=0}^{\infty}\frac{2n+1}{2}f_n\int_{-1}^{1}d\mu\, P_m\left[\frac{n^2(n+1)}{(2n+1)(2n-1)} - \frac{n(n+1)^2}{(2n+1)(2n+3)}\right]P_n$$

$$- \frac{1}{2}\frac{\partial U_i}{\partial x_i}\sum_{n=0}^{\infty}\frac{2n+1}{2}f_n\int_{-1}^{1}d\mu\, P_m\frac{n(n+1)}{2n+1}\frac{n+2}{2n+3}P_{n+2}.$$

The first term contributes only for $n = m+2$, the second for $n = m$, and the third for $n = m-2$, from which we obtain

$$\frac{1-\mu^2}{2}\mu\frac{\partial U_i}{\partial x_i}\frac{\partial f}{\partial \mu} : \frac{1}{2}\frac{\partial U_i}{\partial x_i}\frac{(m+1)(m+2)(m+3)}{(2m+1)(2m+3)}f_{m+2}$$

$$+ \frac{1}{2}\frac{\partial U_i}{\partial x_i}\left[\frac{m^2(m+1)}{(2m+1)(2m-1)} - \frac{m(m+1)^2}{(2m+1)(2m+3)}\right]f_m$$

$$- \frac{1}{2}\frac{\partial U_i}{\partial x_i}\frac{m(m-1)(m-2)}{(2m+1)(2m-1)}f_{m-2}.$$

Since

$$\left[\frac{m^2(m+1)}{(2m+1)(2m-1)} - \frac{m(m+1)^2}{(2m+1)(2m+3)}\right] = \frac{m(m+1)}{(2m-1)(2m+3)},$$

we can express term eight as

$$\frac{1-\mu^2}{2}\mu\frac{\partial U_i}{\partial x_i}\frac{\partial f}{\partial \mu} : \frac{1}{2}\frac{\partial U_i}{\partial x_i}\frac{(m+1)(m+2)(m+3)}{(2m+1)(2m+3)}f_{m+2}$$

$$+\frac{1}{2}\frac{\partial U_i}{\partial x_i}\frac{m(m+1)}{(2m-1)(2m+3)}f_m$$

$$-\frac{1}{2}\frac{\partial U_i}{\partial x_i}\frac{m(m-1)(m-2)}{(2m+1)(2m-1)}f_{m-2}.$$

We can utilize these results to express term nine in (5.4) as

$$-\frac{1-\mu^2}{2}3\mu b_i b_j \frac{\partial U_j}{\partial x_i}\frac{\partial f}{\partial \mu} : -\frac{3}{2}b_i b_j \frac{\partial U_j}{\partial x_i}\frac{(m+1)(m+2)(m+3)}{(2m+1)(2m+3)}f_{m+2}$$

$$-\frac{3}{2}b_i b_j \frac{\partial U_j}{\partial x_i}\frac{m(m+1)}{(2m-1)(2m+3)}f_m$$

$$+\frac{3}{2}b_i b_j \frac{\partial U_j}{\partial x_i}\frac{m(m-1)(m-2)}{(2m+1)(2m-1)}f_{m-2}.$$

The results from evaluating the seventh term yield

$$-\frac{b_i}{c}\frac{DU_i}{Dt}(1-\mu^2)\frac{\partial f}{\partial \mu} : -\frac{b_i}{c}\frac{DU_i}{Dt}\left[\frac{(m+1)(m+2)}{2m+1}f_{m+1} - \frac{m(m-1)}{2m+1}f_{m-1}\right].$$

Finally, let us consider the specific form of the diffusion term in pitch-angle μ,

$$\frac{\partial}{\partial \mu}\left[v\frac{1-\mu^2}{2}\frac{\partial f}{\partial \mu}\right] : \int_{-1}^{1}P_m\frac{\partial}{\partial \mu}\left[v\frac{1-\mu^2}{2}\frac{\partial f}{\partial \mu}\right]d\mu$$

$$=\sum_{n=0}^{\infty}\frac{2n+1}{2}f_n\int_{-1}^{1}d\mu\, P_m\frac{\partial}{\partial \mu}\left[v\frac{1-\mu^2}{2}\frac{\partial P_n}{\partial \mu}\right].$$

The recursion operator

$$\frac{\partial P_n}{\partial \mu} = P'_n = \frac{n}{1-\mu^2}[P_{n-1} - \mu P_n],$$

yields

$$\int_{-1}^{1}P_m\frac{\partial}{\partial \mu}\left[v\frac{1-\mu^2}{2}\frac{\partial f}{\partial \mu}\right]d\mu = \sum_{n=0}^{\infty}\frac{2n+1}{2}f_n$$

$$\times \int_{-1}^{1}P_m\frac{\partial}{\partial \mu}\left[v\frac{1-\mu^2}{2}\frac{n}{1-\mu^2}(P_{n-1} - \mu P_n)\right]d\mu.$$

5.1 Transport Equations for Non-relativistic Particles Scattered by Plasma Fluctuations

Since v is independent of μ, we find

$$\int_{-1}^{1} P_m \frac{\partial}{\partial \mu}\left[v\frac{1-\mu^2}{2}\frac{\partial f}{\partial \mu}\right]d\mu = v\sum_{n=0}^{\infty}\frac{2n+1}{2}\frac{n}{2}f_n\int_{-1}^{1} P_m \frac{\partial}{\partial \mu}(P_{n-1} - \mu P_n)\,d\mu.$$

On using the following relation,

$$\frac{\partial}{\partial \mu}(P_{n-1} - \mu P_n) = \frac{\partial}{\partial \mu}\left(P_{n-1} - \frac{n+1}{2n+1}P_{n+1} - \frac{n}{2n+1}P_{n-1}\right)$$

$$= \frac{n+1}{2n+1}\frac{\partial}{\partial \mu}(P_{n-1} - P_{n+1}),$$

together with the definition

$$\frac{\partial}{\partial \mu}(P_{n-1} - P_{n+1}) = -(2n+1)P_n,$$

we find

$$\frac{\partial}{\partial \mu}(P_{n-1} - \mu P_n) = -(n+1)P_n.$$

We therefore have the result that

$$\int_{-1}^{1} P_m \frac{\partial}{\partial \mu}\left[v\frac{1-\mu^2}{2}\frac{\partial f}{\partial \mu}\right]d\mu = -v\sum_{n=0}^{\infty}\frac{2n+1}{2}\frac{n}{2}f_n(n+1)\int_{-1}^{1} P_m P_n\,d\mu,$$

and since the integral only contributes for $n = m$, we have

$$\int_{-1}^{1} P_m \frac{\partial}{\partial \mu}\left[v\frac{1-\mu^2}{2}\frac{\partial f}{\partial \mu}\right]d\mu = -v\frac{m(m+1)}{2}f_m.$$

This completes the evaluation of each of the terms in (5.4).

By gathering the results above together, the complete transformed focused transport equation (5.4) can now be expressed as an infinite set of partial differential equations in the coefficients f_n of the Legendre polynomials,

$$\frac{\partial f_m}{\partial t} + U_i \frac{\partial f_m}{\partial x_i} + \frac{cb_i}{2m+1}\left[(m+1)\frac{\partial f_{m+1}}{\partial x_i} + m\frac{\partial f_{m-1}}{\partial x_i}\right] + cb_ib_j\frac{\partial U_j}{\partial x_i}\frac{1}{2}\frac{\partial f_m}{\partial c}$$

$$-cb_ib_j\frac{\partial U_j}{\partial x_i}\frac{3}{2}\frac{(m-1)m}{(2m+1)(2m-1)}\frac{\partial f_{m-2}}{\partial c} - cb_ib_j\frac{\partial U_j}{\partial x_i}\frac{3}{2}\left(\frac{(m+1)^2}{2m+3} + \frac{m^2}{2m-1}\right)$$

$$\times \frac{1}{2m+1}\frac{\partial f_m}{\partial c} - cb_ib_j\frac{\partial U_j}{\partial x_i}\frac{3}{2}\frac{(m+2)(m+1)}{(2m+3)(2m+1)}\frac{\partial f_{m+2}}{\partial c} - \frac{c}{2}\frac{\partial U_i}{\partial x_i}\frac{\partial f_m}{\partial c}$$

$$+\frac{c}{2}\frac{\partial U_i}{\partial x_i}\frac{(m-1)m}{(2m+1)(2m-1)}\frac{\partial f_{m-2}}{\partial c} + \frac{c}{2}\frac{\partial U_i}{\partial x_i}\left(\frac{(m+1)^2}{2m+3} + \frac{m^2}{2m-1}\right)\frac{1}{2m+1}\frac{\partial f_m}{\partial c}$$

$$+ \frac{c}{2}\frac{\partial U_i}{\partial x_i}\frac{(m+2)(m+1)}{(2m+3)(2m+1)}\frac{\partial f_{m+2}}{\partial c} - \frac{DU_i}{Dt}\frac{b_i}{2m+1}\left[(m+1)\frac{\partial f_{m+1}}{\partial c} + m\frac{\partial f_{m-1}}{\partial c}\right]$$

$$+ \frac{c}{2}\frac{\partial b_i}{\partial x_i}\left[\frac{(m+1)(m+2)}{2m+1}f_{m+1} - \frac{m(m-1)}{2m+1}f_{m-1}\right]$$

$$+ \frac{1}{2}\frac{\partial U_i}{\partial x_i}\frac{(m+1)(m+2)(m+3)}{(2m+1)(2m+3)}f_{m+2} + \frac{1}{2}\frac{\partial U_i}{\partial x_i}\frac{m(m+1)}{(2m-1)(2m+3)}f_m$$

$$- \frac{1}{2}\frac{\partial U_i}{\partial x_i}\frac{m(m-1)(m-2)}{(2m+1)(2m-1)}f_{m-2} - \frac{3}{2}b_ib_j\frac{\partial U_j}{\partial x_i}\frac{(m+1)(m+2)(m+3)}{(2m+1)(2m+3)}f_{m+2}$$

$$- \frac{3}{2}b_ib_j\frac{\partial U_j}{\partial x_i}\frac{m(m+1)}{(2m-1)(2m+3)}f_m + \frac{3}{2}b_ib_j\frac{\partial U_j}{\partial x_i}\frac{m(m-1)(m-2)}{(2m+1)(2m-1)}f_{m-2}$$

$$- \frac{b_i}{c}\frac{DU_i}{Dt}\left[\frac{(m+1)(m+2)}{2m+1}f_{m+1} - \frac{m(m-1)}{2m+1}f_{m-1}\right] = -\nu\frac{m(m+1)}{2}f_m. \qquad (5.9)$$

The infinite set of partial differential equations (5.9) is equivalent to the focused transport equation (5.4) and therefore as challenging to solve. At each order of the expansion, i.e., the pde for a Legendre coefficient of particular order, it is clearly seen that the equation possesses coefficients of a higher order. This is another expression of the closure problem. Closure is typically affected by simply truncating the Legendre polynomial expansion at a finite number of coefficients. This procedure is somewhat arbitrary and one formally needs to establish that the truncation remains sufficiently close to the full solution. This is typically very difficult in practice, and so is rarely done. An example of the subtleties that can arise was discussed in Chap. 2, Sect. 2.8, where an even or an odd truncation of the Legendre polynomial expansion of a simplified Boltzmann equation yielded fundamentally different solutions, with the even truncation capturing the non-propagating characteristic solution and the odd truncation missing that particular mode.

Let us consider the simplest reduction of the set of equations (5.9) by truncating the infinite set of equations at some arbitrary order with the hope that this does not introduce any unphysical character into the reduced model. Typically, truncations are made at the lowest order possible. For the f_1 approximation (i.e. assume $f_n = 0 \ \forall \ n \geq 2$), we have, on setting $m = 0$,

$$\frac{\partial f_0}{\partial t} + U_i\frac{\partial f_0}{\partial x_i} - \frac{c}{3}\frac{\partial U_i}{\partial x_i}\frac{\partial f_0}{\partial c} = -cb_i\frac{\partial f_1}{\partial x_i} + \frac{DU_i}{Dt}b_i\frac{\partial f_1}{\partial c} - c\frac{\partial b_i}{\partial x_i}f_1 + 2\frac{b_i}{c}\frac{DU_i}{Dt}f_1,$$

$$(5.10)$$

and on setting $m = 1$ and neglecting all terms with indices having $i \geq 2$, we find

$$\frac{\partial f_1}{\partial t} + U_i\frac{\partial f_1}{\partial x_i} + \frac{cb_i}{3}\frac{\partial f_0}{\partial x_i} + \frac{1}{2}cb_ib_j\frac{\partial U_j}{\partial x_i}\frac{\partial f_1}{\partial c} - \frac{9}{10}cb_ib_j\frac{\partial U_j}{\partial x_i}\frac{\partial f_1}{\partial c} - \frac{c}{2}\frac{\partial U_i}{\partial x_i}\frac{\partial f_1}{\partial c}$$

$$+ \frac{9}{10}c\frac{\partial U_i}{\partial x_i}\frac{\partial f_1}{\partial c} - \frac{DU_i}{Dt}\frac{b_i}{3}\frac{\partial f_0}{\partial c} + \frac{1}{5}\frac{\partial U_i}{\partial x_i}f_1 - \frac{3}{5}b_ib_j\frac{\partial U_j}{\partial x_i}f_1 = -\nu f_1.$$

5.1 Transport Equations for Non-relativistic Particles Scattered by Plasma Fluctuations

On rearranging the above expression, we obtain

$$\frac{\partial f_1}{\partial t} + U_i \frac{\partial f_1}{\partial x_i} - \frac{2}{5} c b_i b_j \frac{\partial U_j}{\partial x_i} \frac{\partial f_1}{\partial c} + \frac{2}{5} c \frac{\partial U_i}{\partial x_i} \frac{\partial f_1}{\partial c} + \frac{1}{5} \frac{\partial U_i}{\partial x_i} f_1 - \frac{3}{5} b_i b_j \frac{\partial U_j}{\partial x_i} f_1$$
$$= -\nu f_1 - \frac{c b_i}{3} \frac{\partial f_0}{\partial x_i} + \frac{D U_i}{Dt} \frac{b_i}{3} \frac{\partial f_0}{\partial c}, \tag{5.11}$$

where the f_0 Legendre coefficients are expressed as source terms in the evaluation of the next order Legendre coefficients f_1. To solve Eq. (5.11) for f_1 in terms of the lower order Legendre coefficient f_0, we make the further assumption that the zeroth-order coefficient f_0 is almost isotropic, implying that $f_1 \ll f_0$. The next assumption that we impose is that $\nu = \tau^{-1}$ is large, i.e., rapid scattering of the charged particles (which is consistent with the assumption that the particle distribution is nearly isotropic), so that the term $\nu f_1 \sim O(f_0)$. Subject to these assumptions, Eq. (5.11) can then be solved, yielding

$$\nu f_1 \simeq -\frac{c b_i}{3} \frac{\partial f_0}{\partial x_i} + \frac{D U_i}{Dt} \frac{b_i}{3} \frac{\partial f_0}{\partial c}. \tag{5.12}$$

Suppose first that the background flow possesses no large-scale accelerations or gradients, i.e., $Du_i/Dt = 0$, so that f_1 can be expressed as a diffusion term,

$$f_1 = -\frac{c \tau b_i}{3} \frac{\partial f_0}{\partial x_i}. \tag{5.13}$$

For the case that $DU_i/Dt = 0$, use of (5.13) in (5.10) yields

$$\frac{\partial f_0}{\partial t} + U_i \frac{\partial f_0}{\partial x_i} - \frac{c}{3} \frac{\partial U_i}{\partial x_i} \frac{\partial f_0}{\partial c} = -c b_i \frac{\partial f_1}{\partial x_i} - c \frac{\partial b_i}{\partial x_i} f_1 = -\frac{\partial}{\partial x_i}(b_i c f_1)$$
$$= \frac{\partial}{\partial x_i} \left(b_i \kappa b_j \frac{\partial f_0}{\partial x_j} \right),$$

where we introduced the diffusion coefficient

$$\kappa = \frac{c^2 \tau}{3}.$$

The diffusion term $b_i \kappa b_j$ is a tensor comprising an isotropic part and an anisotropic part,

$$\mathbf{K} = \begin{pmatrix} \kappa_{11} & 0 & 0 \\ 0 & \kappa_{22} & 0 \\ 0 & 0 & \kappa_{33} \end{pmatrix} + \begin{pmatrix} 0 & \kappa_{12} & \kappa_{13} \\ \kappa_{12} & 0 & \kappa_{23} \\ \kappa_{13} & \kappa_{23} & 0 \end{pmatrix} = \begin{pmatrix} \kappa_{11} & \kappa_{12} & \kappa_{13} \\ \kappa_{12} & \kappa_{22} & \kappa_{23} \\ \kappa_{13} & \kappa_{23} & \kappa_{33} \end{pmatrix},$$

where the elements of the tensor are simply $\kappa_{ii} = b_i^2 \kappa$, $\kappa_{ij} = b_i b_j \kappa$ for $i \neq j$ and $\kappa_{ij} = \kappa_{ji}$. Use of the diffusion tensor **K** allows us to express the convective-diffusive or advective-diffusive transport equation as

$$\frac{\partial f_0}{\partial t} + \mathbf{U} \cdot \nabla f_0 - \frac{c}{3} \nabla \cdot \mathbf{U} \frac{\partial f_0}{\partial c} = \nabla \cdot (\mathbf{K} \nabla f_0). \quad (5.14)$$

Subject to the assumptions imposed in deriving Eq. (5.14), this is the standard form of the transport equation for non-relativistic charged particles experiencing scattering in a turbulent magnetized medium. The physical content of the transport equation (5.14) is interesting when considered term-by-term. The second term shows that the scattered particles that comprise the distribution f_0 essentially co-move with the background flow in which the "scatterers" are embedded. The third term is an energy change term in response to the divergence of the background flow. This is seen by considering

$$\frac{\partial f_0}{\partial t} - \frac{c}{3} \nabla \cdot \mathbf{U} \frac{\partial f_0}{\partial c} = 0 \Leftrightarrow \frac{\partial f_0}{\partial t} - \frac{1}{3} \nabla \cdot \mathbf{U} \frac{\partial f_0}{\partial \xi} = 0,$$

where $\xi = \ln c$. The characteristics for this equation are given by

$$\frac{d\xi}{dt} \left(= \frac{1}{c} \frac{dc}{dt} \right) = -\frac{1}{3} \nabla \cdot \mathbf{U};$$

$$\frac{df_0}{dt} = \text{const.},$$

which yields

$$\xi(t) - \xi(t_0) = -\frac{1}{3} \int_{t_0}^{t} \nabla \cdot \mathbf{U} dt,$$

from some initial time t_0 to a time t. If **U** is stationary, then the change in particle velocity is given by

$$\frac{\ln c(t) - \ln c(t_0)}{t - t_0} = -\frac{1}{3} \nabla \cdot \mathbf{U}.$$

According as $\nabla \cdot \mathbf{U}$ is convergent (< 0) or divergent (> 0), particles will gain or lose speed c in the flow. For example, if the particle distribution function upstream of a region of a 1D decelerating flow ($\partial U/\partial x < 0$) is a power law $f \sim c^{-q}$, then the spectrum behind the decelerating flow will be shifted uniformly to the "right" in which each speed $\ln c$ increased by an amount proportional to the velocity gradient. Consequently, the energy of the particle distribution function will increase.

The diffusion term contains much of the physics of the magnetic field structure as well as the scattering properties of the small scale fluctuating field. As a consequence, the term **K** contains much more than simply diffusion. The isotropic

5.1 Transport Equations for Non-relativistic Particles Scattered by Plasma Fluctuations

part of the tensor **K** describes particle diffusion along (parallel) and perpendicular to the magnetic field. The anisotropic terms are generally thought to describe the collective drift of particles due to gradients and curvature in the magnetic field **B** and magnitude |**B**|. However, in a sense shown below, the particle response to the large-scale magnetic field geometry and gradients is present in all the elements of the tensor **K**. This can be seen by expressing

$$-cb_i \frac{\partial f_1}{\partial x_i} - c \frac{\partial b_i}{\partial x_i} f_1 = b_i \frac{\partial}{\partial x_i} \left(\kappa b_j \frac{\partial f_0}{\partial x_j} \right) + \frac{\partial b_i}{\partial x_i} \kappa b_j \frac{\partial f_0}{\partial x_j}$$

$$= \kappa b_i b_j \frac{\partial^2 f_0}{\partial x_i \partial x_j} + \kappa b_i \frac{\partial b_j}{\partial x_i} \frac{\partial f_0}{\partial x_j} + \frac{\partial b_i}{\partial x_i} \kappa b_j \frac{\partial f_0}{\partial x_j}. \tag{5.15}$$

The first term of (5.15) describes the isotropic and the anisotropic diffusive propagation of charged particles. The coefficients of $\partial f_0/\partial x_j$ in the second and third terms of (5.15) are evidently velocity terms that are associated with variations in b_i, i.e., these are drift terms associated either with gradients in **B**, |**B**|, or large-scale curvature of **B**. Note that

$$\frac{\partial b_i}{\partial x_i} = \nabla \cdot \mathbf{b} = \nabla \cdot \left(\frac{\mathbf{B}}{|\mathbf{B}|} \right) = -\frac{\mathbf{B} \cdot \nabla |\mathbf{B}|}{|\mathbf{B}|^2},$$

is non-zero only if |**B**| varies spatially. This term is therefore related to the variation in pitch-angle that a single particle experiences as it propagates along a magnetic field that is converging or diverging. Consequently, the term $\nabla \cdot \mathbf{b} = L^{-1}$ defines the so-called focusing length L, and the collective effect of focusing is therefore embedded in the "diffusion" term of the transport equation (5.14). The terms $\partial b_j/\partial x_i$ when $i \neq j$ include the large-scale curvature in **B** since

$$\frac{\partial b_j}{\partial x_i} = \frac{1}{|\mathbf{B}|} \frac{\partial B_j}{\partial x_i} - \frac{B_j}{|\mathbf{B}|^2} \frac{\partial |\mathbf{B}|}{\partial x_i}.$$

The terms $\partial b_j/\partial x_i$ also describe gradients in the components of **B**.

If we now include the DU_i/Dt convective derivative that was neglected in the solution of first-order correction f_1, i.e., (5.12), the transport equation for f_0 becomes

$$\frac{\partial f_0}{\partial t} + U_i \frac{\partial f_0}{\partial x_i} - \frac{\partial U_i}{\partial x_i} \frac{c}{3} \frac{\partial f_0}{\partial c} = \frac{\partial}{\partial x_i} \left(\frac{c^2 \tau}{3} b_i b_j \frac{\partial f_0}{\partial x_j} \right) - \frac{\partial}{\partial x_i} \left(\frac{c\tau}{3} b_i b_j \frac{DU_i}{Dt} \frac{\partial f_0}{\partial c} \right).$$

Use of the definition $\kappa = c^2\tau/3$ and the diffusion tensor **K** allows us to express the transport equation in the presence of large-scale flow gradients and accelerations as

$$\frac{\partial f_0}{\partial t} + \mathbf{U} \cdot \nabla f_0 - \nabla \cdot \mathbf{u} \frac{c}{3} \frac{\partial f_0}{\partial c} + \nabla \cdot \left(\mathbf{K} \frac{D\mathbf{U}}{Dt} \frac{1}{c} \frac{\partial f_0}{\partial c} \right) = \nabla \cdot (\mathbf{K} \nabla f_0). \tag{5.16}$$

The convective transport equation (5.14) and its extensions to relativistic charged particles is one of the most intensively studied equations in space physics and astrophysics as it is the basis for almost all work on energetic charged particle transport, ranging from galactic cosmic rays to solar energetic particles.

5.2 Transport Equation for Relativistic Charged Particles

5.2.1 Derivation of the Focussed Transport Equation

Consider now the extension of the previous two sections to include relativistic charged particles propagating in a non-relativistic background plasma flow with infinite conductivity.[1] It is assumed from the outset that the charged particles experience resonant scattering due to turbulent fluctuations in the background magnetic field. The fluctuations have typically been assumed to be magnetohydrodynamic waves, typically Alfvén waves, which tends to ensure that the scattered particles are trapped by the waves and stream with them. The waves define a frame of reference, the "wave frame," which propagates through the inertial or observer's (rest) frame and this is the frame in which the scattering is executed. In general, the wave frame is non-inertial, since, if we assume that the waves propagate at the local Alfvén speed V_A and they experience convection at the background plasma flow velocity $\mathbf{U}$, the wave frame velocity, $\mathbf{V}_A + \mathbf{U}$ may vary with space and time. This frame as expressed here also assumes that all the waves propagate uniformly in one direction which may not be appropriate. To avoid these complications, we shall assume that the background plasma flow speed sufficiently exceeds the Alfvén speed that we can neglect V_A. This is certainly true in the solar wind where $V_A \simeq 50$ km/s compared to the solar wind radial flow speed of 350–700 km/s.

The collisionless Vlasov equation that is valid for both relativistic and non-relativistic particles may be written as

$$\frac{d}{dt} f(\mathbf{x}, \mathbf{p}, t) \equiv \frac{\partial f}{\partial t} + v_i \frac{\partial f}{\partial x_i} + \frac{dp_i}{dt} \frac{\partial f}{\partial p_i} = 0, \qquad (5.17)$$

where $f(\mathbf{x}, \mathbf{p}, t)$ is the distribution function in the rest frame and $d\mathbf{p}/dt$ is the force on the charged relativistic particle. In the wave frame, scattering of the particles does not change the momentum or energy of the particles, so we need to transform (5.17) into the wave frame. The transformations that we need are listed in the footnote.[2]

---

[1] As noted earlier, the transport equation was derived by Skilling (1971). His treatment is very brief and the development given here is guided by an excellent set of notes developed originally by Dr's G.M. Webb and J.A. le Roux, to whom I am indebted for sharing them with me.

[2] We summarize the various Lorentz transformations that are needed in the derivation of the focussed transport equation. A four-vector has three spatial components and one time component, $(x_0, x_1, x_2, x_3) = (x_0, x^a) = x^\alpha$, where small Roman superscripts denote spatial coordinates of the four-vector and Greek superscripts denote all four components. The length of a four-

5.2 Transport Equation for Relativistic Charged Particles

We need to derive

$$\frac{d}{dt'} f(\mathbf{x'}, \mathbf{p'}, t') = 0,$$

where the distribution function and the variables correspond to the wave frame. Nevertheless, we will observe the cosmic rays in the rest or observer's frame, and this will therefore introduce a set of mixed coordinates as was done above. Exploiting the Lorentz invariance of the distribution function, $f(\mathbf{x}, \mathbf{p}, t) = f(\mathbf{x'}, \mathbf{p'}, t')$, we have

---

vector is $x^\alpha x_\alpha = x_1^2 + x_2^2 + x_3^2 - x_0^2 = x^a x^a - x_0^2$, and is invariant between coordinate systems. The contraction of any two four-vectors is invariant between coordinate systems. The Lorentz transformation matrix (see Jackson 1975, Sect. 11.7) enables one to transform one tensor to another. When the Lorentz matrix operates on a four-vector, it yields

$$x'_0 = \gamma(x_0 - \beta^a x^a);$$

$$x^{a'} = x^a + \beta^a \left(\frac{\beta^b x^b}{\beta^2} (\gamma - 1) - \gamma x_0 \right); \gamma = \frac{1}{\sqrt{1 - U^2/c^2}}; \quad \beta^a = U^a/c,$$

where the transformation of the four vector is between reference frames in which the primed variable has velocity U^a relative to the non-primed variable. c denotes the speed of light. The corresponding inverse Lorentz matrix can of course be used. Typical four vectors are time-space $(ct, x^a) = x^\alpha$, and the energy and momentum of a particle $(E/c, p^a) = (mc, p^a) = \gamma m_0(c, v^a)$, where m_0 is the rest mass of the particle. Since m_0 is constant, $\gamma(c, v^a)$ is a four-vector. Defining the proper time of a particle τ of a particle as $dt/d\tau = \gamma$ allows the four-velocity to be expressed as $dx^\alpha/d\tau$. The various transformations that we need are as follows:

$$t = \gamma \left(t' + \frac{\mathbf{x'} \cdot \mathbf{U}}{c^2} \right); \quad t' = \gamma \left(t - \frac{\mathbf{x} \cdot \mathbf{U}}{c^2} \right);$$

$$\mathbf{p} = \mathbf{p'} + m' \mathbf{U} \left[\frac{\mathbf{v'} \cdot \mathbf{U}}{c^2} (\gamma - 1) + \gamma \right];$$

$$\left(1 - \frac{\mathbf{v} \cdot \mathbf{U}}{c^2} \right) \gamma \mathbf{v'} = \mathbf{v} + \mathbf{U} \left[\frac{\mathbf{v} \cdot \mathbf{U}}{c^2} (\gamma - 1) - \gamma \right];$$

$$\left(1 + \frac{\mathbf{v'} \cdot \mathbf{U}}{c^2} \right) \gamma \mathbf{v} = \mathbf{v'} + \mathbf{U} \left[\frac{\mathbf{v'} \cdot \mathbf{U}}{c^2} (\gamma - 1) + \gamma \right];$$

$$\mathbf{E'} = \gamma (\mathbf{E} + \mathbf{U} \times \mathbf{B}) + (1 - \gamma) \frac{\mathbf{E} \cdot \mathbf{U}}{c^2} \mathbf{U};$$

$$\mathbf{B'} = \gamma \left(\mathbf{B} - \frac{1}{c^2} \mathbf{U} \times \mathbf{E} \right) + (1 - \gamma) \frac{\mathbf{B} \cdot \mathbf{U}}{v^2} \mathbf{U}.$$

Note that if $|\mathbf{U}|/c \ll 1$,

$$\gamma = (1 - U^2/c^2)^{-1/2} \simeq 1 + (U^2/2c)^2,$$

so that $\gamma \simeq 1$ is valid to the first order in U/c.

$$\frac{d}{dt} f(\mathbf{x}, \mathbf{p}, t) \frac{dt}{dt'} = 0.$$

The Lorentz transformation for time between the observer and wave frames yields to first-order in $|\mathbf{U}|/c$

$$t \simeq t' + \frac{\mathbf{x}' \cdot \mathbf{U}}{c^2}.$$

Consequently, we have

$$\frac{dt}{dt'} \simeq \left(1 + \frac{\mathbf{v}' \cdot \mathbf{U}}{c^2}\right) \Longrightarrow \left(1 + \frac{\mathbf{v}' \cdot \mathbf{U}}{c^2}\right) \frac{df}{dt}(\mathbf{x}, \mathbf{p}, t) = 0.$$

We now need to introduce a transformation so that the particle momentum is measured in the wave frame. This requires that the various partial derivatives in the Vlasov equation are transformed from the observer's frame to the wave frame i.e., $(\mathbf{x}, \mathbf{p}, t) \mapsto (\mathbf{x}, \mathbf{p}', t)$. This requires the use of the inverse Lorentz transformation for particle momentum (Footnote), which to first order in $|\mathbf{U}|/c$ yields $\gamma \simeq 1$ and

$$\mathbf{p}' = \mathbf{p} - m'\mathbf{U},$$

where $m' = \gamma' m_0$ and $\gamma' = 1/\sqrt{1 - v'^2/c^2}$ for the relativistic particle in the observer's frame. Considering the time derivative yields

$$\frac{\partial}{\partial t} = \frac{\partial}{\partial t} + \frac{\partial p'_i}{\partial t} \frac{\partial}{\partial p'_i} = \frac{\partial}{\partial t} + \frac{\partial}{\partial t}(p_i - m'U_i) \frac{\partial}{\partial p'_i} = \frac{\partial}{\partial t} - m' \frac{\partial U_i}{\partial t} \frac{\partial}{\partial p'_i}.$$

The spatial derivative transforms as

$$\frac{\partial}{\partial x_i} = \frac{\partial}{\partial x_i} + \frac{\partial p'_j}{\partial x_i} \frac{\partial}{\partial p'_j} = \frac{\partial}{\partial x_i} - m' \frac{\partial U_j}{\partial x_i} \frac{\partial}{\partial p'_j}.$$

Finally, instead of the inverse transform, we use $\mathbf{p}' = \mathbf{p} - m\mathbf{U}$ to obtain

$$\frac{\partial}{\partial p_i} = \frac{\partial p'_j}{\partial p_i} \frac{\partial}{\partial p'_j} = \frac{\partial}{\partial p_i}(p_j - mU_j) \frac{\partial}{\partial p'_j} = \delta_{ij} \frac{\partial}{\partial p'_j} - U_j \frac{\partial m}{\partial p_i} \frac{\partial}{\partial p'_j}.$$

Introducing the basis vector for spherical coordinates allows us to express

$$\frac{\partial m}{\partial p_i} = \hat{e}_{pi} \frac{\partial m}{\partial p},$$

and since

$$m = \gamma m_0 = m_0 \left(1 - \frac{v^2}{c^2}\right)^{-1/2} = m_0 \left(1 - \frac{p^2}{m^2 c^2}\right)^{-1/2} \Rightarrow m = \left(1 + \frac{p^2}{m_0^2 c^2}\right)^{1/2},$$

5.2 Transport Equation for Relativistic Charged Particles

we have

$$\frac{dm}{dp} = \frac{p}{mc^2}.$$

We then obtain

$$\frac{\partial}{\partial p_i} = \delta_{ij}\frac{\partial}{\partial p'_j} - \frac{v_i U_j}{c^2}\frac{\partial}{\partial p'_j}.$$

On retaining only terms of $O(U/c)$, we obtain

$$\left(1 + \frac{\mathbf{v}' \cdot \mathbf{U}}{c^2}\right)\frac{\partial f}{\partial t} = \left(1 + \frac{\mathbf{v}' \cdot \mathbf{U}}{c^2}\right)\left(\frac{\partial f}{\partial t} - m\frac{\partial U_i}{\partial t}\frac{\partial f}{\partial p'_i}\right)$$

$$\simeq \left(1 + \frac{\mathbf{v}' \cdot \mathbf{U}}{c^2}\right)\frac{\partial f}{\partial t} - m\frac{\partial U_i}{\partial t}\frac{\partial f}{\partial p'_i}.$$

Consider now the convective term, $\left(1 + \mathbf{v}' \cdot \mathbf{U}/c^2\right) v_i \partial f/\partial x_i$. To the first order in U/c, using the Lorentz transformation for the velocity, and $\gamma \simeq 1$ gives

$$\mathbf{v} = \frac{\mathbf{v}' + \mathbf{U}}{1 + \mathbf{v}' \cdot \mathbf{U}/c^2}.$$

This then yields

$$\left(1 + \frac{\mathbf{v}' \cdot \mathbf{U}}{c^2}\right)\frac{v'_i + U_i}{1 + \mathbf{v}' \cdot \mathbf{U}/c^2}\left(\frac{\partial f}{\partial x_i} - m'\frac{\partial U_j}{\partial x_i}\frac{\partial f}{\partial p'_j}\right) = (v'_i + U_i)\frac{\partial f}{\partial x_i}$$

$$- m'\left(v'_i + U_i\right)\frac{\partial U_j}{\partial x_i}\frac{\partial f}{\partial \partial'_j}.$$

Consider now the momentum change term $\left(1 + \mathbf{v}' \cdot \mathbf{U}/c^2\right)(dp_i/dt)\partial f/\partial p_i$. We assume that the momentum change is due to electromagnetic fields only. Thus, we have the Lorentz force

$$\frac{dp_i}{dt} = q\left(E_i + \varepsilon_{ijk}v_j B_k\right),$$

where q is the particle charge, $\mathbf{B}$ the external magnetic field, $\mathbf{E}$ the electric field, and ε_{ijk} is the Levi-Civita tensor. The first order Lorentz transformation for $\mathbf{E}$ is simply

$$\mathbf{E}' \simeq \mathbf{E} + \mathbf{U} \times \mathbf{B} \iff \mathbf{E} \simeq \mathbf{E}' - \mathbf{U} \times \mathbf{B},$$

which yields

$$\frac{dp_i}{dt} = q\left(E'_i + \varepsilon_{ijk}(v_j - U_j)B_k\right).$$

To address the transformation of the velocity, the Lorentz transformation yields $\gamma \simeq 1$ and

$$\left(1 - \frac{\mathbf{v} \cdot \mathbf{U}}{c^2}\right) \mathbf{v}' = \mathbf{v} - \mathbf{U},$$

at $O(U/v)$, from which we find

$$\frac{dp_i}{dt} = q \left[E'_i + \left(1 - \frac{\mathbf{v} \cdot \mathbf{U}}{c^2}\right) \varepsilon_{ijk} v'_j B_k \right],$$

so that

$$\left(1 + \frac{\mathbf{v}' \cdot \mathbf{U}}{c^2}\right) \frac{dp_i}{dt} \frac{\partial f}{\partial p_i} = q \left(1 + \frac{\mathbf{v}' \cdot \mathbf{U}}{c^2}\right) E'_i \frac{\partial f}{\partial p_i}$$

$$+ q \left(1 + \frac{\mathbf{v}' \cdot \mathbf{U}}{c^2}\right) \left(1 - \frac{\mathbf{v} \cdot \mathbf{U}}{c^2}\right) \varepsilon_{ijk} v'_j B_k \frac{\partial f}{\partial p_i}.$$

The Lorentz transformation for time and its inverse yield

$$\frac{dt'}{dt} = \gamma \left(1 - \frac{\mathbf{v} \cdot \mathbf{U}}{c^2}\right); \quad \frac{dt}{dt'} = \gamma \left(1 + \frac{\mathbf{v}' \cdot \mathbf{U}}{c^2}\right),$$

from which we obtain

$$\gamma^2 \left(1 - \frac{\mathbf{v} \cdot \mathbf{U}}{c^2}\right) \left(1 + \frac{\mathbf{v}' \cdot \mathbf{U}}{c^2}\right) = 1,$$

or

$$\left(1 - \frac{\mathbf{v} \cdot \mathbf{U}}{c^2}\right) \left(1 + \frac{\mathbf{v}' \cdot \mathbf{U}}{c^2}\right) = 1$$

in the limit $U/c \ll 1$. We may therefore derive

$$\left(1 + \frac{\mathbf{v}' \cdot \mathbf{U}}{c^2}\right) \frac{dp_i}{dt} \frac{\partial f}{\partial p_i} = q \left[\left(1 + \frac{\mathbf{v}' \cdot \mathbf{U}}{c^2}\right) E'_i + \varepsilon_{ijk} v'_j B_k \right] \left(\delta_{ij} - \frac{v_i U_j}{c^2}\right) \frac{\partial f}{\partial p'_j}.$$

Now consider the Lorentz transformation of the magnetic field. To first order, we have

$$\mathbf{B}' = \mathbf{B} - \frac{1}{c^2} \mathbf{U} \times \mathbf{E},$$

but since $\mathbf{E} = -\mathbf{U} \times \mathbf{B}$, $\mathbf{B}' = \mathbf{B} + \mathbf{U} \times (\mathbf{U} \times \mathbf{B})/c^2$, this implies that

$$\mathbf{B}' = \mathbf{B},$$

5.2 Transport Equation for Relativistic Charged Particles

at this order. This, together with

$$\frac{dp'_i}{dt} = q\left(E'_i + \varepsilon_{ijk} v'_j B'_k\right),$$

allows us to write

$$\left(1 + \frac{\mathbf{v}' \cdot \mathbf{U}}{c^2}\right) \frac{dp_i}{dt} \frac{\partial f}{\partial p_i} = q \frac{\mathbf{v}' \cdot \mathbf{U}}{c^2} E'_i \left(\delta_{ij} - \frac{v_i U_j}{c^2}\right) \frac{\partial f}{\partial p'_j} + \frac{dp'_i}{dt}\left(\delta_{ij} - \frac{v_i U_j}{c^2}\right)\frac{\partial f}{\partial p'_j}.$$

Consider the term $q\varepsilon_{ijk} v'_j B'_k (v_i U_j/c^2) \partial f/\partial p'_j$. The first order velocity transformation yields

$$\mathbf{v} = \frac{\mathbf{v}' + \mathbf{U}}{1 + \mathbf{v}' \cdot \mathbf{U}/c^2},$$

so that

$$q\varepsilon_{ijk} v'_j B'_k \frac{v_i U_j}{c^2}\frac{\partial f}{\partial p'_j} = \frac{q}{1+\mathbf{v}'\cdot\mathbf{U}/c^2} \varepsilon_{ijk} v'_j B'_k \frac{v'_i + U_i}{c^2} U_j \frac{\partial f}{\partial p'_j}$$

$$\simeq \frac{q}{1+\mathbf{v}'\cdot\mathbf{U}/c^2}\frac{1}{c^2} \varepsilon_{ijk} v'_j B'_k v'_i U_j \frac{\partial f}{\partial p'_j}$$

$$= \frac{q}{1+\mathbf{v}'\cdot\mathbf{U}/c^2}\frac{1}{c^2}(\mathbf{v}'\times\mathbf{B}')\cdot\mathbf{v}' U_j \frac{\partial f}{\partial p'_j} = 0,$$

after neglecting the $U_i U_j / c^2$ term in the second line. The term $q(\mathbf{v}' \cdot \mathbf{U}/c^2) E'_i v_i (U_j/c^2) \partial f/\partial p'_j$ is $O\left((U/c)^2\right)$ and so is neglected. We therefore obtain

$$\left(1 + \frac{\mathbf{v}' \cdot \mathbf{U}}{c^2}\right) \frac{dp_i}{dt} \frac{\partial f}{\partial p_i} \simeq \left[q\frac{\mathbf{v}' \cdot \mathbf{U}}{c^2} E'_i + \frac{dp'_i}{dt}\right] \frac{\partial f}{\partial p'_i},$$

where

$$\frac{dp'_i}{dt} = q\left(E'_i + \varepsilon_{ijk} v'_j B'_k\right).$$

On combining the results above, we obtain the Vlasov equation in mixed coordinates,

$$\left(1 + \frac{\mathbf{v}'\cdot\mathbf{U}}{c^2}\right)\frac{\partial f}{\partial t} + (\mathbf{v}' + \mathbf{U})\cdot\nabla f + q\frac{\mathbf{v}'\cdot\mathbf{U}}{c^2}E'_i\frac{\partial f}{\partial p'_i}$$

$$-\left[m'\left(\frac{\partial U_i}{\partial t} + U_j \frac{\partial U_i}{\partial x_j}\right) + p'_j \frac{\partial U_i}{\partial x_j} - \frac{dp'_i}{dt}\right]\frac{\partial f}{\partial p'_i} = 0, \qquad (5.18)$$

with $f(\mathbf{x}, \mathbf{p}, t)$. However, since the coordinates $(\mathbf{x}, \mathbf{p}', t)$ are in the mixed coordinate system, we need to introduce the transformation $f(\mathbf{x}, \mathbf{p}, t) \mapsto f'(\mathbf{x}, \mathbf{p}', t)$. Recall that

$$f(\mathbf{x}, \mathbf{p}, t) d^3x\, d^3p = f'(\mathbf{x}, \mathbf{p}', t) d^3x'\, d^3p',$$

and that $d^3x = \gamma d^3x'$. Consider the transformation of the volume element in momentum space. On using $\mathbf{p} = \mathbf{p}' + m'\mathbf{U}$, $dm'/dp = p/(m'c^2)$, and $dp/dp_i = p_i/p$, we have

$$d^3\mathbf{p} = dp_x dp_y dp_z$$

$$= \left(dp'_x + U_x \frac{p'}{m'c^2} dp'\right)\left(dp'_y + U_y \frac{p'}{m'c^2} dp'\right)\left(dp'_z + U_z \frac{p'}{m'c^2} dp'\right)$$

$$= dp'_x dp'_y dp'_z \left(1 + \frac{p'_x U_x}{m'c^2}\right)\left(1 + \frac{p'_y U_y}{m'c^2}\right)\left(1 + \frac{p'_z U_z}{m'c^2}\right)$$

$$= dp'_x dp'_y dp'_z \left(1 + \frac{\mathbf{p}' \cdot \mathbf{U}}{m'c^2}\right) + O\left(\frac{U^2}{c^2}\right)$$

$$\simeq dp'_x dp'_y dp'_z \left(1 + \frac{\mathbf{v}' \cdot \mathbf{U}}{c^2}\right).$$

Thus, we have the transformation

$$f(\mathbf{x}, \mathbf{p}, t) = \frac{f'(\mathbf{x}, \mathbf{p}', t)}{\gamma(1 + \mathbf{v}' \cdot \mathbf{U}/c^2)},$$

which to first order in U/c, $\gamma \simeq 1$ yields

$$f(\mathbf{x}, \mathbf{p}, t) = \frac{f'(\mathbf{x}, \mathbf{p}', t)}{(1 + \mathbf{v}' \cdot \mathbf{U}/c^2)} \equiv f''(\mathbf{x}, \mathbf{p}', t).$$

On setting $f(\mathbf{x}, \mathbf{p}, t) = f''(\mathbf{x}, \mathbf{p}', t)$ in (5.18), we have the final form of the transformed equation,

$$\left(1 + \frac{\mathbf{v}' \cdot \mathbf{U}}{c^2}\right)\frac{\partial f''}{\partial t} + (\mathbf{v}' + \mathbf{U}) \cdot \nabla f'' + q\frac{\mathbf{v}' \cdot \mathbf{U}}{c^2} E'_i \frac{\partial f''}{\partial p'_i}$$

$$- \left[m'\left(\frac{\partial U_i}{\partial t} + U_j \frac{\partial U_i}{\partial x_j}\right) + p'_j \frac{\partial U_i}{\partial x_j}\right]\frac{\partial f''}{\partial p'_i} + \frac{\partial}{\partial p'_i}(F'_i f'') = 0. \quad (5.19)$$

5.2 Transport Equation for Relativistic Charged Particles

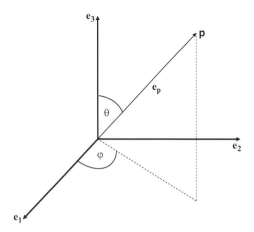

Fig. 5.1 The coordinates for a particle gyrating about a mean magnetic field **B** oriented along the z-axis. The particle momentum is given by the vector **p**, the pitch-angle by θ, and the gyrophase by ϕ. The directional vector $\mathbf{b} \equiv \mathbf{B}/|\mathbf{B}| = \mathbf{e}_3$

In deriving (5.19), we used

$$\frac{\partial}{\partial p'_i}\left(\frac{dp'_i}{dt}\right) = \frac{\partial}{\partial p'_i}\left(q\varepsilon_{ijk}v'_j B'_k\right) = q\varepsilon_{ijk}\frac{\partial v'_j}{\partial p'_i}B'_k$$

$$= q\varepsilon_{ijk}B'_k e'_{pi}\frac{\partial}{\partial p'}\left(e'_{pj}v'\right)$$

$$= q\varepsilon_{ijk}e'_{pi}e'_{pj}\frac{\partial v'}{\partial p'}B'_k$$

$$= qe'_{pi}(\varepsilon_{ijk}e'_{pj}B'_k)\frac{1}{m'\gamma'^2}$$

$$= \frac{q}{m'\gamma'^2}\mathbf{e}'_p \cdot \left(\mathbf{e}'_p \times \mathbf{B}'\right) = 0.$$

Just as we did in the derivation of the focussed transport equation for non-relativistic particles, we shall assume that the particle distribution function is nearly gyrotropic, making $f(\mathbf{x}, \mathbf{v}, t) \simeq f(\mathbf{x}, v, \mu, t)$ where the particle pitch angle is $\mu \equiv \cos\theta$ as before. For the sake of notational convenience, we henceforth drop the "prime" on the variables and distribution function. The averaging procedure proceeds in much the same way as before. For completeness, we provide some of the details in the derivation although using a slightly more general notation. The local geometry of a charged particle gyrating about the mean magnetic field **B** is illustrated in Fig. 5.1. The coordinates (x_1, x_2, x_3) refer to a magnetic field system and $\mathbf{e}_3 = \mathbf{b} \equiv \mathbf{B}/|\mathbf{B}|$. Since the magnetic field is not assumed to be uniform, the unit vectors $(\mathbf{e}_1, \mathbf{e}_2, \mathbf{e}_3 = \mathbf{b})$ are functions of $\mathbf{x}$. As before, $\mu = \cos\theta = \mathbf{e}_p \cdot \mathbf{b}$.

Recall that the momentum can be expressed in spherical coordinates as

$$\frac{\partial f}{\partial \mathbf{p}} = \mathbf{e}_p\frac{\partial f}{\partial p} + \mathbf{e}_\theta\frac{1}{p}\frac{\partial f}{\partial \theta} + \mathbf{e}_\phi\frac{1}{p\sin\theta}\frac{\partial f}{\partial \phi}$$

where

$$\mathbf{p} = p(\sin\theta\cos\phi\mathbf{e}_1 + \sin\theta\sin\phi\mathbf{e}_2 + \cos\theta\mathbf{b});$$

$$\mathbf{e}_p = \frac{\partial \mathbf{p}}{\partial p} = \sin\theta\cos\phi\mathbf{e}_1 + \sin\theta\sin\phi\mathbf{e}_2 + \cos\theta\mathbf{b};$$

$$\mathbf{e}_\theta = \frac{1}{p}\frac{\partial \mathbf{p}}{\partial \theta} = \cos\theta\cos\phi\mathbf{e}_1 + \cos\theta\sin\phi\mathbf{e}_2 - \sin\theta\mathbf{b};$$

$$\mathbf{e}_\phi = \frac{1}{p\sin\theta}\frac{\partial \mathbf{p}}{\partial \phi} = -\sin\phi\mathbf{e}_1 + \cos\phi\mathbf{e}_2.$$

As before, we require the following integrals,

$$\langle\mathbf{e}_p\rangle = \frac{1}{2\pi}\int_0^{2\pi} \mathbf{e}_p d\phi = \cos\theta\mathbf{b};$$

$$\langle\mathbf{e}_p\mathbf{e}_p\rangle = \frac{1}{2\pi}\int_0^{2\pi} \mathbf{e}_p\mathbf{e}_p d\phi = \frac{1}{2}\sin^2\theta\,[\mathbf{I} - \mathbf{bb}] + \cos^2\theta\mathbf{bb},$$

after using $\langle\cos^2\phi\rangle = \langle\sin^2\phi\rangle = 1/2$.

Consider the time derivative

$$\frac{\partial f}{\partial t} = \frac{\partial f}{\partial t} + \frac{\partial \mu}{\partial t}\frac{\partial f}{\partial \mu}$$

$$= \frac{\partial f}{\partial t} + \frac{\partial}{\partial t}(\mathbf{e}_p\cdot\mathbf{b})\frac{\partial f}{\partial \mu} = \frac{\partial f}{\partial t}$$

after using $b_i\partial b_i/\partial t = 0$ as before. Evidently, $\langle\partial f/\partial t\rangle = \partial f/\partial t$. Now,

$$\left\langle\frac{\mathbf{p}\cdot\mathbf{U}}{mc^2}\right\rangle\frac{\partial f}{\partial t} = \frac{1}{2\pi}\int_0^{2\pi}\frac{p_iU_i}{mc^2}\left[\frac{\partial f}{\partial t} + e_{pj}\frac{\partial b_j}{\partial t}\frac{\partial f}{\partial \mu}\right]d\phi$$

$$= \frac{U_i}{mc^2}\frac{1}{2\pi}\int_0^{2\pi} p_i\frac{\partial f}{\partial t}d\phi + \frac{U_i}{mc^2}\frac{\partial b_j}{\partial t}\frac{1}{2\pi}\int_0^{2\pi} p_ie_{pj}\frac{\partial f}{\partial \mu}d\phi$$

$$= \frac{pU_i}{mc^2}\frac{\partial f}{\partial t}\frac{1}{2\pi}\int_0^{2\pi} e_{pi}d\phi + \frac{pU_i}{mc^2}\frac{\partial b_j}{\partial t}\frac{\partial f}{\partial \mu}\frac{1}{2\pi}\int_0^{2\pi} e_{pi}e_{pj}d\phi$$

$$= \frac{p\mu}{mc^2}U_ib_i\frac{\partial f}{\partial t} + \frac{pU_i}{mc^2}\frac{\partial b_j}{\partial t}\frac{\partial f}{\partial \mu}\left[\frac{1}{2}\sin^2\theta(\delta_{ij} - b_ib_j) + \cos^2\theta b_ib_j\right]$$

$$= \frac{v\mu}{c}\left(\frac{\mathbf{U}}{c}\cdot\mathbf{b}\right)\frac{\partial f}{\partial t} + \frac{1}{2}\frac{v(1-\mu^2)}{c}\left(\frac{\mathbf{U}}{c}\cdot\frac{\partial \mathbf{b}}{\partial t}\right)\frac{\partial f}{\partial \mu}.$$

5.2 Transport Equation for Relativistic Charged Particles

On considering the convective term,

$$\frac{\partial f}{\partial x_i} = \frac{\partial f}{\partial x_i} + e_{pj}\frac{\partial b_j}{\partial x_i}\frac{\partial f}{\partial \mu},$$

we derive the gyrophase averaged expression

$$\left\langle \left(U_i + \frac{p_i}{m}\right)\frac{\partial f}{\partial x_i}\right\rangle = U_i\frac{\partial f}{\partial x_i} + \frac{p}{m}\frac{\partial f}{\partial x_i}\langle e_{pi}\rangle + U_i\frac{\partial b_j}{\partial x_i}\frac{\partial f}{\partial \mu}\langle e_{pj}\rangle + \frac{p}{m}\frac{\partial b_j}{\partial x_i}\frac{\partial f}{\partial \mu}\langle e_{pi}e_{pj}\rangle$$

$$= (U_i + v\mu b_i)\frac{\partial f}{\partial x_i} + \frac{1}{2}v(1-\mu^2)\frac{\partial b_i}{\partial x_i}\frac{\partial f}{\partial \mu}.$$

On expressing

$$\frac{\partial f}{\partial p_i} = \frac{\partial p}{\partial p_i}\frac{\partial f}{\partial p} + \frac{\partial \mu}{\partial p_i}\frac{\partial f}{\partial \mu}$$

$$= e_{pi}\frac{\partial f}{\partial p} + \frac{\partial}{\partial p_i}(e_{pj}b_j)\frac{\partial f}{\partial \mu}$$

$$= e_{pi}\frac{\partial f}{\partial p} + b_j\frac{\partial}{\partial p_i}\left(\frac{p_j}{p}\right)\frac{\partial f}{\partial \mu}$$

$$= e_{pi}\frac{\partial f}{\partial p} + \frac{b_j}{p}\left(\delta_{ij} - \frac{p_j}{p}\frac{\partial p}{\partial p_i}\right)\frac{\partial f}{\partial \mu}$$

$$= e_{pi}\frac{\partial f}{\partial p} + \frac{b_j}{p}(\delta_{ij} - e_{pi}e_{pj})\frac{\partial f}{\partial \mu},$$

we may consider

$$\left\langle -m\frac{dU_i}{dt}\frac{\partial f}{\partial p_i}\right\rangle = -m\frac{dU_i}{dt}\frac{\partial f}{\partial p}\langle e_{pi}\rangle - m\frac{dU_i}{dt}\frac{b_j}{p}\delta_{ij}\frac{\partial f}{\partial \mu} + m\frac{dU_i}{dt}\frac{b_j}{p}\frac{\partial f}{\partial \mu}\langle e_{pi}e_{pj}\rangle$$

$$= -m\mu\frac{dU_i}{dt}b_i\frac{\partial f}{\partial p} - \frac{m}{p}(1-\mu^2)\frac{dU_i}{dt}b_i\frac{\partial f}{\partial \mu}.$$

We can similarly evaluate

$$\left\langle -p_k\frac{\partial U_i}{\partial x_k}\frac{\partial f}{\partial p_i}\right\rangle = -p\left[\frac{1}{2}(1-\mu^2)\frac{\partial U_i}{\partial x_i} + \frac{1}{2}(3\mu^2-1)b_ib_j\frac{\partial U_i}{\partial x_j}\right]\frac{\partial f}{\partial p}$$

$$+ \frac{1}{2}(1-\mu^2)\left[\frac{\partial U_i}{\partial x_i} - 3b_ib_j\frac{\partial U_i}{\partial x_j}\right]\mu\frac{\partial f}{\partial \mu}.$$

Finally,

$$\frac{q}{m}(\mathbf{p}\times\mathbf{B})\cdot\frac{\partial f}{\partial \mathbf{p}} = \frac{q}{m}(\mathbf{p}\times\mathbf{B})\cdot\left[\mathbf{e}_p\frac{\partial f}{\partial p} + \left(\frac{\mathbf{b}}{p} - \mu\mathbf{e}_p\right)\frac{\partial f}{\partial \mu}\right] = 0,$$

since $(\mathbf{p}\times\mathbf{B})\cdot\mathbf{b} = 0$ and $(\mathbf{p}\times\mathbf{B})\cdot\mathbf{e}_p = 0$.

On combining the results above, we obtain the focussed transport equation or, equivalently, the Boltzmann equation for a gyrotropic particle distribution,

$$\left(1 + \frac{v\mu}{c}\frac{\mathbf{U}\cdot\mathbf{b}}{c}\right)\frac{\partial f}{\partial t} + (\mathbf{U} + v\mu\mathbf{b})\cdot\nabla f + \frac{1-\mu^2}{2}\left[\frac{v}{c}\left(\frac{\mathbf{U}}{c}\cdot\frac{\partial\mathbf{b}}{\partial t}\right) + v\nabla\cdot\mathbf{b}\right.$$
$$\left. - \frac{2m}{p}\left(\frac{d\mathbf{U}}{dt}\cdot\mathbf{b}\right) + \mu\left(\nabla\cdot\mathbf{U} - 3\mathbf{bb}:\nabla\mathbf{U}\right)\right]\frac{\partial f}{\partial\mu} - \left[\frac{\mu m}{p}\left(\frac{d\mathbf{U}}{dt}\cdot\mathbf{b}\right)\right.$$
$$\left. + \frac{1}{2}(1-\mu^2)\nabla\cdot\mathbf{U} + \frac{1}{2}(3\mu^2 - 1)\mathbf{bb}:\nabla\mathbf{U}\right]p\frac{\partial f}{\partial p} = \left(\frac{\delta f}{\delta t}\right)_s. \quad (5.20)$$

The righthand term is the scattering term, due charged particles scattering in pitch-angle due to the stochastically fluctuating magnetic field. Certainly for parallel propagation, the scattering fluctuations are typically assumed to be Alfvén waves. The scattering of charged particles conserves particle energy in the wave frame. In the transformation from the observer's frame (the rest frame) to the wave frame, the macroscopic electric fields are transformed away by the background velocity $\mathbf{U}$ because the plasma is infinitely conductive. Electric fields associated with the waves disappear in a frame moving with the waves. In the absence of electric fields, charged particles can only experience scattering in pitch angle. Energy is not, however, conserved in the observer's frame.

On assuming that $d\mathbf{U}/dt = 0$ and neglecting terms $O(U/c)$, we recover the usual form of the focussed transport equation,

$$\frac{\partial f}{\partial t} + (\mathbf{U} + v\mu\mathbf{b})\cdot\nabla f + \left[\frac{1-3\mu^2}{2}(\mathbf{bb}:\nabla\mathbf{U}) - \frac{1}{2}(1-\mu^2)\nabla\cdot\mathbf{U}\right]p\frac{\partial f}{\partial p}$$
$$+ \frac{1-\mu^2}{2}\left[v\nabla\cdot\mathbf{b} + \mu\nabla\cdot\mathbf{U} - 3\mu\mathbf{bb}:\nabla\mathbf{U}\right]\frac{\partial f}{\partial\mu} = \left(\frac{\delta f}{\delta t}\right)_s. \quad (5.21)$$

The focussed transport equation (5.21) can be reduced to the convective-diffusive equation if the distribution function $f(\mathbf{x}, p, \mu, t) \simeq f(\mathbf{x}, p, t)$ i.e., if the scattering experienced by the particle is sufficiently strong that the distribution is nearly isotropic. The analysis of Sect. 2 carries over directly with "c" being replaced by "p", and the general convective-diffusive transport equation is given by

$$\frac{\partial f}{\partial t} + \mathbf{U}\cdot\nabla f - \frac{p}{3}\nabla\cdot\mathbf{U}\frac{\partial f}{\partial p} = \nabla(\mathbf{K}\nabla f). \quad (5.22)$$

This is the standard form of the transport equation for relativistic charged particles experiencing scattering in a non-relativistic turbulent plasma.

Exercises

1. Derive the following averaging relations:

$$\langle \mathbf{e}_p \rangle = \frac{1}{2\pi} \int_0^{2\pi} \mathbf{e}_p d\phi = \cos\theta \mathbf{b};$$

$$\langle \mathbf{e}_p \mathbf{e}_p \rangle = \frac{1}{2\pi} \int_0^{2\pi} \mathbf{e}_p \mathbf{e}_p d\phi = \frac{1}{2}\sin^2\theta\,[\mathbf{I} - \mathbf{bb}] + \cos^2\theta\,\mathbf{bb}.$$

2. Complete the derivation of

$$\left\langle \left(U_i + \frac{p_i}{m}\right)\frac{\partial f}{\partial x_i} \right\rangle = (\mathbf{U} + v\mu\mathbf{b})\cdot\nabla f + \frac{1}{2}v(1-\mu^2)\nabla\cdot\mathbf{b}\frac{\partial f}{\partial \mu}.$$

3. Show that

$$\frac{\partial}{\partial p_i}(e_{pj}b_j) = \frac{b_j}{p}(\delta_{ij} - e_{pi}e_{pj}).$$

4. Complete the derivation of

$$\left\langle -p_k \frac{\partial U_i}{\partial x_k}\frac{\partial f}{\partial p_i} \right\rangle = -p\left[\frac{1}{2}(1-\mu^2)\frac{\partial U_i}{\partial x_i} + \frac{1}{2}(3\mu^2 - 1)b_i b_j \frac{\partial U_i}{\partial x_j}\right]\frac{\partial f}{\partial p}$$

$$+ \frac{1}{2}(1-\mu^2)\left[\frac{\partial U_i}{\partial x_i} - 3b_i b_j \frac{\partial U_i}{\partial x_j}\right]\mu\frac{\partial f}{\partial \mu}.$$

5.3 The Magnetic Correlation Tensor

As will be discussed in detail below, the magnetic correlation tensor plays a central role in determining the transport properties of particles experiencing pitch-angle scattering by turbulent magnetic field fluctuations. A very detailed discussion of different forms of the magnetic correlation tensor has been presented by Shalchi (2009).[3] The general form of the two-point, two-time magnetic correlation tensor has the form

$$R_{ij}(\mathbf{r},t,\mathbf{r}\prime,t_0) = \langle \delta B_i(\mathbf{r},t), \delta B_j(\mathbf{r}',t_0)\rangle,$$

where r' denotes a different spatial location and $\langle \cdot \rangle$ an ensemble average. It is convenient to consider the correlation tensor using a Fourier representation

---

[3] See also Tautz and Lerche (2011).

$$\delta B_i(\mathbf{r}, t) = \int \delta B_i(\mathbf{k}, t) e^{i\mathbf{k}\cdot\mathbf{r}} d^3k,$$

from which we find

$$R_{ij}(\mathbf{r}, t, \mathbf{r}', t_0) = \int d^3k \int d^3k' \langle \delta B_i(\mathbf{k}, t) \delta B_j(\mathbf{k}', t_0) \rangle e^{i\mathbf{k}\cdot\mathbf{r} + i\mathbf{k}'\cdot\mathbf{r}'}. \quad (5.23)$$

As is typically assumed, we suppose that the magnetic turbulence is homogeneous, so that the correlation function depends only on the separation $|\mathbf{r} - \mathbf{r}'|$ between two points. Then we can express $\langle \delta B_i(\mathbf{k}, t), \delta B_j(\mathbf{k}', t_0) \rangle$ as $\langle \delta B_i(\mathbf{k}, t), \delta B_j(\mathbf{k}', t_0) \rangle \delta(\mathbf{k} + \mathbf{k}')$, which allows us to integrate (5.23) as

$$R_{ij} = \int d^3k \langle \delta B_i(\mathbf{k}, t), \delta B_j(-\mathbf{k}, t_0) \rangle e^{i\mathbf{k}\cdot(\mathbf{r}-\mathbf{r}')}.$$

From the definition of the Fourier transform, $\delta B_j(-\mathbf{k}) = \delta B_j^*(\mathbf{k})$ where * denotes the complex conjugate. This allows us to introduce the usual definition of the correlation tensor,

$$P_{ij}(\mathbf{k}, t, t_0) = \langle \delta B_i(\mathbf{k}, t) \delta B_j^*(\mathbf{k}, t_0) \rangle,$$

and the correlation tensor $P_{ij}(\mathbf{k}, t, t_0)$ is expressed in wave number space. The correlation tensor (5.23) then reduces to

$$R_{ij}(\mathbf{r}, t, \mathbf{r}', t_0) = \int d^3k P_{ij}(\mathbf{k}, t, t_0) e^{i\mathbf{k}\cdot(\mathbf{r}-\mathbf{r}')}.$$

On setting $t_0 = 0$ and $\mathbf{r}' = 0$, we have

$$P_{ij}(\mathbf{k}, t) = \langle \delta B_i(\mathbf{k}, t) \delta B_j^*(\mathbf{k}, 0) \rangle,$$

with

$$R_{ij}(\mathbf{r}, t) = \int d^3k P_{ij}(\mathbf{k}, t) e^{i\mathbf{k}\cdot\mathbf{r}}. \quad (5.24)$$

Although we restrict ourselves to stationary turbulence, we note that the inclusion of temporal effects in the correlation tensor is typically accomplished by assuming that the correlation tensor has a separable form in the spatial and temporal components,

$$P_{ij}(\mathbf{k}, t) = P_{ij}(\mathbf{k}, 0) \Gamma(\mathbf{k}, t),$$

where $\Gamma(\mathbf{k}, t)$ is a dynamical correlation function and $P_{ij}(\mathbf{k}, 0) \equiv P_{ij}(\mathbf{k})$ is the magnetostatic correlation tensor.

5.3 The Magnetic Correlation Tensor

For completeness, we first consider isotropic turbulence. The general form of an isotropic rank-2 tensor is[4]

$$P_{ij}(\mathbf{k}) = A(k)\delta_{ij} + B(k)k_i k_j + C(k)\sum_k \varepsilon_{ijk} k_k.$$

Recall that ε_{ijk} is the Levi-Civita or unit alternating tensor and has values $\varepsilon_{ijk} = 0$ if any of i, j, and k are repeated, $\varepsilon_{ijk} = +1$ or -1 when i, j, and k are all different and in cyclic or acyclic order respectively.

Since $\nabla \cdot \delta \mathbf{B} = 0$,

$$\sum_i k_i \delta B_i(\mathbf{k}) = 0,$$

which yields

$$\sum_{i,j} \langle k_i \delta B_i k_j \delta B_j^* \rangle = \sum_{i,j} k_i k_j P_{ij} = 0.$$

If we substitute the general form P_{ij} of an isotropic rank-2 tensor, it therefore follows immediately that for magnetic turbulence

$$0 = A(k)\sum_{i,j} k_i k_j \delta_{ij} + B(k)\sum_{i,j} k_i^2 k_j^2 + C(k)\sum_{i,j,k} \varepsilon_{ijk} k_i k_j k_k$$

$$= A(k)k^2 + B(k)k^4,$$

and hence that

$$B(k) = -\frac{A(k)}{k^2}.$$

The general form of the magnetic isotropic tensor is therefore

$$P_{ij}(\mathbf{k}) = A(k)\left(\delta_{ij} - \frac{k_i k_j}{k^2}\right) + C(k)\sum_k \varepsilon_{ijk} k_k.$$

Since

$$P_{ij}(\mathbf{k}) = \langle \delta B_i \delta B_j^* \rangle = \langle \delta B_i^* \delta B_j \rangle^* = \langle \delta B_j \delta B_i^* \rangle^* = P_{ji}^*(\mathbf{k}),$$

---

[4]Batchelor (1953)

we have

$$P_{ji}^* = A^*(k)\left(\delta_{ji} - \frac{k_j k_i}{k^2}\right) + C^*(k)\sum_k \varepsilon_{jik} k_k$$

$$= A^*(k)\left(\delta_{ij} - \frac{k_i k_j}{k^2}\right) - C^*(k)\sum_k \varepsilon_{ijk} k_k$$

$$= P_{ij}(\mathbf{k}) = A(k)\left(\delta_{ij} - \frac{k_i k_j}{k^2}\right) + C(k)\sum_k \varepsilon_{ijk} k_k,$$

after using $\varepsilon_{jik} = -\varepsilon_{ijk}$. We therefore have $A(k) = A^*(k)$, i.e., $A(k)$ is real, and $C(k) = -C^*(k)$, implying that $C(k)$ is imaginary. Quite generally, we can express

$$C(k) = iA(k)\frac{\sigma(k)}{k},$$

to obtain

$$P_{ij}(\mathbf{k}) = A(k)\left[\delta_{ij} - \frac{k_i k_j}{k^2} + i\sigma(k)\sum_k \varepsilon_{ijk}\frac{k_k}{k}\right], \quad (5.25)$$

where $A(k)$ and $\sigma(k)$ are real, and $\sigma(k)$ is known as the *magnetic helicity*. Appropriate models for $A(k)$ and $\sigma(k)$ must be given.

Let us reconsider now the correlation tensor in the presence of magnetic turbulence that is axisymmetric with respect to a preferred direction; typically the z-axis along which the uniform mean magnetic field is assumed to be oriented. In this case, it can be shown (not done here, see Matthaeus and Smith (1981)) that the isotropic form of the correlation tensor also holds for axisymmetric turbulence,

$$P_{ij}(\mathbf{k}) = A(k_\parallel, k_\perp)\left[\delta_{ij} - \frac{k_i k_j}{k^2} + i\sigma(k_\parallel, k_\perp)\sum_k \varepsilon_{ijk}\frac{k_k}{k}\right].$$

In most applications to cosmic ray or energetic particle transport, the magnetic helicity term is neglected, as is the parallel component of the turbulent magnetic field δB_z. In this case, the correlation tensor reduces to

$$P_{ij}(\mathbf{k}) = A(k_\parallel, k_\perp)\left[\delta_{ij} - \frac{k_i k_j}{k^2}\right], \quad (5.26)$$

where $i, j = x, y$ and $P_{iz} = 0 = P_{zj}$.

To complete the correlation tensor for use in a transport equation describing particle scattering in a turbulent magnetic field, we need to specify both the geometry of the magnetic turbulence and the spectrum of the magnetic field fluctuations. This

5.3 The Magnetic Correlation Tensor

will allow us to model the function $A(k_\parallel, k_\perp)$. Three possible geometries, besides the isotropic case discussed already, are possible in the interplanetary (and possibly interstellar) environment. The first is the *slab model*, which is a one-dimensional model in that the turbulent magnetic field depends only on the z-coordinate

$$\delta B_i^{slab}(\mathbf{r}) = \delta B_i^{slab}(z),$$

allowing us to express the function

$$A^{slab}(k_\parallel, k_\perp) = g^{slab}(k_\parallel) \frac{\delta(k_\perp)}{k_\perp}.$$

For the slab model, the wave vectors are parallel to the mean magnetic field, i.e., $\mathbf{k} \parallel \mathbf{B}_0$.

Alternatively, we can consider a 2D or *perpendicular turbulence model* in which the turbulent field is a function of the perpendicular coordinates (x, y) only, i.e.,

$$\delta B_i^{2D}(\mathbf{r}) = \delta B_i^{2D}(x, y),$$

so that

$$A^{2D}(k_\parallel, k_\perp) = g^{2D}(k_\perp) \frac{\delta(k_\parallel)}{k_\perp}.$$

In this case, the wave vectors are orthogonal to the mean magnetic field, $\mathbf{k} \perp \mathbf{B}_0$, and therefore lie in a 2D plane perpendicular to the mean field.

Finally, one can construct a *two-component model* that corresponds to a superposition of the slab and 2D models. This model is quasi-3D and

$$\delta B_i^{comp}(\mathbf{r}) = \delta B_i^{2D}(x, y) + \delta B_i^{slab}(z).$$

Because we have

$$\langle \delta B_i^{slab}(z) \delta B_i^{*,2D}(x, y) \rangle = 0,$$

the correlation tensor has the form

$$P_{ij}^{comp}(\mathbf{k}) = P_{ij}^{slab}(\mathbf{k}) + P_{ij}^{2D}(\mathbf{k}).$$

In addition to the underlying geometry of the assumed interplanetary or interstellar turbulence, we need to specify the wave number dependence of $A(k_\parallel, k_\perp)$, i.e., the wave number spectrum. For the slab model, this requires that we prescribe $g^{slab}(k_\parallel)$ and similarly $g^{2D}(k_\perp)$ for the 2D model. A typical spectrum observed in the solar wind has three distinct regions: (i) an *energy containing range* at small wave numbers (i.e., large scales), and is typically of the form k^{-1}. The energy

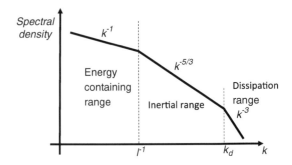

Fig. 5.2 Schematic of the wave number spectrum observed typically in the solar wind, illustrating the energy containing range, the inertial range, and the dissipation range. The bendover scale ℓ^{-1} and the dissipation scale k_d are identified

range is defined by a *bendover* or *turnover* scale such that $g^{slab/2D}(k_{slab/2D} \leq \ell^{-1}_{slab/2D})$ = energy range of the spectrum, depending on whether the turbulence is of the slab or 2D kind. (ii) At larger wave number scales, energy in turbulent fluctuations is transferred locally from larger to smaller scales in a self-similar manner. This part of the spectrum is called the *inertial range* and typically has the form $k^{-5/3}$, which is the Kolmogorov form of the spectrum.[5] For the inertial range, we introduce a dissipation wave number $k_{d,slab/2D}$ and defined the spectrum by $g^{slab/2D}(\ell^{-1}_{slab/2D} \leq k_{slab/2D} \leq k_{d,slab/2D})$ = inertial range of the spectrum. (iii) Finally, for large wave numbers or small scales, the turbulence loses energy through dissipation, and so this part of the spectrum is called the *dissipation range*, and is much steeper than the rest of the spectrum, typically k^{-3}. The dissipation range may be defined as $g^{slab/2D}(k_{d,slab/2D} \leq k_{slab/2D})$ = dissipation range of the spectrum (see Fig. 5.2 for a schematic illustration and Fig. 5.3 for several examples observed in the solar wind).

In most studies of energetic particle transport, the dissipation range plays very little role and is therefore neglected typically. The energy and inertial ranges are however critical in determining particle transport properties and a useful analytic form of the wave number spectrum for magnetic (and other) fluctuations is

$$g^i(k_i) \sim \left(1 + k_i^2 \ell_i^2\right)^{-\nu}, \quad i \equiv \text{slab or 2D}.$$

This form of the spectrum contains both the energy range (modeled as a constant) and an inertial range with slope $k^{-2\nu}$ defined by a bendover scale ℓ_i.

An important quantity used to characterize turbulence and closely related to the bendover scale is the correlation length, defined by the following integral,

$$\ell_{c,ij}\left\langle \delta B_j^2 \right\rangle = \int_0^\infty R_{jj}(\mathbf{r}, 0) dr_i.$$

The correlation length represents the characteristic length scale for the spatial decorrelation of turbulence. Hence, $\ell_{c,ij}\delta B_j^2$ is simply the area under the correlation function R_{ii}. Clearly, the correlation length depends intimately on the nature of the turbulence and wave number spectrum through the correlation function.

---

[5]Kolmogorov (1941).

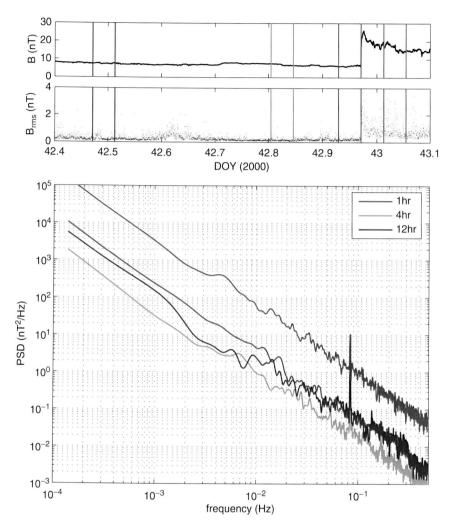

Fig. 5.3 Example of spectra upstream and downstream of a perpendicular interplanetary shock wave (Zank et al. 2006)

Consider now the correlation function for slab turbulence, assuming that $\Gamma(\mathbf{r}, t) = 1$, i.e., magnetostatic turbulence. Turbulent magnetic fluctuations vary only along the direction of the mean magnetic field z, so

$$R_{ij}^{slab} = \langle \delta B_i(z) \delta B_j^*(0) \rangle,$$

assuming $z(0) = 0$ because of homogeneous turbulence. On using the form of the axisymmetric magnetic correlation tensor, and the results from the geometric form of $A(k_\parallel, k_\perp)$, we find

$$P_{ij}(\mathbf{k}) = g^{slab}(k_\parallel) \frac{\delta(k_\perp)}{k_\perp} \left(\delta_{ij} - \frac{k_i k_j}{k^2} \right)$$

$$= g^{slab}(k_\parallel) \frac{\delta(k_\perp)}{k_\perp} \delta_{ij} \quad \text{if } i, j = x, y,$$

and $P_{iz} = 0 = P_{zj}$. If we assume the general form of the turbulence spectrum above, we can express g^{slab} as

$$g^{slab}(k_\parallel) = \frac{C(\nu)}{2\pi} \ell_{slab} \langle \delta B^2_{slab} \rangle \left(1 + k_\parallel^2 \ell^2_{slab} \right)^{-\nu}, \tag{5.27}$$

where the normalization constant has to be determined. Thus, using cylindrical coordinates $k_x = k_\perp \cos\theta$, $k_y = k_\perp \sin\theta$, $k_z = k_\parallel$ to express the wave vector, we find

$$\langle \delta B^2_{slab} \rangle = \langle \delta B^2_x \rangle + \langle \delta B^2_y \rangle = R_{xx}(0) + R_{yy}(0) = \int d^3k \left[P_{xx}(\mathbf{k}) + P_{yy}(\mathbf{k}) \right]$$

$$= 2 \int_0^{2\pi} \int_0^\infty \int_{-\infty}^\infty g^{slab}(k_\parallel) \frac{\delta(k_\perp)}{k_\perp} k_\perp d\theta dk_\perp dk_\parallel$$

$$= 8\pi \int_0^\infty g^{slab}(k_\parallel) dk_\parallel.$$

On using (5.27), we find

$$C^{-1}(\nu) = 4\ell_{slab} \int_0^\infty \left(1 + k_\parallel^2 \ell^2_{slab} \right)^{-\nu} dk_\parallel,$$

$$= 2 \int_0^\infty t^{-1/2} (1-t)^{-\nu} dt,$$

after using the change of variables $t = k_\parallel^2 \ell^2_{slab}$. This integral is the beta function (related to the gamma function $\Gamma(x)$) defined by $B(x, y) \equiv \int_0^\infty t^{x-1}/(1+t)^{x+y} dt$, $x > 0$, $y > 0$, and $B(x, y) = \Gamma(x)\Gamma(y)/\Gamma(x+y)$. Thus, setting $x = \frac{1}{2}$, $y = \nu - \frac{1}{2}$ yields

$$C(\nu) = \frac{1}{2\sqrt{\pi}} \frac{\Gamma(\nu)}{\Gamma(\nu - 1/2)},$$

since $\Gamma(1/2) = \sqrt{\pi}$.

The slab correlation function can now be calculated using (5.24)

$$R^{slab}_{xx}(z) = \langle \delta B_x(z) \delta B^*_x(0) \rangle = \int d^3k P^{slab}_{xx} \cos(k_\parallel z)$$

$$= 4\pi \int_0^\infty g^{slab}(k_\parallel) \cos(k_\parallel z) dk_\parallel$$

$$= 2C(\nu) \langle \delta B^2_{slab} \rangle \int_0^\infty (1+x^2)^{-\nu} \cos(ax) dx,$$

5.3 The Magnetic Correlation Tensor

where $x \equiv k_\parallel \ell_{slab}$ and $a \equiv z/\ell_{slab}$. The last integral is of a standard tabulated form

$$\int_0^\infty (1+x^2)^{-\nu} \cos(ax) dx = \frac{\sqrt{\pi}}{\Gamma(\nu)} \left(\frac{2}{a}\right)^{1/2-\nu} K_{\nu-1/2}(a),$$

where $K_\beta(z)$ is the modified Bessel function of imaginary argument. The perpendicular correlation function $R_\perp = R_{xx} + R_{yy}$ can therefore be expressed as

$$R_\perp^{slab} = \frac{2\langle \delta B_{slab}^2 \rangle}{\Gamma(\nu - 1/2)} \left(\frac{2\ell_{slab}}{z}\right)^{1/2-\nu} K_{\nu-1/2}\left(\frac{z}{\ell_{slab}}\right). \tag{5.28}$$

Shalchi provides two useful asymptotic forms[6] for the slab correlation function in the limits $z \ll \ell_{slab}$ and $z \gg \ell_{slab}$, respectively

$$K_{\nu-1/2}(z \ll \ell_{slab}) \simeq \frac{1}{2}\Gamma(\nu - 1/2)\left(\frac{2\ell_{slab}}{z}\right)^{\nu-1/2},$$

$$\implies R_\perp^{slab}(z \ll \ell_{slab}) = \langle \delta B_{slab}^2 \rangle \quad \text{if } \nu > 1/2;$$

$$K_{\nu-1/2}(z \gg \ell_{slab}) \simeq \sqrt{\frac{\pi \ell_{slab}}{2z}} e^{-z/\ell_{slab}},$$

$$\implies R_\perp^{slab}(z \ll \ell_{slab}) = \frac{\sqrt{\pi}}{\Gamma(\nu - 1/2)} \langle \delta B_{slab}^2 \rangle \left(\frac{2\ell_{slab}}{z}\right)^{1-\nu} e^{-z/\ell_{slab}}.$$

The bendover scale ℓ_{slab} is the characteristic length scale for the spatial decorrelation of the turbulence for the exponentially decaying correlation function in the limit $z \gg \ell_{slab}$.

The slab correlation length can also be computed, and this illustrates the relationship between $\ell_{c,slab}$ and the bendover scale length ℓ_{slab}. Recall from the definition of $\ell_{c,slab}$

$$\ell_{c,slab} \langle \delta B_{slab}^2 \rangle = \int_0^\infty R_\perp^{slab}(z) dz$$

$$= 2\pi \int_{-\infty}^\infty dk_\parallel g^{slab}(k_\parallel) \int_{-\infty}^\infty dz e^{ik_\parallel z}$$

$$= (2\pi)^2 \int_{-\infty}^\infty dk_\parallel g^{slab}(k_\parallel) \delta(k_\parallel)$$

$$= (2\pi)^2 g^{slab}(0) = 2\pi C(\nu) \ell_{slab} \langle \delta B_{slab}^2 \rangle,$$

---

[6]Useful limits of these and many other related functions are tabulated in Abramowitz and Stegun (1974). For this case, A. Shalchi used the formulae (9.6.9) and (9.7.2).

since $\int dz e^{ik_\| z} = 2\pi\delta(k_\|)$. Thus the slab correlation length and the bendover scale are related via

$$\ell_{c,slab} = 2\pi C(\nu)\ell_{slab},$$

which if we assume a Kolmogorov power law for the inertial range, $\nu = 5/6$, we have $C(5/6) = 0.1188$ and hence $\ell_{c,slab} \simeq 0.75\ell_{slab}$.

The 2D magnetostatic correlation function is a little more laborious to compute. Since $\delta B_i(\mathbf{r}) = \delta B_i(x, y)$, the 2D correlation tensor is given by

$$R_{ij}^{2D}(x, y) = \langle \delta B_i(x, y)\delta B_j(0, 0)\rangle,$$

or

$$R_{xx}(x, y) = \int d^3k P_{xx}(\mathbf{k})e^{i\mathbf{k}\cdot\mathbf{r}} = \int d^3k P_{xx}(\mathbf{k})e^{ik_x x + ik_y y},$$

and we have

$$P_{ij}^{2D}(\mathbf{k}) = g^{2D}(k_\perp)\frac{\delta(k_\|)}{k_\perp}\left(\delta_{ij} - \frac{k_i k_j}{k^2}\right) \quad \text{if } i, j = x, y,$$

$$\text{or} \quad = 0 \quad \text{if } i \text{ or } j = z.$$

For the wave spectrum, we assume the same normalized form as for the slab case except that we introduce the 2D counterparts ℓ_{2D} and $\langle\delta B_{2D}^2\rangle$,

$$g^{2D}(k_\perp) = \frac{C(\nu)}{2\pi}\ell_{2D}\langle\delta B_{2D}^2\rangle\left(1 + k_\perp^2 \ell_{2D}^2\right)^{-\nu}.$$

On introducing cylindrical coordinates for the wave vector and position

$$k_x = k_\perp \cos\Psi, \quad k_y = k_\perp \sin\Psi;$$
$$x = r\cos\Phi, \quad y = r\sin\Phi,$$

we find

$$R_{xx}^{2D}(x, y) = \int_0^{2\pi}\int_0^\infty\int_{-\infty}^\infty g^{2D}(k_\perp)\frac{\delta(k_\|)}{k_\perp}\left(1 - \frac{k_\perp^2}{k^2}\cos^2\Psi\right)e^{i\mathbf{k}\cdot\mathbf{r}}k_\perp d\Psi dk_\perp dk_\|$$

$$= \int_0^\infty dk_\perp g^{2D}(k_\perp)\int_0^{2\pi} d\Psi \sin^2\Psi \exp[ik_\perp r(\cos\Phi\cos\Psi + \sin\Phi\sin\Psi)]$$

$$= \int_0^\infty dk_\perp g^{2D}(k_\perp)\int_0^{2\pi} d\Psi \sin^2\Psi \exp[ik_\perp r\cos(\Phi - \Psi)].$$

5.3 The Magnetic Correlation Tensor

A standard simplification of these integrals makes use of a series expansion in terms of Bessel functions,

$$e^{ix\sin\alpha} = \sum_{n=-\infty}^{\infty} J_n(x) e^{in\alpha}, \quad e^{ix\cos\alpha} = \sum_{n=-\infty}^{\infty} J_n(x) e^{in(\alpha+\pi/2)},$$

which allows us to obtain

$$R_{xx}^{2D}(x,y) = \int_0^\infty dk_\perp g^{2D}(k_\perp) \sum_{n=-\infty}^{\infty} J_n(k_\perp r) \int_0^{2\pi} d\Psi \sin^2\Psi e^{-in\Psi} e^{in(\Phi+\pi/2)}.$$

The corresponding expression for R_{yy} is given by

$$R_{yy}^{2D}(x,y) = \int_0^\infty dk_\perp g^{2D}(k_\perp) \sum_{n=-\infty}^{\infty} J_n(k_\perp r) \int_0^{2\pi} d\Psi \cos^2\Psi e^{-in\Psi} e^{in(\Phi+\pi/2)},$$

meaning that

$$R_\perp^{2D}(x,y) = \int_0^\infty dk_\perp g^{2D}(k_\perp) \sum_{n=-\infty}^{\infty} J_n(k_\perp r) \int_0^{2\pi} d\Psi e^{-in\Psi} e^{in(\Phi+\pi/2)}.$$

Since

$$\int_0^{2\pi} d\Psi e^{\pm in\Psi} = 2\pi \delta_{n0},$$

the 2D perpendicular correlation function reduces to

$$R_\perp^{2D}(x,y) = 2\pi \int_0^\infty dk_\perp g^{2D}(k_\perp) J_0(k_\perp r),$$

which can be further expressed as (using as before $x \equiv k_\perp \ell_{2D}$ and $a \equiv r/\ell_{2D}$)

$$R_\perp^{2D}(r) = 4C(\nu) \langle \delta B_{2D}^2 \rangle \int_0^\infty (1+x^2)^{-\nu} J_0(ax) dx. \tag{5.29}$$

As before, it is instructive to consider the limits $a = 0$ and $a \to \infty$. The former limit yields ($J_0(0) = 1$)

$$\int_0^\infty (1+x^2)^{-\nu} dx = (4C(\nu))^{-1} \Rightarrow R_\perp^{2D}(r=0) = \langle \delta B_{2D}^2 \rangle.$$

The latter limit yields ($\int_0^\infty J_0(y)dy = 1$)

$$\int_0^\infty (1+x^2)^{-\nu} J_0(ax)dx = \frac{1}{a}\int_0^\infty \left(1+\frac{y^2}{a^2}\right)^{-\nu} J_0(y)dy$$

$$\simeq \frac{1}{a}\int_0^\infty J_0(y)dy = \frac{1}{a};$$

$$\Rightarrow R_\perp^{2D}(r \gg \ell_{2D}) = 4C(\nu)\langle\delta B_{2D}^2\rangle \frac{\ell_{2D}}{r}.$$

Note that the spatial decorrelation length for the turbulence is determined by the 2D bendover scale ℓ_{2D}. Notice too that although the same forms of the wave number spectrum were used for both the slab and 2D cases, the correlation functions are nonetheless different, with the 2D correlation function decaying more slowly with increasing distance compared to the slab case (which falls off exponentially).

As before, we can relate the 2D correlation length $\ell_{c,2D}$ to the bendover scale ℓ_{2D}. In this case, we need to introduce a minimum wave number, $x_{min} \equiv \ell_{2D}/L_{2D}$, to avoid a divergent integral,

$$\ell_{c,2D} = \frac{1}{\langle\delta B_{2D}^2\rangle}\int_0^\infty R_\perp(r)dr$$

$$= 4C(\nu)\int_{x_{min}}^\infty dx(1+x^2)^{-\nu}\int_0^\infty drJ_0\left(\frac{xr}{\ell_{2D}}\right)$$

$$= 4C(\nu)\ell_{2D}\int_{x_{min}}^\infty \frac{dx}{x}(1+x^2)^{-\nu}$$

$$\simeq 4C(\nu)\ell_{2D}\left(\int_{x_{min}}^1 \frac{dx}{x} + \int_1^\infty x^{-2\nu-1}dx\right)$$

$$\simeq 4C(\nu)\ell_{2D}\left(\frac{1}{2\nu} + \ln\frac{L_{2D}}{\ell_{2D}}\right).$$

The wave spectrum used here is normalized correctly only if $L_{2D} \gg \ell_{2D}$, and in the limit of an infinitely large box, $L_{2D} \to \infty$, the correlation length is infinite.

5.4 Quasi-linear Transport Theory of Charged Particle Transport: Derivation of the Scattering Tensor

We have so far prescribed a very simple diffusion in pitch-angle expression to describe the scattering of particles by in situ magnetic fluctuations. In this and the next section, we derive expressions that describe the scattering of energetic particles in low-frequency magnetic turbulence.

5.4 Quasi-linear Transport Theory of Charged Particle Transport... 227

Since we consider particles that can have high energies, we begin with the momentum form of the Vlasov or collisionless Boltzmann equation

$$\frac{\partial f}{\partial t} + \frac{\mathbf{p}}{m} \cdot \nabla f + q\left(\mathbf{E} + \frac{\mathbf{p} \times \mathbf{B}}{m}\right) \cdot \frac{\partial f}{\partial \mathbf{p}} = 0, \qquad (5.30)$$

for particles of mass m and charge q. Following le Roux et al. (2004)[7] we use a quasi-linear approach to derive a Fokker-Planck kinetic transport equation for the diffusion of charged particles experiencing scattering in pitch-angle and momentum space due to the presence of Alfvénic/slab and quasi-2D turbulence in the solar wind. Quasi-linear theory proceeds essentially by assuming that charged particle gyro-orbits are only weakly perturbed by electromagnetic fluctuations. Typically, there are three ways to proceed. One can proceed from the formalism discussed in the derivation of the Fokker-Planck equation from the Chapman-Kolmogorov equation, assuming a Markovian process, and evaluate the diffusion coefficients directly. A second approach, which we follow here, is to directly expand Eq. (5.30) to determine the diffusion coefficients. A third approach is to work directly from the Newton-Lorentz equations for particle motion in a fluctuating electromagnetic field and directly compute momentum and spatial diffusion coefficients from the Taylor-Green-Kubo (TGK) forms,[8]

$$D_{\mu\mu}(\mu) \equiv \int_0^\infty dt \langle \dot{\mu}(t) \dot{\mu}(t) \rangle;$$

$$D_{ij}(\mu) \equiv \int_0^\infty dt \langle v_i(t) v_j(t) \rangle,$$

where μ is the cosine of the particle pitch angle and $\mathbf{v}$ is the particle velocity.

Several assumptions are made explicitly to ensure the validity of the quasi-linear approximation. The first is that the electromagnetic fluctuations are of small amplitude. This ensures that particles follow approximately undisturbed helical orbits on a particle correlation time τ_c^p, which is the characteristic time for a particle to gyrate on an undisturbed trajectory before being disturbed by incoherent or random fluctuations. This obviously means that the particle correlation time is much less than the characteristic time scale for particle pitch-angle scattering τ_μ i.e., $\tau_c^p \ll \tau_\mu$. The time scale over which particle orbits are significantly distorted by pitch-angle scattering is therefore much longer than the particle correlation time scale on which a coherent helical orbit is maintained.

In Eq. (5.30), we may expand the electromagnetic fields, $\mathbf{E}$ and $\mathbf{B}$, the flow velocity $\mathbf{u}$, and the distribution f into mean and fluctuating parts using a *mean field decomposition*, i.e., a field or scalar Q is may be decomposed as $Q = Q_0 + \delta Q$

---

[7] See also le Roux and Webb (2007).
[8] See Shalchi (2009) for a general discussion of this approach.

such that the ensemble average $\langle Q \rangle = Q_0$ and $\langle \delta Q \rangle = 0$. It does not necessarily follow that $\delta Q \ll Q_0$, although in quasi-linear theory, this assumption is made to eliminate second-order and higher correlations. Hence,

$$\mathbf{E} = \mathbf{E}_0 + \delta \mathbf{E}, \quad \langle \delta \mathbf{E} \rangle = 0;$$
$$\mathbf{B} = \mathbf{B}_0 + \delta \mathbf{B}, \quad \langle \delta \mathbf{B} \rangle = 0;$$
$$\mathbf{u} = \mathbf{u}_0 + \delta \mathbf{u}, \quad \langle \delta \mathbf{u} \rangle = 0;$$
$$f = f_0 + \delta f, \quad \langle \delta f \rangle = 0.$$

The fields are assumed to vary smoothly on the large scale L, and randomly varying fluctuations occur on the smaller correlation length scale $\ell_c \ll L$. The power spectrum of fluctuations ranges from scales on the order of the correlation length to smaller than the particle gyroradius r_g. In the analysis here, we assume an infinitely extended wave number power spectrum for simplicity, rather than include the details of the dissipation range part of the spectrum. The total electric field, in the MHD approximation, is

$$\mathbf{E} = -\mathbf{u} \times \mathbf{B},$$

where $\mathbf{u}$ and $\mathbf{B}$ are measured in the observer's frame. Applying the small amplitude assumption to the mean field decomposition of the electric field $\mathbf{E}$ yields

$$\mathbf{E}_0 = -\mathbf{u}_0 \times \mathbf{B}_0, \quad \text{and} \quad \delta \mathbf{E} = -\mathbf{u}_0 \times \delta \mathbf{B} - \delta \mathbf{u} \times \mathbf{B}_0,$$

after neglecting quadratically small terms ($\delta \mathbf{u} \ll \mathbf{u}_0$, and $\delta \mathbf{B} \ll \mathbf{B}_0$). We will neglect the induced turbulent electric field $\delta \mathbf{E}$ (although see le Roux et al. for the case where this term is retained). We will make the assumption that the particle distribution is co-moving with the background plasma frame, so that the mean motional electric field term is zero, $\mathbf{E}_0 = 0$.

The mean field decomposition above is substituted into the collisionless Boltzmann equation (5.30). The ensemble averaged form of this equation is then subtracted from the full, unaveraged transport equation (5.30), yielding a transport equation for the rapidly fluctuating variable δf. This equation contains the differences of second-order terms and their corresponding ensemble averages. Since we assume from the outset that $\delta f \ll f_0$, $\delta \mathbf{B} \ll \mathbf{B}_0$, the quadratic terms are small and can be neglected (Exercise). The linearized equation for the correction δf is

$$\frac{\partial}{\partial t} \delta f + \frac{\mathbf{p}}{m} \cdot \nabla \delta f + (\mathbf{p} \times \mathbf{\Omega}) \cdot \frac{\partial \delta f}{\partial \mathbf{p}} = -q \frac{\mathbf{p} \times \delta \mathbf{B}}{m} \cdot \frac{\partial f_0}{\partial \mathbf{p}}, \tag{5.31}$$

where $\mathbf{\Omega} = q\mathbf{B}_0/m$ is the particle gyrofrequency. The corresponding mean-field equation for the distribution function f_0 is given by

$$\frac{\partial f_0}{\partial t} + \frac{\mathbf{p}}{m} \cdot \nabla f_0 + (\mathbf{p} \times \mathbf{\Omega}) \cdot \frac{\partial f_0}{\partial \mathbf{p}} = -q \left\langle \frac{\mathbf{p} \times \delta \mathbf{B}}{m} \cdot \frac{\partial \delta f}{\partial \mathbf{p}} \right\rangle, \tag{5.32}$$

5.4 Quasi-linear Transport Theory of Charged Particle Transport...

where the right-hand nonlinear term describes the perturbing effect of the fluctuating magnetic field on the scattered particle distribution. As we illustrate below, this term introduces a diffusion coefficient in pitch-angle space. The closure of (5.32) can be affected by solving the quasi-linear equation (5.31) for δf, and then evaluating the ensemble-averaged term in (5.32).

Consider a homogeneous, infinitely extended plasma system with Cartesian coordinates (x, y, z) with the z-axis aligned with the mean magnetic field $\mathbf{B}_0 = B_0 \hat{\mathbf{z}}$. Since the turbulence is comprised of slab turbulence with wave vectors along the mean magnetic field and 2D turbulence with fluctuations and wave vectors transverse to $\mathbf{B}_0$, we have

$$\delta \mathbf{B}(x, y, z) = \delta B_x(x, y)\hat{\mathbf{x}} + \delta B_y(x, y)\hat{\mathbf{y}} + \delta B_x(z)\hat{\mathbf{x}} + \delta B_y(z)\hat{\mathbf{y}},$$

where the 2D component $\delta B_{x/y}(x, y)$ describes the magnetic field fluctuations that convect with the background flow. The second set of terms $\delta B_{x/y}(z)$ comprises the slab or Alfvénic component. For notational convenience, we express magnetic field variations as $\delta B_{x/y}$ and this includes both the slab and 2D components.

The Cartesian form of the momentum coordinates $\mathbf{p} = (p_x, p_y, p_z)$ in the mean-field aligned co-moving coordinate system (p_z is along the mean-field direction) can be expressed in terms of spherical coordinates, so that $\mathbf{p} = p(\sin\theta\cos\phi, \sin\theta\sin\phi, \cos\theta)$, where p is the particle momentum magnitude, θ the particle pitch-angle, and ϕ the particle phase angle. Consider the right-hand side of (5.31),

$$(\mathbf{p} \times \delta\mathbf{B}) \cdot \nabla_p f_0 = p(\delta B_z \sin\theta\sin\phi - \delta B_y \cos\theta, \delta B_x \cos\theta - \delta B_z \sin\theta\cos\phi,$$
$$\delta B_y \sin\theta\cos\phi - \delta B_x \sin\theta\sin\phi) \cdot \nabla_p f_0$$
$$= (\mathbf{p} \times \delta\mathbf{B})_x \frac{\partial f_0}{\partial p_x} + (\mathbf{p} \times \delta\mathbf{B})_y \frac{\partial f_0}{\partial p_y} + (\mathbf{p} \times \delta\mathbf{B})_z \frac{\partial f_0}{\partial p_z}.$$

On using the results,

$$\frac{\partial}{\partial p_x} = \sin\theta\cos\phi \frac{\partial}{\partial p} + \cos\theta\cos\phi \frac{1}{p}\frac{\partial}{\partial \theta} - \frac{\sin\phi}{p\sin\theta}\frac{\partial}{\partial \phi};$$
$$\frac{\partial}{\partial p_y} = \sin\theta\sin\phi \frac{\partial}{\partial p} + \cos\theta\sin\phi \frac{1}{p}\frac{\partial}{\partial \theta} + \frac{\cos\phi}{p\sin\theta}\frac{\partial}{\partial \phi}; \quad (5.33)$$
$$\frac{\partial}{\partial p_z} = \cos\theta \frac{\partial}{\partial p} - \sin\theta \frac{1}{p}\frac{\partial}{\partial \theta},$$

we find that

$$-\frac{q}{m}(\mathbf{p} \times \delta\mathbf{B} \cdot \nabla_p f_0) = -\frac{|\Omega|}{B}(\delta B_x \sin\phi - \delta B_y \cos\phi)\frac{\partial f_0}{\partial \theta},$$

and the coefficient of $\partial f_0/\partial p$ is identically zero. In deriving this result, we have invoked the further assumption that the ensemble averaged distribution function is gyrotropic i.e., is independent of the particle phase angle ϕ. Thus, effects such as diffusion perpendicular to the mean magnetic field and gradient and curvature drifts are neglected in this description of particle transport. This is equivalent to assuming that the particle gyroradius $r_g \ll \ell_c$, the correlation length of the turbulent fluctuations. Equivalently, this requires that the particle gyroperiod $\tau_g = \Omega^{-1} \ll \tau_c^p$.

The evolution equation for δf is a first-order quasi-linear equation and therefore can be solved using the method of characteristics. Accordingly, we have the following set of seven ordinary differential equations to solve,

$$\frac{d}{dt}\delta f = -\frac{|\Omega|}{B}\left(\delta B_x \sin\phi - \delta B_y \cos\phi\right)\frac{\partial f_0}{\partial \theta}; \tag{5.34}$$

$$\frac{d\mathbf{r}}{dt} = \frac{\mathbf{p}}{m}; \tag{5.35}$$

$$\frac{d\mathbf{p}}{dt} = \mathbf{p} \times \mathbf{\Omega}. \tag{5.36}$$

For particles located initially at $\mathbf{r}_0 = \mathbf{r}(t_0) = (x_0, y_0, z_0)$ with momentum p_0 and phase angle ϕ_0, we can solve the above odes to obtain

$$\phi(t') = \phi_0 - \Omega(t' - t_0); \quad x(t') = x_0 - r_g\left(\sin\phi(t') - \sin\phi_0\right);$$
$$y(t') = y_0 + r_g\left(\cos\phi(t') - \cos\phi_0\right); \quad z(t') = z_0 + v\cos\theta(t' - t_0);$$
$$\delta f(\mathbf{r}, \mathbf{p}, t) = \int_{t_0}^{t}\left(-\frac{|\Omega|}{B}\left(\delta B_x \sin\phi' - \delta B_y \cos\phi'\right)\frac{\partial f_0'}{\partial \theta}\right)dt' + \delta f(\mathbf{r}_0, \mathbf{p}_0, t_0), \tag{5.37}$$

where $r_g = v\sin\theta/\Omega$ is the particle gyroradius, and $\phi' \equiv \phi(t')$, $\delta B_i(\mathbf{r}(t'), t')$, and $f_0 = f_0(\mathbf{r}(t'), \mathbf{p}(t'), t')$. The particles evidently follow undisturbed helical orbits along $\mathbf{B}_0$ since p and θ are unchanged during the interaction period, this being less than the characteristic time scale for particles to interact with small-amplitude turbulence, viz. τ_c^p. Consequently, τ_c^p must be restricted so that $t' - t_0$ remains sufficiently small that $\delta f \ll f_0$.

The above expressions can be rewritten in terms of the time difference $\Delta t \equiv t - t'$, where t denotes the observation time and t' is the time during which the particle executes a helical trajectory. Hence, $\Delta t \in [t - t_0, 0]$ so this substitution implies that we follow the particle trajectory backward in time. Rewriting the solution for δf yields

$$\delta f(\mathbf{r}, \mathbf{p}, t) = \int_0^{t-t_0}\left(-\frac{\Omega}{B}\left(\delta B_x \sin\phi - \delta B_y \cos\phi\right)\frac{\partial f_0}{\partial \theta}\right)d(\Delta t) + \delta f(\mathbf{r}_0, \mathbf{p}_0, t_0),$$

5.4 Quasi-linear Transport Theory of Charged Particle Transport... 231

where $\phi = \phi(t-\Delta t)$, $\delta B_i(\mathbf{r}(t-\Delta t), t-\Delta t)$, and $f_0 = f_0(\mathbf{r}(t-\Delta t), \mathbf{p}(t-\Delta t), t-\Delta t)$. The expressions for the undisturbed particle orbits are now independent of the initial values, and are given by

$$\phi(t-\Delta t) = \phi(t) + \Omega(\Delta t); \quad x(t-\Delta t) = x(t) - r_g(\sin\phi(t-\Delta t) - \sin\phi(t));$$

$$y(t-\Delta t) = y(t) + r_g(\cos\phi(t-\Delta t) - \cos\phi(t)); \quad z(t-\Delta t)$$

$$= z(t) + -v\cos\theta(\Delta t).$$

Note that $t - t_0 \gg \tau_c^p$ and thus $|\mathbf{r} - \mathbf{r}_0| \gg \ell_c$. If $\lambda_\parallel$ denotes the parallel mean free path for the spatial diffusion of particles, then the assumption of small amplitude turbulence implies that $\ell_c \ll \lambda_\parallel$. The overall ordering of scales is therefore $r_g \ll \ell_c \ll \lambda_\parallel \ll L$.

Having obtained the solution δf, we can evaluate the ensemble-averaged collision term on the right-hand-side of (5.32). Introducing

$$\Psi_1 \equiv -\cos\phi\,\delta B_y + \sin\phi\,\delta B_x; \qquad \Psi_2 \equiv \sin\phi\,\delta B_y + \cos\phi\,\delta B_x,$$

we have

$$\frac{q}{m}\left\langle \mathbf{p}\times\delta\mathbf{B}\cdot\frac{\partial}{\partial\mathbf{p}}\delta f\right\rangle = \frac{|\Omega|}{B}\left\langle (-\cos\phi\,\delta B_y + \sin\phi\,\delta B_x)\frac{\partial}{\partial\theta}\delta f\right\rangle$$

$$+\frac{|\Omega|}{B}\left\langle \frac{\cos\theta}{\sin\theta}(\sin\phi\,\delta B_y + \cos\phi\,\delta B_x)\frac{\partial}{\partial\phi}\delta f\right\rangle$$

$$= \frac{|\Omega|}{B}\frac{\partial}{\partial\theta}\langle\delta f\,\Psi_1\rangle + \frac{|\Omega|}{B}\frac{\cos\theta}{\sin\theta}\frac{\partial}{\partial\phi}\langle\delta f\,\Psi_2\rangle$$

$$-\frac{|\Omega|}{B}\frac{\cos\theta}{\sin\theta}\left\langle\delta f\,\frac{\partial\Psi_2}{\partial\phi}\right\rangle$$

$$= \frac{|\Omega|}{B}\left(\frac{\partial}{\partial\theta}\langle\delta f\,\Psi_1\rangle + \frac{\cos\theta}{\sin\theta}\langle\delta f\,\Psi_1\rangle\right)$$

$$+\frac{|\Omega|}{B}\frac{\cos\theta}{\sin\theta}\frac{\partial}{\partial\phi}\langle\delta f\,\Psi_2\rangle$$

$$= \frac{|\Omega|}{B}\frac{1}{\sin\theta}\frac{\partial}{\partial\theta}(\sin\theta\,\langle\delta f\,\Psi_1\rangle) + \frac{|\Omega|}{B}\frac{\cos\theta}{\sin\theta}\frac{\partial}{\partial\phi}\langle\delta f\,\Psi_2\rangle,$$

after using $\partial\Psi_2/\partial\phi = -\Psi_1$. Since f_0 is independent of gyrophase, we neglect the last term. Thus, in spherical coordinates, we have the relation

$$-q\left\langle\frac{\mathbf{p}\times\delta\mathbf{B}}{m}\cdot\frac{\partial\delta f}{\partial\mathbf{p}}\right\rangle$$

$$= -\frac{1}{\sin\theta}\frac{\partial}{\partial\theta}\left(\frac{|\Omega|}{B}\sin\theta\,\langle(\delta B_x(\mathbf{r},t)\sin\phi(t) - \delta B_y(\mathbf{r},t)\cos\phi(t))\,\delta f\rangle\right).$$

On substituting for δf and using the relations,

$$\cos\phi(t) = \cos(\phi(t-\Delta t) - \Omega(\Delta t))$$
$$= \cos\phi(t-\Delta t)\cos(\Omega\Delta t) + \sin\phi(t-\Delta t)\sin(\Omega\Delta t);$$
$$\sin\phi(t) = \sin(\phi(t-\Delta t) - \Omega(\Delta t))$$
$$= \sin\phi(t-\Delta t)\cos(\Omega\Delta t) - \cos\phi(t-\Delta t)\sin(\Omega\Delta t).$$

we obtain a diffusion equation in particle pitch angle,

$$-q\left\langle \frac{\mathbf{p}\times\delta\mathbf{B}}{m} \cdot \frac{\partial \delta f}{\partial \mathbf{p}} \right\rangle = \frac{1}{\sin\theta}\frac{\partial}{\partial\theta}\left(D_{\theta\theta}\frac{\partial f_0}{\partial\theta}\right), \tag{5.38}$$

where the diffusion coefficient $D_{\theta\theta}$ is given by

$$D_{\theta\theta}(\mathbf{r},t) = \left(\frac{\Omega}{B}\right)^2 \sin\theta \int_0^\infty \left(R_{yy}c^2 - (R_{xy}-R_{yx})cs + R_{xx}s^2\right)$$
$$\times \cos(\Omega\Delta t)d(\Delta t)$$
$$+ \left(\frac{\Omega}{B_0}\right)^2 \sin\theta \int_0^\infty \left(R_{yy}cs + R_{xy}c^2 - R_{yx}s^2 - R_{xx}cs\right)\sin(\Omega\Delta t)d(\Delta t).$$

Here, $c \equiv \cos\phi(t-\Delta t) = \cos(\phi(t)+\Omega\Delta t)$ and $s = \sin\phi(t-\Delta t) = \sin(\phi(t)+\Omega\Delta t)$, and R_{ij} is the two-point, two-time correlation function for the magnetic fluctuations along the unperturbed particle orbit, i.e.,

$$R_{ij}(\Delta\mathbf{r}(\Delta t), \Delta t) \equiv \langle \delta B_i(0)\delta B_j(\Delta\mathbf{r}(\Delta t), \Delta t)\rangle.$$

We then have

$$R_{ij}(\mathbf{r},\mathbf{r}(t-\Delta t),t,t-\Delta t) = \langle \delta B_i(\mathbf{r},t), \delta B_j(\mathbf{r}(t-\Delta t),t-\Delta t)\rangle,$$

where the components of $\mathbf{r}(t-\Delta t)$ are determined above.

Observe that in deriving the diffusion form of the particle transport equation, we moved the pitch-angle derivative of the distribution function f_0 from under the integral in the expression for δf. There is an important implication embedded in the time scales associated with the ordering of particle scattering and diffusion, $\tau_c^p \ll \tau_\mu$. This ordering implies that $R_{ij} \to 0$ on a much shorter time scale than the time scale on which the particle orbit deviates from an undisturbed trajectory, implying that the integrand contributes only over the time τ_c^p rather than τ_μ to the time integration. Since the gyrotropic-independent distribution function f_0 varies on a time scale comparable to the pitch-angle diffusion time τ_μ, derivatives of f_0 can be taken out from under the integral. The second implication is that we can then extend the integral describing pitch-angle diffusion to ∞ ($t_0 \to -\infty$ in the expression for δf).

5.4 Quasi-linear Transport Theory of Charged Particle Transport...

The turbulence responsible for scattering the particles has been assumed to be axisymmetric about the mean magnetic field $\mathbf{B}_0 = B_0 \hat{\mathbf{z}}$. The axisymmetry condition for the correlation matrix $\mathbf{R}(\delta \mathbf{r})$ under an arbitrary rotation ϕ' about $\mathbf{B}_0 = B_0 \hat{\mathbf{z}}$ is expressed by

$$\mathbf{R}(\delta \mathbf{r}) = \mathbf{O} \mathbf{R}(\mathbf{O}^T \delta \mathbf{r}) \mathbf{O}^T,$$

where both the left- and right-hand sides are independent of ϕ', and $\mathbf{O}$ is the rotation matrix

$$\mathbf{O} = \begin{pmatrix} \cos \phi' & \sin \phi' & 0 \\ -\sin \phi' & \cos \phi' & 0 \\ 0 & 0 & 1 \end{pmatrix},$$

and $\mathbf{O}^T$ is the transpose. Hence, the elements of the left and right matrices

$$\mathbf{R}(\delta \mathbf{r}) = \begin{pmatrix} R_{xx} & R_{xy} & R_{xz} \\ R_{yx} & R_{yy} & R_{yz} \\ R_{zx} & R_{zy} & R_{zz} \end{pmatrix} = \mathbf{O} \mathbf{R}(\mathbf{O}^T \delta \mathbf{r}) \mathbf{O}^T$$

$$= \begin{pmatrix} R_{xx}c^2 + R_{xy}sc + R_{yx}sc + R_{yy}s^2 & -R_{xx}sc + R_{xy}c^2 - R_{yx}s^2 + R_{yy}sc & R_{xz}c + R_{yz}s \\ -R_{xx}sc - R_{xy}s^2 + R_{yx}c^2 + R_{yy}sc & R_{xx}s^2 - R_{xy}sc + R_{yx}sc + R_{yy}c^2 & -R_{xz}s + R_{yz}c \\ R_{zx}c + R_{zy}s & -R_{zx}s + R_{zy}c & R_{zz} \end{pmatrix},$$

(5.39)

are independent of ϕ'. Inspection of the axisymmetric matrix conditions show that the integrands of the diffusion coefficient $D_{\theta\theta}$ are therefore independent of ϕ'. Consequently, using $\phi' = \phi(t - \Delta t)$, we have

$$\delta x \equiv x(t - \Delta t) - x(t)$$
$$= r_g [\sin \phi(t) - \sin \phi(t - \Delta t)]$$
$$= r_g [\sin \phi(t - \Delta t)(\cos(\Omega \Delta t) - 1) - \cos \phi(t - \Delta t) \sin(\Omega \Delta t)]$$
$$= -r_g \sin(\Omega \Delta t),$$

etc. if we set $\phi(t - \Delta t) = 0$. This therefore yields

$$\delta \mathbf{r} = \left[-r_g \sin(\Omega \Delta t), r_g (1 - \cos(\Omega \Delta t)), -v \cos \theta (\Delta t) \right],$$

from which we find

$$\left(\mathbf{O}^T \delta \mathbf{r} \right)_x = -r_g [\cos \phi(t - \Delta t) \sin \Omega \Delta t + \sin \phi(t - \Delta t)(1 - \cos \Omega \Delta t)];$$
$$\left(\mathbf{O}^T \delta \mathbf{r} \right)_y = r_g [-\sin \phi(t - \Delta t) \sin \Omega \Delta t + \cos \phi(t - \Delta t)(1 - \cos \Omega \Delta t)];$$
$$\left(\mathbf{O}^T \delta \mathbf{r} \right)_y = -v \cos \theta \Delta t,$$

which corresponds to the unperturbed helical trajectories derived by substituting the trigonometric expansions for $\cos\phi(t)$ etc. as done above. Thus, for axisymmetric turbulence, the R_{ij} terms in the pitch-angle diffusion coefficient are independent of $\phi(t-\Delta t)$, so we may without loss of generality set $\phi(t-\Delta t) = \pi/2$, significantly simplifying the expression for the diffusion coefficient,

$$D_{\theta\theta} = \sin\theta \left(\frac{\Omega}{B_0}\right)^2 \int_0^\infty \left[\cos(\Omega\Delta t)R_{xx} - \sin(\Omega\Delta t)R_{yx}\right] d(\Delta t). \tag{5.40}$$

The integral (5.40), divided by B_0^2, is essentially the particle decorrelation time. In addition, setting $\phi(t-\Delta t) = \pi/2$ allows the arguments of the two-point, two-time correlation functions to be expressed as

$$x(t-\Delta t) = x(t) + r_g[\cos(\Omega\Delta t) - 1]; \quad y(t-\Delta t) = y(t) - r_g\sin(\Omega\Delta t);$$
$$z(t-\Delta t) = z(t) - v\cos(\theta\Delta t).$$

By introducing a mean magnetic field $\mathbf{B}_0 = B_0\hat{\mathbf{z}}$ into Eq. (5.32), and using $\mu = \cos\theta$, the cosine of the particle pitch-angle, we obtain the simplest 1D form of the collisionless transport equation as

$$\frac{\partial f_0}{\partial t} + \mu v \frac{\partial f_0}{\partial z} = \frac{\partial}{\partial\mu}\left(D_{\mu\mu}\frac{\partial f_0}{\partial\mu}\right), \tag{5.41}$$

where the Fokker-Planck diffusion coefficient in pitch-angle space is given by

$$D_{\mu\mu} = (1-\mu^2)\left(\frac{\Omega}{B_0}\right)^2 \int_0^\infty \left(\cos(\Omega\Delta t)R_{xx} - \sin(\Omega\Delta t)R_{yx}\right) d(\Delta t). \tag{5.42}$$

For slab turbulence, the pitch-angle scattering diffusion coefficient can be further simplified since $R_{yx} = 0$, yielding the standard expression

$$D_{\mu\mu} = (1-\mu^2)\left(\frac{\Omega}{B}\right)^2 \int_0^\infty R_{xx}^{slab}\cos(\Omega\Delta t)d(\Delta t). \tag{5.43}$$

Using the results of the previous section, we may evaluate $D_{\mu\mu}$ for slab turbulence. Recall that

$$R_{xx} = \int d^3k P_{xx}^{slab}(\mathbf{k})e^{i\mathbf{k}\cdot\mathbf{r}}$$

$$= \int_0^{2\pi}\int_0^\infty\int_{-\infty}^\infty g^{slab}(k_\parallel)\frac{\delta(k_\perp)}{k_\perp}e^{i\mathbf{k}\cdot\mathbf{r}}k_\perp d\theta dk_\perp dk_\parallel$$

$$= 4\pi \int_0^\infty g^{slab}(k_\parallel)e^{ik_\parallel z}dk_\parallel$$

$$= 4\pi \int_0^\infty g^{slab}(k_\parallel)e^{ik_\parallel \mu v}dk_\parallel,$$

where we used $z = -v\cos\theta t$. On replacing Δt by t in (5.43), we have

$$D_{\mu\mu} = 4\pi(1-\mu^2)\left(\frac{\Omega}{B}\right)^2 \int_0^\infty \int_0^\infty g^{slab}(k_\|)e^{-i(k_\|\mu v - \Omega)t}dk_\| dt$$

$$= 4\pi^2(1-\mu^2)\left(\frac{\Omega}{B}\right)^2 \int_0^\infty g^{slab}(k_\|)\delta(k_\|\mu v - \Omega)dk_\|$$

$$= 4\pi^2(1-\mu^2)\left(\frac{\Omega}{B}\right)^2 g^{slab}\left(k_\| = \frac{\Omega}{\mu v}\right).$$

Thus, for slab turbulence, energetic charged particles diffuse in pitch angle due to their scattering with waves that satisfy the resonance condition $\mu v k_\| = \Omega$.

Exercises

1. Rewrite the Vlasov equation (5.30) using a mean field expansion for the electromagnetic variables, assuming that the particle distribution function is co-moving with the plasma (thus ensuring that $\mathbf{E}_0 = 0$), and neglecting the fluctuating electric field term. Hence derive (5.31) and (5.32).
2. Derive the relations (5.33) and hence show that

$$-\frac{q}{m}\left(\mathbf{p}\times\delta\mathbf{B}\cdot\nabla_p f_0\right) = -\frac{\Omega}{B}\left(\delta B_x \sin\phi - \delta B_y \cos\phi\right)\frac{\partial f_0}{\partial\theta}.$$

5.5 Diffusion Perpendicular to the Mean Magnetic Field: The Nonlinear Guiding Center Theory

To determine the transport of energetic particles perpendicular to a mean magnetic field is not possible within a gyrophase averaged formulation of the Fokker-Planck equation. Instead, we can compute directly the perpendicular spatial diffusion coefficient $\kappa_\perp$ from the Fokker-Planck coefficients. Recall that the mean square displacement is given by

$$\langle(\Delta x)^2\rangle = \langle(x(t) - x(0))^2\rangle,$$

for an averaging operator $\langle\ldots\rangle$. Several forms of diffusion can be described if we suppose that the following temporal scaling holds for the spatial variance

$$\langle(\Delta x)^2\rangle \sim t^\sigma.$$

The following regimes are typically identified:

1. $0 < \sigma < 1$: subdiffusion;
2. $\sigma = 1$: regular Markovian diffusion;

3. $1 < \sigma < 2$: superdiffusion, and
4. $\sigma = 2$: free streaming or ballistic particle motion.

There have been suggestions that energetic particles can be subdiffusive, and at early times in an impulsive solar energetic particle event, particles are typically free streaming. Over long time scales, however, particle motion is more typically diffusive.

The diffusion coefficient is defined as

$$\kappa_{xx} = \lim_{t \to \infty} \frac{\langle (\Delta x)^2 \rangle}{2t},$$

where we assume that x is normal to the mean magnetic field. To estimate the spatial variance, we appeal to the Taylor-Green-Kubo (TGK) formalism. In general, the variance is given by

$$\langle (\Delta x)^2 \rangle (t) = \left\langle \left(\int_0^t v_x(\tau) d\tau \right)^2 \right\rangle$$

$$= \int_0^t d\tau \int_0^t d\xi \langle v_x(\tau) v_x(\xi) \rangle$$

$$= \int_0^t d\tau \int_0^\tau d\xi \langle v_x(\tau) v_x(\xi) \rangle + \int_0^t d\tau \int_\tau^t d\xi \langle v_x(\tau) v_x(\xi) \rangle.$$

On assuming temporal homogeneity, i.e., that the velocity correlation depends only on the time difference, then we choose

$$\langle v_x(\tau) v_x(\xi) \rangle = \langle v_x(\tau - \xi) v_x(0) \rangle$$

for the first integral, and

$$\langle v_x(\tau) v_x(\xi) \rangle = \langle v_x(\xi - \tau) v_x(0) \rangle$$

for the second, to obtain

$$\langle (\Delta x)^2 \rangle (t) = \int_0^t d\tau \int_0^\tau d\xi \langle v_x(\tau - \xi) v_x(0) \rangle + \int_0^t d\tau \int_\tau^t d\xi \langle v_x(\xi - \tau) v_x(0) \rangle$$

$$= \int_0^t d\tau \int_0^\tau d\xi \langle v_x(\xi) v_x(0) \rangle + \int_0^t d\tau \int_0^{t-\tau} d\xi \langle v_x(\xi) v_x(0) \rangle,$$

after using the transformations $\tau - \xi \to \xi$ and $\xi - \tau \to \xi$ in the respective integrals. These integrals can be simplified using partial integration and applying Leibnitz' rule to obtain

5.5 Diffusion Perpendicular to the Mean Magnetic Field...

$$\langle(\Delta x)^2\rangle(t) = \tau \int_0^\tau d\xi \langle v_x(\xi)v_0(0)\rangle\Big|_0^t - \int_0^t d\tau\tau\langle v_x(\tau)v_x(0)\rangle$$

$$+ \tau \int_0^{t-\tau} d\xi \langle v_x(\xi)v_x(0)\rangle\Big|_0^t + \int_0^t d\tau\tau\langle v_x(t-\tau)v_x(0)\rangle$$

$$= t \int_0^t d\xi \langle v_x(\xi)v_x(0)\rangle - \int_0^t d\tau\tau\langle v_x(\tau)v_x(0)\rangle$$

$$+ \int_0^t d\tau\tau\langle v_x(t-\tau)v_x(0)\rangle$$

$$= \int_0^t d\tau(t-\tau)\langle v_x(\tau)v_x(0)\rangle + \int_0^t d\tau\tau\langle v_x(t-\tau)v_x(0)\rangle$$

$$= 2\int_0^t d\tau(t-\tau)\langle v_x(\tau)v_x(0)\rangle.$$

The running diffusion coefficient $d_{xx}(t)$ is defined as

$$d_{xx}(t) = \frac{1}{2}\frac{d}{dt}\langle(\Delta x)^2\rangle(t)$$

$$= \frac{1}{2}\frac{d}{dt}2\int_0^t (t-\tau)\langle v_x(\tau)v_x(0)\rangle$$

$$= \int_0^t d\tau\langle v_x(\tau)v_x(0)\rangle.$$

The limit $d_{xx}(t \to \infty)$ defines diffusive particle transport, therefore

$$\kappa_{xx} = \int_0^\infty d\tau\langle v_x(\tau)v_x(0)\rangle,$$

which is the Kubo formula for the diffusion coefficient.

A detailed discussion of guiding center motion of energetic charged particles can be found in many plasma text books and so is not repeated here. Instead, if we assume that background magnetic field is varying slowly, that for any of the slab, 2D, or composite turbulence models discussed above, the guiding center velocity (assuming $\mathbf{B} = B_0\hat{\mathbf{z}} + \delta\mathbf{B}$) is given by

$$v_x^g(t) \simeq v_z(t)\frac{\delta B_x}{B_0}; \quad v_y^g \simeq v_z(t)\frac{\delta B_y}{B_0}.$$

Note that the assumption of slab, 2D, or composite turbulence models implies that $\delta B_z = 0$. Particle motion is thus a superposition of the particle gyromotion and the stochastic motion of the particle's guiding center, which follows the random motion of magnetic field lines. The gyromotion can be neglected when computing a diffusion coefficient.

The first systematic derivation of the perpendicular diffusion coefficient was proposed by Matthaeus et al. (2003) and is called the nonlinear guiding center (NLGC) theory. Improvements and extensions to the original model have been made[9] but the original development is very instructive in its simplicity. To ensure agreement with numerical simulations of particles experiencing scattering in low frequency turbulence, we introduce a parameter a (typically taken to be $1/3$) that allows for slight deviations from purely guiding center motion, and take

$$v_x^g = av_z \frac{\delta B_x}{B_0}.$$

This is reasonable since the magnetic field can occasionally experience variation on scales that are not necessarily slowly varying. The TGK expression for the perpendicular diffusion coefficient is

$$\kappa_{xx} = \int_0^\infty dt \, \langle v_x^g(t) v_x^g(0) \rangle$$
$$= \frac{a^2}{B_0^2} \int_0^\infty dt \, \langle v_z(t) \delta B_x(t) v_z(0) \delta B_x^*(0) \rangle.$$

The fourth-order correlation introduces a closure problem. This is frequently resolved by the assumption that the fourth-order correlation can be replaced by the product of second-order correlations (motivated by the example of Gaussian statistics), which yields

$$\kappa_{xx} = \frac{a^2}{B_0^2} \int_0^\infty dt \, \langle v_z(t) v_z(0) \rangle \langle \delta B_x(t) \delta B_x^*(0) \rangle.$$

Since the particle velocity along the field is mediated by pitch-angle scattering, we may suppose that particle distribution becomes approximately isotropic on diffusion time scales and that there is a decorrelation time scale associated with the parallel velocity. The decorrelation time will be related to the parallel mean free path, so we can use an exponential model to describe the two-point velocity correlation function,

$$\langle v_z(t) v_z(0) \rangle = \frac{v^2}{3} e^{-vt/\lambda_\parallel}.$$

---

[9]Well summarized by Shalchi (2009)

5.5 Diffusion Perpendicular to the Mean Magnetic Field...

The magnetic correlation function $R_{xx}(t) = \langle \delta B_x(t) \delta B_x^*(0) \rangle$ can be expressed as a Fourier transform

$$\delta B_x(\mathbf{x}, t) = \int d^3k\, \delta B_x(\mathbf{k}, t) e^{i\mathbf{k}\cdot\mathbf{x}} \Rightarrow R_{xx}(t) = \int d^3k \left\langle \delta B_x(t) \delta B_x^*(0) e^{i\mathbf{k}\cdot\Delta\mathbf{x}} \right\rangle,$$

under the assumption of homogeneous turbulence.

At this point, it is still unclear how to further decompose the ensemble averaged integrand in the magnetic correlation function. Corrsin (1959) suggested that at long diffusion times, the probability distribution of particle displacements and the probability distribution of the Eulerian velocity field would become statistically independent of each other – this is Corrsin's independence hypothesis. At large values of the diffusion time, the independence hypothesis asserts that the joint average in R_{xx} can be expressed as the product of two separate averages, i.e.,

$$\left\langle \delta B_x(t) \delta B_x^*(0) e^{i\mathbf{k}\cdot\Delta\mathbf{x}} \right\rangle = \left\langle \delta B_x(t) \delta B_x^*(0) \right\rangle \left\langle e^{i\mathbf{k}\cdot\Delta\mathbf{x}} \right\rangle.$$

Applying Corrsin's independence hypothesis then yields

$$R_{xx}(t) = \int d^3k\, P_{xx}(\mathbf{k}, t) \left\langle e^{i\mathbf{k}\cdot\Delta\mathbf{x}} \right\rangle,$$

requiring only that we estimate the characteristic function $\left\langle e^{i\mathbf{k}\cdot\Delta\mathbf{x}} \right\rangle$. The simplest approximation is to assume a Gaussian distribution of the particles, so that

$$\left\langle e^{i\mathbf{k}\cdot\Delta\mathbf{x}} \right\rangle = \exp\left[-\frac{1}{2}\langle(\Delta x)^2\rangle k_x^2 - \frac{1}{2}\langle(\Delta y)^2\rangle k_y^2 - \frac{1}{2}\langle(\Delta z)^2\rangle k_z^2\right].$$

Since we are considering time scales that correspond to large values of the diffusion time, we can approximate the parallel and perpendicular transport as diffusion, so that $\langle(\Delta x)^2\rangle = 2t\kappa_{xx}$ for example, yielding

$$\left\langle e^{i\mathbf{k}\cdot\Delta\mathbf{x}} \right\rangle = \exp\left[-\kappa_{xx} k_x^2 t - \kappa_{yy} k_y^2 t - \kappa_{zz} k_z^2 t\right].$$

Subject to these six assumptions, we obtain a nonlinear integral equation for the perpendicular diffusion coefficient

$$\kappa_{xx} = \frac{a^2}{B_0^2} \int d^3k \int_0^\infty dt\, P_{xx}(\mathbf{k}, t) \exp\left[-vt/\lambda_\| - \kappa_{xx} k_x^2 t - \kappa_{yy} k_y^2 t - \kappa_{zz} k_z^2 t\right].$$

On expressing the correlation tensor $P_{xx}(\mathbf{k}, t)$ as the product of a stationary tensor $P_{xx}(\mathbf{k})$ and a dynamical correlation tensor $\Gamma(\mathbf{k}, t)$, i.e., $P_{xx}(\mathbf{k}, t) = \Gamma(\mathbf{k}, t) P_{xx}(\mathbf{k})$, and assuming the exponential form,

$$\Gamma(\mathbf{k}, t) = e^{-\gamma(\mathbf{k})t},$$

allows the time integral to be solved

$$\kappa_{xx} = \frac{a^2 v^2}{3B_0^2} \int d^3k \frac{P_{xx}}{v/\lambda_\| + \kappa_{xx}k_x^2 + \kappa_{yy}k_y^2 + \kappa_{zz}k_z^2 + \gamma(\mathbf{k})}. \quad (5.44)$$

The nonlinear integral equation (5.44) is the central result of the NLGC theory, describing the diffusion of energetic particles perpendicular to the mean magnetic field where $\delta B_z = 0$. The particle transport results from a combination of pitch-angle scattering along the magnetic field while the underlying magnetic field is experiencing random diffusive motion. The superposition of parallel transport and random magnetic field transport of the particle guiding center leads to a nonlinear diffusion of particle normal to the large-scale magnetic field. As indicated, more sophisticated treatments of the NLGC theory have been developed since. The nonlinear integral equation (5.44) can be solved approximately and analytically for the slab, 2D, and composite turbulence models in the magnetostatic limit.[10]

5.6 Hydrodynamic Description of Energetic Particles

In deriving the cosmic ray transport equation, we have assumed that the underlying energetic particle distribution function is isotropic to zeroth order. We further assumed that the energetic particle number density and momentum is sufficiently small that the background flow in which the "scattering centers" (Alfvén waves or MHD turbulence) are convected is not altered by the energetic particle population, nor is the convection electric field. Energetic particles therefore behave essentially as massless particles that may possess a significant internal energy, which will be expressed through an isotropic or scalar pressure, say P_c, and energy density E_c, and an energy flux $\mathbf{F}_c$.[11] In this case, the general system of MHD equations will be modified by the inclusion of the cosmic rays, through

$$\frac{\partial n}{\partial t} + \nabla \cdot (n\mathbf{u}) = 0; \quad (5.45)$$

$$\frac{\partial \mathbf{G}}{\partial t} + \nabla \cdot \mathbf{\Pi} = 0; \quad (5.46)$$

$$\frac{\partial W}{\partial t} + \nabla \cdot \mathbf{S} = 0, \quad (5.47)$$

---

[10] Zank et al. (2004) and Shalchi et al. (2004).
[11] Webb (1983) and Zank (1988).

5.6 Hydrodynamic Description of Energetic Particles

where

$$\mathbf{G} = n\mathbf{u};$$
$$\Pi_{ij} = (P_g + P_c)\delta_{ij} + nu_i u_j - T_{ij};$$
$$T_{ij} = \frac{1}{\mu} B_i B_j - \frac{1}{2\mu} B^2 \delta_{ij};$$
$$W = E_g + E_c + \frac{mnu^2}{2} + \frac{1}{2\mu} B^2;$$
$$\mathbf{S} = \frac{1}{\mu}\mathbf{E} \times \mathbf{B} + \left(E_g + P_g + \frac{mnu^2}{2}\right)\mathbf{u} + \mathbf{F}_c;$$
$$E_g = \frac{P_g}{\gamma_g - 1}.$$

The terms $\mathbf{G}$, $\mathbf{\Pi}$, T_{ij}, W, and $\mathbf{S}$ denote respectively the momentum density, the momentum flux, the Maxwell stress tensor, the total energy density, and the total energy flux. The physical quantities n, $\mathbf{u}$, P_g, E_g, and γ_g are the (background) plasma number density, bulk flow velocity, thermal gas pressure, thermal gas energy density, and adiabatic index of the background plasma (typically $= 5/3$) respectively. As usual, the electric field, magnetic field, and magnetic permeability are denoted respectively by $\mathbf{E}$, $\mathbf{B}$, and μ. The energetic particle or cosmic ray energy density, pressure, and energy flux are given by E_c, P_c, and $\mathbf{F}_c$. To close the system of equations (5.44)–(5.47), we appeal to the transport equation derived above (and neglecting drifts)

$$\frac{\partial f}{\partial t} + \mathbf{u} \cdot \nabla f - \frac{p}{3}\nabla \cdot \mathbf{u}\frac{\partial f}{\partial p} = \nabla \cdot (\kappa \cdot \nabla f),$$

and use the following moments,

$$P_c = \frac{4\pi}{3} \int_0^\infty p^3 v f(p, \mathbf{x}, t) dp; \tag{5.48}$$

$$E_c = 4\pi \int_0^\infty p^2 T(p) f(p, \mathbf{x}, t) dp, \tag{5.49}$$

where $T(p)$ is the particle kinetic energy. By taking the energy moment of the transport equation, we obtain a "hydrodynamic" equation for the transport of cosmic rays,

$$\frac{\partial E_c}{\partial t} + \nabla \cdot \mathbf{F}_c - \mathbf{u} \cdot \nabla P_c = 0;$$
$$\mathbf{F}_c = \mathbf{u}(E_c + P_c) - \bar{\kappa} \cdot \nabla E_c,$$

where

$$\bar{\kappa}(\mathbf{x}) = \frac{\int_0^\infty p^2 T(p)\kappa(\mathbf{x}, p) \cdot \nabla f \, dp}{\int_0^\infty p^2 T(p) \nabla f \, dp},$$

is an effective mean diffusion coefficient. The internal energy density is related to the energetic particle pressure through an adiabatic exponent γ_c (that is $4/3$ for relativistic particles and $5/3$ for non-relativistic particles) by the closure relation

$$P_c = (\gamma_c - 1)E_c.$$

To complete the system of equations, the electromagnetic equations must satisfy Maxwell's equations,

$$\frac{\partial \mathbf{B}}{\partial t} = -\nabla \times \mathbf{E}, \quad \nabla \cdot \mathbf{B} = 0, \quad \mathbf{J} = \frac{1}{\mu}\nabla \times \mathbf{B}, \tag{5.50}$$

where $\mathbf{J}$ is the current density. For a highly conducting MHD fluid, the electric field is given by Ohm's law,

$$\mathbf{E} = -\mathbf{u} \times \mathbf{B}.$$

Although a little laborious, the elimination of the cosmic ray terms in the total energy equation (5.47) and using Poynting's theorem to eliminate the electromagnetic terms yields the simpler adiabatic equation for the thermal gas pressure,

$$\left(\frac{\partial}{\partial t} + \mathbf{u} \cdot \nabla\right) P_g = -\gamma_g P_g \nabla \cdot \mathbf{u}, \tag{5.51}$$

which can be used whenever the background flow is smooth (i.e., in the absence of shock waves or other discontinuities in the flow).

The so-called "two-fluid" system of equations incorporating cosmic rays has been used to investigate the structure of shocks in the presence of an energetic particle population, including the evolution of shocks associated with supernova remnants. These equations provide a relatively tractable approach to the inclusion of nonlinearities and their modification by cosmic rays. To illustrate the effect that cosmic rays or other energetic particles can have on the background plasma, we will consider briefly the propagation of linear and nonlinear waves in an astrophysical plasma mediated by cosmic rays.

Suppose that wave propagation is 1D and is represented by a wave vector $\mathbf{k} = k\hat{\mathbf{x}}$ in the Cartesian coordinates (x, y, z) and that $\partial/\partial y = \partial/\partial z = 0$. By writing $\mathbf{u} = (u_x, u_y, u_z)$ and $\mathbf{B} = (B_x, B_y, B_z)$, equations (5.44)–(5.47) reduce to

$$\frac{\partial \rho}{\partial t} + \frac{\partial}{\partial x}(\rho u_x) = 0;$$

$$\rho \frac{du_x}{dt} = -\frac{\partial}{\partial x}(P_g + P_c) - \frac{1}{2\mu}\frac{\partial}{\partial x}\left(B_y^2 + B_z^2\right);$$

5.6 Hydrodynamic Description of Energetic Particles

$$\rho \frac{du_y}{dt} = \frac{B_x}{\mu} \frac{\partial B_y}{\partial x}; \quad \rho \frac{du_z}{dt} = \frac{B_x}{\mu} \frac{\partial B_z}{\partial x}; \quad B_x = \text{const.};$$

$$\frac{\partial B_y}{\partial t} = \frac{\partial}{\partial x}\left(u_y B_x - u_x B_y\right);$$

$$\frac{\partial B_z}{\partial t} = \frac{\partial}{\partial x}\left(u_z B_x - u_x B_z\right);$$

$$\frac{dP_g}{dt} + \gamma_g P_g \frac{\partial u_x}{\partial x} = 0;$$

$$\frac{dP_c}{dt} + \gamma_c P_c \frac{\partial u_x}{\partial x} - \kappa \frac{\partial^2 P_c}{\partial x^2} = 0,$$

where κ is taken to be spatially constant and

$$\frac{d}{dt} \equiv \frac{\partial}{\partial t} + u_x \frac{\partial}{\partial x}$$

is the Lagrangian time derivative along the flow. Note that the induction equations ensure that the magnetic field is frozen into the fluid flow $\mathbf{u}$. Evidently, the cosmic rays introduce a characteristic length scale into the problem, $L = \kappa/u$, which we shall exploit shortly. By linearizing about the uniform equilibrium state $\mathbf{u} = 0$, $\rho = \rho_0$, $P_g = P_{g0}$, P_{c0}, $\mathbf{B} = (B_x, B_{y0}, B_{z0})$ and seeking solutions $\propto \exp[i(\omega t - kx)]$, with $V_p \equiv \omega/k$ the wave phase speed, we can derive the dispersion relation (Exercise)

$$(V_p^2 - V_x^2)\left[V_p^5 - i\kappa k V_p^4 - (V_A^2 + a_*^2)V_p^3 + i\kappa k(V_A^2 + a_{g0}^2)V_p^2 \right.$$
$$\left. + a_*^2 V_x^2 V_p - i\kappa k a_{g0}^2 V_x^2\right] = 0, \tag{5.52}$$

where

$$V_A^2 = \frac{B_0^2}{\mu \rho_0}, \quad V_x = V_A \cos\phi, \quad a_{g0}^2 = \frac{\gamma_g P_{g0}}{\rho_0},$$

$$a_{c0}^2 = \frac{\gamma_c P_{c0}}{\rho_0}, \quad a_*^2 = a_{g0}^2 + a_{c0}^2.$$

Here, V_A and V_x are the Alfvén speeds, ϕ the angle between the magnetic field and the wave vector $\mathbf{k}$, and a_{g0}, a_{c0}, and a_* are the sound speeds of the thermal gas, the cosmic ray gas, and the mixture respectively. Obviously, since the Alfvén wave is incompressible, cosmic rays do not effect their propagation characteristics and we have from (5.52),

$$V_p = \pm V_x.$$

By contrast, the fifth-order dispersion relation within the square brackets of (5.52) contains the spatial dispersion coefficient κ. By setting $\kappa = 0$ in (5.52), we see that the reduced dispersion relation possesses exactly the same biquadratic form as that of MHD, viz.,

$$V_p^4 - (V_A^2 + a_*^2)V_p^2 + a_*^2 V_x^2 = 0,$$

together with a non-propagating mode $V_p = 0$, and hence the fast and slow magnetosonic modes,

$$V_p^2 = V_{f,s}^2 = \frac{1}{2}\left[V_A^2 + a_*^2 \pm \sqrt{(V_A^2 + a_*^2)^2 - 4a_*^2 V_x^2}\right].$$

However, the critical difference between these wave modes and those derived in the usual MHD theory is the presence of the *mixed sound speed* $a_* = \sqrt{(\gamma_g P_{g0} + \gamma_c P_{c0})/\rho_0}$ indicating that the cosmic rays couple to the background plasma and alter the phase speed for these wave modes. If we instead consider the *long wavelength* limit $\kappa k/V_p \ll 1$, we find that the long wavelength magnetosonic modes satisfy (Exercise)

$$V_p = V_{f,s} + i\kappa k\beta + O\left((\kappa k)^2\right),$$

where

$$\beta = \frac{a_{c0}^2 \left(V_{f,s}^2 - V_x^2\right)}{2\left[(V_A^2 + a_*^2)V_{f,s}^2 - 2a_*^2 V_x^2\right]}.$$

Short wavelength modes propagate at the usual magnetosonic speeds defined by the thermal plasma but are nonetheless damped by cosmic rays (Exercise).

Following our general theme of weak shock structure, we may consider the nonlinear propagation of either long wavelength or short wavelength modes in the two-fluid cosmic ray model as a simplification of the full shock structure problem. We use the method of multiple scales, which is closely related to the reductive perturbation method utilized already although a little more systematic. To this end, we exploit the length scale L introduced by the spatial diffusion coefficient κ i.e., $L \sim \kappa/V_p$ for a characteristic speed V_p. We introduce a time scale T such that the relationship

$$\frac{V_p T}{L} = 1,$$

and the following normalizations,

$$x = L\bar{x}, \quad t = T\bar{t}, \quad \bar{\mathbf{B}} = \mathbf{B}/B_0, \quad \bar{P}_{g,c} = P_{g,c}/P_{g0,c0},$$

$$\bar{\rho} = \rho/\rho_0, \quad \mathbf{u} = V_p \bar{\mathbf{u}}.$$

5.6 Hydrodynamic Description of Energetic Particles

This allows us to rewrite the 1D system of equations in the non-dimensional form,

$$\frac{\partial \rho}{\partial t} + \frac{\partial}{\partial x}(\rho u_x) = 0;$$

$$\rho \frac{du_x}{dt} = -\frac{\bar{a}_{g0}^2}{\gamma_g}\frac{\partial P_g}{\partial x} - \frac{\bar{a}_{c0}^2}{\gamma_c}\frac{\partial P_c}{\partial x} - \frac{\bar{V}_A^2}{2}\frac{\partial}{\partial x}\left(B_y^2 + B_z^2\right);$$

$$\rho \frac{du_y}{dt} = \bar{V}_A^2 \frac{B_x}{\mu}\frac{\partial B_y}{\partial x}; \quad \rho \frac{du_z}{dt} = \bar{V}_A^2 \frac{B_x}{\mu}\frac{\partial B_z}{\partial x}; \quad B_x = \text{const.};$$

$$\frac{\partial B_y}{\partial t} = \frac{\partial}{\partial x}\left(u_y B_x - u_x B_y\right);$$

$$\frac{\partial B_z}{\partial t} = \frac{\partial}{\partial x}\left(u_z B_x - u_x B_z\right);$$

$$\frac{dP_g}{dt} + \gamma_g P_g \frac{\partial u_x}{\partial x} = 0;$$

$$\frac{dP_c}{dt} + \gamma_c P_c \frac{\partial u_x}{\partial x} - \nu \frac{\partial^2 P_c}{\partial x^2} = 0.$$

For convenience, the bars are omitted over the various quantities with the exception of the sound speeds and Alfvén speed,

$$\bar{a}_{g0,c0} = a_{g0,c0}/V_p, \quad \bar{V}_a = V_A/V_p,$$

and the long wavelength parameter is defined as

$$\nu = \kappa/(V_p L).$$

The dependent variables are expanded as an asymptotic series in a small parameter ε by using the fast and slow variables

$$\xi = x - t, \quad \tau = \varepsilon t,$$

together with the expansions

$$\rho = 1 + \varepsilon \rho^1 + \cdots, \quad u_x = \varepsilon u_x^1 + \cdots, \quad u_z = \varepsilon u_z^1 + \cdots, \quad B_z = B_z^0 + \varepsilon B_z^1 + \cdots,$$

$$P_g = 1 + \varepsilon P_g^1 + \cdots, \quad P_c = 1 + \varepsilon P_c^1 + \cdots,$$

where it is convenient to assume $u_y = 0$ and $B_y = 0$. The derivatives are calculated as

$$\frac{\partial}{\partial x} = \frac{\partial}{\partial \xi}, \quad \frac{\partial}{\partial t} = \varepsilon \frac{\partial}{\partial \tau} + \frac{\partial}{\partial \xi}.$$

The lowest order system of equations is given by

$$-\frac{\partial \rho^1}{\partial \xi} + \frac{\partial u_x^1}{\partial \xi} = 0;$$

$$\frac{\partial u_x^1}{\partial \xi} = \frac{\bar{a}_{g0}^2}{\gamma_g}\frac{\partial P_g^1}{\partial \xi} + \frac{\bar{a}_{c0}^2}{\gamma_c}\frac{\partial P_c^1}{\partial \xi} + \bar{V}_A^2 B_z^0 \frac{\partial B_z^1}{\partial \xi};$$

$$-\frac{\partial u_z^1}{\partial \xi} = \bar{V}_A^2 B_x \frac{\partial B_z^1}{\partial \xi};$$

$$-\frac{\partial B_z^1}{\partial \xi} = B_x \frac{\partial u_z^1}{\partial \xi} - B_z^0 \frac{\partial u_x^1}{\partial \xi};$$

$$-\frac{\partial P_g^1}{\partial \xi} + \gamma_g \frac{\partial u_x^1}{\partial \xi} = 0;$$

$$-\frac{\partial P_c^1}{\partial \xi} + \gamma_c \frac{\partial u_x^1}{\partial \xi} = 0.$$

To integrate this system of equations, we assume that the plasma is in a uniform state upstream, so that to $O(\varepsilon)$ we obtain

$$\rho^1 = u_x^1;$$

$$u_x^1 = \frac{\bar{a}_{g0}^2}{\gamma_g} P_g^1 + \frac{\bar{a}_{c0}^2}{\gamma_c} P_c^1 + \bar{V}_A^2 B_z^0 B_z^1;$$

$$u_z^1 = -\bar{V}_A^2 B_x^0 B_z^1;$$

$$B_z^1 = -B_x u_z^1 + B_z^0 u_x^1;$$

$$P_g^1 = \gamma_g u_x^1; \quad P_c^1 = \gamma_c u_x^1,$$

so that we have the eigenvector solutions

$$\left(\rho^1, u_x^1, u_z^1, B_z^1, P_g^1, P_c^1\right) = u_x^1 \left(1, 1, -\frac{\bar{V}_A^2 B_x B_z^0}{1-\bar{V}_x^2}, \gamma_g, \gamma_c\right),$$

provided that the relation

$$1 = \bar{a}_{g0}^2 + \bar{a}_{c0}^2 - \frac{\bar{V}_A^2 B_z^{02}}{1-\bar{V}_x^2}$$

5.6 Hydrodynamic Description of Energetic Particles

holds. This is of course nothing more than the normalized form of the dispersion relation for magnetosonic waves with the cosmic ray pressure and thermal gas pressure contributing i.e., the long wavelength limit of the dispersion relation

$$V_p^4 - (a_*^2 + V_A^2)V_p^2 + a_*^2 V_x^2 = 0.$$

The slow time dependence is determined from the $O(\varepsilon^2)$ set of equations,

$$-\frac{\partial \rho^2}{\partial \xi} + \frac{\partial u_x^2}{\partial \xi} = -\frac{\partial u_x^1}{\partial \tau} - 2u_x^1 \frac{\partial u_x^1}{\partial \xi};$$

$$-\frac{\partial u_x^2}{\partial \xi} + \frac{\bar{a}_{g0}^2}{\gamma_g}\frac{\partial P_g^2}{\partial \xi} + \frac{\bar{a}_{c0}^2}{\gamma_c}\frac{\partial P_c^2}{\partial \xi} + \bar{V}_A^2 B_z^0 \frac{\partial B_z^2}{\partial \xi} = -\frac{\partial u_x^1}{\partial \tau} - \frac{\bar{V}_A^2 B_z^{0^2}}{(1-\bar{V}_x^2)^2} u_x^1 \frac{\partial u_x^1}{\partial \xi};$$

$$\frac{\partial u_z^2}{\partial \xi} + \bar{V}_A^2 B_x \frac{\partial B_z^2}{\partial \xi} = -\frac{\bar{V}_A^2 B_x B_z^0}{1-\bar{V}_x^2}\frac{\partial u_x^1}{\partial \tau};$$

$$-\frac{\partial B_z^2}{\partial \xi} - B_x \frac{\partial u_z^2}{\partial \xi} + B_z^0 \frac{\partial u_x^2}{\partial \xi} = -\frac{B_z^0}{1-\bar{V}_x^2}\frac{\partial u_x^1}{\partial \tau} - 2\frac{B_z^1}{1-\bar{V}_x^2}\frac{\partial u_x^1}{\partial \xi};$$

$$-\gamma_g \frac{\partial u_x^2}{\partial \xi} + \frac{\partial P_g^2}{\partial \xi} = \gamma_g \frac{\partial u_x^1}{\partial \tau} + \gamma_g(\gamma_g + 1)u_x^1 \frac{\partial u_x^1}{\partial \xi};$$

$$-\gamma_c \frac{\partial u_x^2}{\partial \xi} + \frac{\partial P_c^2}{\partial \xi} = \gamma_c \frac{\partial u_x^1}{\partial \tau} + \gamma_c(\gamma_c + 1)u_x^1 \frac{\partial u_x^1}{\partial \xi} - \gamma_c K \frac{\partial^2 u_x^1}{\partial \xi^2}.$$

The above nonlinear system can reduced quite easily to a single nonlinear evolution equation by eliminating all second-order variables in favor of u_x^2 and then using the dispersion relation repeatedly (Exercise). Doing so yields the Burgers' equation for the nonlinear evolution of the fast mode magnetosonic mode in the presence of an energetic particle population, now presented in a non-normalized form,

$$\frac{\partial u_x^1}{\partial \tau} + \alpha u_x^1 \frac{\partial u_x^1}{\partial \xi} = \lambda \frac{\partial^2 u_x^1}{\partial \xi^2}, \qquad (5.53)$$

where

$$\alpha = \frac{\left[(\gamma_g + 1)a_{g0}^2 + (\gamma_c + 1)a_{c0}^2\right](V_p^2 - V_x^2) + 3(V_A^2 V_p^2 - a_*^2 V_x^2)}{2\left[(a_*^2 + V_A^2)V_p^2 - 2a_*^2 V_x^2\right]};$$

$$\lambda = \frac{\kappa a_{c0}^2 (V_p^2 - V_x^2)}{2\left[(a_*^2 + V_A^2)V_p^2 - 2a_*^2 V_x^2\right]}.$$

As discussed previously, the Burgers' equation can be used to describe weak shocks, and the role of cosmic rays or energetic particles in providing the shock dissipation is revealed through the second-order term in (5.53).

For shocks parallel to the ambient magnetic field $V_x = V_A$ and the dispersion relation reduces to

$$(V_p^2 - V_A^2)(V_p^2 - a_*^2) = 0.$$

Hence for the sound wave, $V_p = \pm a_*$ and the coefficients of Burgers' equation (5.53) reduce to

$$\alpha_\| = \frac{(\gamma_g + 1)a_{g0}^2 + (\gamma_c + 1)a_{c0}^2}{2a_*^2}, \quad \lambda_\| = \frac{\kappa a_{c0}^2}{2a_*^2}.$$

For perpendicular shocks, $V_x = 0$, so the coefficients of Burgers' equation (5.53) become

$$\alpha_\perp = \frac{(\gamma_g + 1)a_{g0}^2 + (\gamma_c + 1)a_{c0}^2 + 3V_A^2}{2(a_*^2 + V_A^2)}, \quad \lambda_\| = \frac{\kappa a_{c0}^2}{2(a_*^2 + V_A^2)}.$$

The preceding analysis was restricted to waves with wavelengths that are greatly in excess of the diffusive length scale. We can consider short wavelength modes for which the parameter $\nu \gg 1$. These short wavelength modes are in fact of considerable interest in the context of the stability of astrophysical shocks mediated by cosmic rays since they can destabilize the foreshock.[12] Following the analysis presented above allows for the investigation of linear and nonlinear short wavelength modes in a homogenous flow. We now suppose that

$$\frac{1}{\nu} = \frac{\varepsilon}{K}.$$

Such a scaling leads again to the integrated $O(\varepsilon)$ system of equations above, except that $P_c^1 = 0$, and the normalized dispersion relation reduces to

$$1 = \bar{a}_{g0}^2 - \frac{\bar{V}_A^2 B_z^{0\,2}}{1 - \bar{V}_x^2}.$$

The $O(\varepsilon^2)$ system of equations is obtained similarly, except that now

$$\frac{\partial P_c^2}{\partial \xi} = \frac{\gamma_c}{K} u_x^1.$$

The nonlinear equation of evolution for short wavelength fluctuations is then given by

$$\frac{\partial u_x^1}{\partial \tau} + \alpha u_x^1 \frac{\partial u_x^1}{\partial \xi} = -\mu u_x^1, \tag{5.54}$$

where

---

[12] Drury and Falle (1986), Zank and McKenzie (1987), and Zank et al. (1990).

5.6 Hydrodynamic Description of Energetic Particles

$$\alpha = \frac{(\gamma_g + 1)a_{g0}^2(V_p^2 - V_x^2) + 3(V_A^2 V_p^2 - a_{g0}^2 V_x^2)}{2\left[(a_{g0}^2 + V_A^2)V_p^2 - 2a_{g0}^2 V_x^2\right]};$$

$$\mu = \frac{a_{c0}^2(V_p^2 - V_x^2)}{2\kappa\left[(a_{g0}^2 + V_A^2)V_p^2 - 2a_{g0}^2 V_x^2\right]}.$$

Unlike Burgers' equation, the nonlinear equation (5.54) does not have a dissipative term to balance the wave steepening, but it does contain a damping term μ that is inversely proportional to the diffusion coefficient κ and proportional to the cosmic ray pressure. Thus, cosmic rays do not mediate the propagation speed of waves that have wavelengths shorter than the diffusive length scale, but they do act to damp these modes. Short wavelength modes therefore damp as they steepen, which can be seen from the general solution to (5.54),

$$u_x^1 = e^{-\mu\tau} f(\xi - \alpha u_x^1 \tau),$$

for arbitrary initial data $f(x, t = 0)$.

Exercises

1. Derive the dispersion relation (5.52) for linear wave modes in a cosmic ray mediated plasma.
2. By considering the long wavelength limit of the dispersion relation (5.52), show that the fast and slow magnetosonic modes are damped by cosmic rays since the waves propagate approximately according to

$$V_p = V_{f,s} + i\kappa k\beta + O\left((\kappa k)^2\right).$$

Show that in the opposite limit, *short wavelength* modes decouple from the cosmic rays in that they propagate at the thermal magnetosonic speed, but are nonetheless damped by cosmic rays since

$$V_p = V_{f,s} + i\frac{\mu}{2\kappa k},$$

where $V_{f,s}$ is the fast/slow magnetosonic speed for the thermal plasma (i.e., the dispersion relation contains only the thermal pressure P_{g0} with no contribution from P_{c0}), and

$$\mu = \frac{a_{c0}^2(V_p^2 - V_x^2)}{(a_{g0}^2 + V_A^2)V_{f,s}^2 - 2a_{g0}^2 V_x^2}.$$

3. Derive Burgers' equation (5.53) from the $O(\varepsilon^2)$ expansion of the magnetized two-fluid equations.

5.7 Application 1: Diffusive Shock Acceleration

It is quite straightforward to see that a particle gains energy by interacting once with a shock, most easily seen for a superluminal shock perpendicular to the magnetic field. In this case, we can suppose that a particle conserves its first adiabatic moment,

$$\frac{p_{\perp,1}^2}{B_1} = \frac{p_{\perp,2}^2}{B_2},$$

where the subscripts 1,2 denote upstream and downstream of the shock. At a perpendicular shock, the jump in magnetic field B_2/B_1 is equal to the shock compression ratio, showing that the perpendicular momentum of an energetic particle can be increased by a factor of 2 or less. This is not a particularly large energy gain, and the effect is of course annulled by the expansion of the downstream medium to the original density. Since the process is purely kinematic and reversible, the energetic particle spectrum is essentially the preacceleration spectrum shifted in energy. The situation is quite different when diffusive effects are included since the number of times that a particle interacts with a shock then becomes a random variable and some particles, by interacting many times with the shock, achieve very high energies. The stochastic character of particles interacting with the shock diffusively corresponds to an increase in entropy for the energetic particle distribution (as it does for the thermal background plasma), with the result that the accelerated particle spectrum is relatively independent of the details of the preacceleration spectrum. We discuss the macroscopic approach to the diffusive acceleration of energetic particles at a shock based on the transport equation that we have derived above. This approach was pioneered by Krymsky (1977), Axford et al. (1977), and Blandford and Ostriker (1978), and is well reviewed by Drury (1983) and Forman and Webb (1985).

The shock is taken to be an infinite plane separating a uniform upstream and downstream state, and we choose a frame in which the shock front is stationary. We shall suppose that all quantities depend only on the x spatial coordinate (a 1D problem) and that the flow velocity is steady, given by

$$u(x) = \begin{cases} u_1 & x < 0 \\ u_2 & x > 0 \end{cases},$$

where u_1 and u_2 are the upstream and downstream constant velocities. To determine the boundary conditions that the energetic particle distribution must satisfy at the shock, we require first that the particle number density must be conserved across the shock i.e., particles are neither created nor lost at the shock, so that

$$[f] = f|_{0-}^{0+} = 0, \tag{5.55}$$

5.7 Application 1: Diffusive Shock Acceleration

where $x = 0-$ and $x = 0+$ denote locations infinitesimally close to the shock on the upstream and downstream side respectively. The second condition (the transport equation governing particle transport is second-order) that we require is that the normal component of the particle current is continuous if there is no source at the surface, and changes by an amount equal to the particle injection rate at the surface. To determine the current, observe that the transport equation

$$\frac{\partial f}{\partial t} + \mathbf{u} \cdot \nabla f - \frac{p}{3} \nabla \cdot \mathbf{u} \frac{\partial f}{\partial p} = \nabla \cdot (\kappa \cdot \nabla f),$$

can be expressed as

$$\frac{\partial f}{\partial t} + \nabla \cdot \left[-\kappa \cdot \nabla f - \frac{p}{3} \frac{\partial f}{\partial p} \mathbf{u} \right] + \frac{p}{3} \mathbf{u} \cdot \nabla \frac{\partial f}{\partial p} + \mathbf{u} \cdot \nabla f = 0$$

$$\Rightarrow \frac{\partial f}{\partial t} + \nabla \cdot \mathbf{S} + \frac{1}{p^2} \frac{\partial}{\partial p} \left[\frac{p^3}{3} \mathbf{u} \cdot \nabla f \right] = 0, \qquad (5.56)$$

where

$$\mathbf{S} = -\kappa \cdot \nabla f - \frac{p}{3} \frac{\partial f}{\partial p} \mathbf{u}$$

is the energetic particle streaming in space and $\mathbf{J}_p = (p/3)\mathbf{u} \cdot \nabla f$ is the streaming in momentum space. Equation (5.56) expresses the transport equation in fully conservative form in phase space, averaged over ϕ and with the distribution function close to isotropy. Because cosmic rays are highly mobile ($v \gg u$), the omnidirectional density f cannot change abruptly, hence the normal component of the net streaming $\mathbf{S}$ must be the same on both sides of any surface of discontinuity. On assuming a steady state and integrating across a sharp discontinuity, we obtain the second boundary condition that energetic particles must satisfy across a shock,

$$[\mathbf{S}] = \mathbf{S} \cdot \mathbf{n}|_{0-}^{0+} = \frac{Q(p)}{4\pi p^2} \Leftrightarrow -\left[\kappa \cdot \nabla f + \frac{p}{3} \frac{\partial f}{\partial p} \mathbf{u}\right] \cdot \mathbf{n} \bigg|_{0-}^{0+} = \frac{Q(p)}{4\pi p^2}. \qquad (5.57)$$

Here, $\mathbf{n}$ is the shock normal, and $Q(p)$ is the particle injection rate at the shock. This form of the boundary conditions includes the effects of shock drift acceleration. Note that the transport equation and the derived boundary conditions are appropriate to relativistic particles i.e., only in the limit that the velocity W (where W is the speed of the scattering frame or the observer's frame relative to the frame in which the electric field vanishes) is much less than the particle velocity v ($W \ll v$), as well as particle drift (through the antisymmetric part of the spatial diffusion tensor κ). That the boundary conditions apply in the limit that $W/v \ll 1$ implies that the boundary conditions (5.55) and (5.57) are valid only for particles of speed $v \gg u_1 \sec \theta_{Bn}$, where θ_{Bn} is the angle between the upstream magnetic field and the

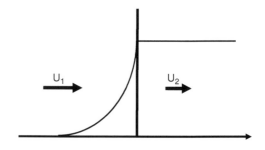

Fig. 5.4 General form of the solution (5.58) illustrating the spatial exponential growth of the distribution function upstream of the shock and the constant ambient value far upstream

shock normal. Furthermore, the transport equation was derived in the limit of near isotropy in the scattering frame, meaning that the particle distribution upstream and downstream of the shock must remain close to isotropy. These conclusions can be weakened slightly for the non-relativisitic form of the transport equation derived above, but isotropy remains a critical assumption. This latter condition is not always met at shocks where energetic particle distributions are often observed to be highly anisotropic.

Consider the 1D transport equation with a constant upstream and downstream velocity and solve the transport equation on either side of the shock, imposing continuity of $f(x, p)$ as $x \to \pm\infty$. The transport equation becomes

$$u_i \frac{df_i}{dx} - \frac{d}{dx}\left(\kappa(x, p)\frac{df_i}{dx}\right) = 0,$$

where $i = 1, 2$ (upstream, downstream) and $\kappa(x, p)$ is the diffusion coefficient parallel to the shock normal. The general solution is

$$f_i(x, p) = A_i(p) + B_i(p) \exp \int_0^x \frac{u}{\kappa(s, p)} ds;$$

$$f_i(x, p) = f(\pm\infty, p) + [f(0, p) - f(\pm\infty, p)]\frac{e^{b(x)} - e^{b(\pm\infty)}}{1 - e^{b(\pm\infty)}},$$

where $b(x) \equiv \int_0^x (u/\kappa) dx$. If $b(\pm\infty)$ are unbounded, the spatial dependence is then given by

$$f(x, p) = f(-\infty, p) + [f(0, p) - f(-\infty, p)] \exp \int_0^x \frac{u}{\kappa(s, p)} ds \quad x < 0;$$

$$= f(0, p) \quad x > 0. \quad (5.58)$$

The general solution (5.58) is illustrated in Fig. 5.4. The general solution $f(x, p)$ has a possible constant background of upstream particles $f(-\infty, p)$ plus an accelerated population that increases toward the shock on a diffusive scale length $\kappa(x, p)/u_1$ but remains constant downstream.

5.7 Application 1: Diffusive Shock Acceleration

The momentum spectrum of the energetic particle population is determined by the streaming boundary condition (5.57) at the shock,

$$-u_2 \frac{p}{3} \frac{df(0, p)}{dp} + u_1 \frac{p}{3} \frac{df(0, p)}{dp} + u_1 [f(0, p) - f(-\infty, p)] = \frac{Q(p)}{4\pi p^2},$$

where we have used the result $u_1 [f(0, p) - f(-\infty, p)] = \kappa \partial f/\partial x$ and have allowed for the injection of $Q(p)$ particles at the shock per unit momentum per cm^2 s at the shock. This then yields the ordinary differential equation in momentum

$$p \frac{df}{dp}(0, p) + \frac{3u_1}{u_1 - u_2} f(0, p) = \frac{3}{u_1 - u_2} \left[u_1 f(-\infty, p) + \frac{Q(p)}{4\pi p^2} \right],$$

illustrating that the source of the energetic particles is the background particle population $f(-\infty, p)$ convected through the shock and locally injected particles. Which particle population is more important depends on the relative flux and the characteristic energies. On solving the equation for the particle spectrum, we obtain the central result of diffusive shock acceleration theory,

$$f(0, p) = \frac{3}{u_1 - u_2} p^{-q} \int_{p_{inj}}^{p} (p')^q \left[u_1 f(-\infty, p') + \frac{Q(p')}{4\pi p'^2} \right] \frac{dp'}{p'}, \quad (5.59)$$

where $q = 3r/(r - 1)$ and $r = u_1/u_2$ is the shock compression ratio. Here, p_{inj} is the injection momentum. The upper limit on particle momentum is particularly important if time-dependent particle acceleration is considered, such as at interplanetary shock waves where the shock propagation time and evolution need to be considered carefully since this places constraints on the time available for a particle to experience acceleration.[13] Time dependent diffusive shock acceleration is discussed below. The spectrum of particles at energies well above the source energy is therefore a power law $\propto p^{-q}$. The characteristic compression ratio for a strong shock is $r = 4$ for a gas with adiabatic index $\gamma_g = 5/3$, implying that $q = 4$, which is very close to the index of 4.3 inferred for the source of galactic cosmic rays. For weak shocks, the power law is steeper, indicating fewer high energy particles.

A very important point to note is that the spectral slope of the accelerated particle spectrum is independent of the details of the scattering process i.e., the diffusion coefficient, depending only the kinematics of the flow. The reason a power law results is because the momentum gained by the particle on each shock interaction is proportional to the momentum it already has and to the probability of its escaping from the acceleration region. This is very nicely discussed by Bell (1978) from a microscopic perspective.

In (5.59), the accelerated particle spectrum p^{-q} is formed from the spectrum of sources at lower momenta $p' < p$. If no source of particles is present for momenta above some p_a, then $f(0, p) \propto p^{-q}$ for all $p > p_a$. If the spectrum of the source

---

[13] Zank et al. (2000).

is steeper than p^{-q}, then at large p, the accelerated spectrum will still approach the p^{-q} power law, but if the source is flatter (harder) than p^{-q}, the reaccelerated spectrum at high energies will have the same slope as the source i.e., the new spectrum will not reflect the characteristics of the last acceleration. In general then, a shock with $q = 3r/(r-1)$ will produce a power law spectrum with p^{-q} if the source spectrum is mono-energetic or has a spectral slope steeper than q, but if the source is harder such that $q' < q$, the spectrum tends to $p^{-q'}$ at large energies (Exercise).

The basic time scale associated with diffusive shock acceleration is of the order of κ/u^2. The importance of the acceleration time scale has to do with the maximum energy to which a particle can be accelerated by a shock wave. Observationally, galactic cosmic rays possess a source spectrum that is a power law $\sim p^{-4.3}$ over many decades up until about 10^{14} eV/nucleon, at which point the spectrum begins to steepen (the *knee*). The maximum energy to which a galactic cosmic ray can be accelerated is related presumably to either the time available to accelerate the particle (the lifetime of shock wave responsible for particle acceleration) or to the size of the acceleration region (both of which are possibly related). Similarly, energetic particles accelerated in *solar energetic particle (SEP)* events have a maximum energy. To estimate the maximum energy, whether at a supernova drive shock wave or at an interplanetary shock requires that we know the particle acceleration time scale, and that this then be related to, for example, the characteristic time scale associated with the shock wave.[14] To make the estimate for the time scale of diffusive shock acceleration more precise, we consider a steady planar shock at which a steady mono-energetic source of particles at the shock is turned on at $t = 0$.[15] We then seek time dependent solutions of the cosmic ray transport equation across a discontinuous shock with $f(t=0, x, p) = 0$ and source $Q\delta(p - p_0)$ at the shock, located at $x = 0$. On introducing the Laplace transform

$$g(s, x, p) = \int_0^\infty e^{-st} f(t, x, p) dt,$$

the transport equation upstream ($i = 1$) and downstream ($i = 2$) of the shock becomes

$$sg + u_i \frac{dg}{dx} = \kappa_i \frac{d^2 g}{dx^2},$$

assuming for simplicity that κ is independent of x. The solutions that satisfy the boundary condition

$$g \to 0 \quad \text{as} \quad x \to \pm\infty \quad \text{are} \quad g \propto \exp(\beta_i x),$$

---

[14] Zank et al. (2000).
[15] Axford (1981).

5.7 Application 1: Diffusive Shock Acceleration

where

$$\beta_i = \frac{u_i}{2\kappa_i} \left[1 - (-1)^i \left(1 + \frac{4\kappa_i s}{u_i^2} \right)^{1/2} \right].$$

The boundary conditions at the shock are given by

$$[f] = 0; \quad \left[\kappa \frac{\partial f}{\partial x} + \frac{p}{3} u \frac{\partial f}{\partial p} \right] = -n\delta(p - p_0),$$

where the square brackets denote as usual a jump in the enclosed quantity. On writing $g_0(s, p) = g(s, 0, p)$ for the Laplace transform of the spectrum at the shock, we find that

$$\kappa_1 \beta_1 g_0 - \kappa_2 \beta_2 g_0 + \frac{u_1 - u_2}{3} p \frac{dg_0}{dp} = \frac{1}{s} n\delta(p - p_0).$$

On letting $A_i = \sqrt{1 + 4\kappa_i s/u_i^2} - 1$, we can rewrite this as

$$\frac{1}{2}(u_1 A_1 + u_2 A_2) g_0 + u_1 g_0 + \frac{u_1 - u_2}{3} p \frac{dg_0}{dp} = \frac{1}{s} n\delta(p - p_0),$$

which has the solution

$$g_0(s, p) = \frac{3n}{s(u_1 - u_2)} \left(\frac{p}{p_0} \right)^{-q} \exp\left[-\int_{p_0}^{p} \frac{3}{2} \frac{u_1 A_1 + u_2 A_2}{u_1 - u_2} \frac{dp'}{p'} \right].$$

By formally inverting the transform, the time-dependent spectrum of accelerated particles at the shock is given by

$$f_0(t, 0, p) = \frac{1}{2\pi i} \int_{-i\infty}^{i\infty} g_0(s, p) e^{ts} ds.$$

To obtain the asymptotic behavior at large times, we consider the contribution of the simple pole at $s = 0$, which gives the steady spectrum,

$$f_0(\infty, 0, p) = f_0(\infty, p) = \frac{3n}{u_1 - u_2} \left(\frac{p}{p_0} \right)^{-q}, \quad p \geq p_0, \quad q = \frac{3r}{r - 1},$$

in agreement with the steady-state result. Obviously,

$$f_0(t, p_0) = \frac{3n}{u_1 - u_2} = f_0(\infty, p_0).$$

At a general time $t > 0$ and momentum $p > p_0$, we can express the spectrum formally as

$$f_0(t, p) = f_0(\infty, p) \int_0^t \phi(t') dt',$$

where

$$\phi(t) = \frac{1}{2\pi i} \int_{-i\infty}^{i\infty} \exp[ts - h(s)] \, ds,$$

and

$$h(s) = \frac{3}{2} \int_{p_0}^{p} \frac{u_1 A_1 + u_2 A_2}{u_1 - u_2} \frac{dp}{p}.$$

The function $\phi(t, p_0, p)$ is the probability distribution function for the time taken to accelerate a particle from momentum p_0 to p. In fact, since

$$\int_0^\infty \phi(t) \exp(-ts) dt = \exp[-h(s)],$$

and $h(0) = 0$, we have that

$$\int_0^\infty \phi(t) dt = 1,$$

indicating that the distribution is properly normalized. Hence, $\exp[-h(s)]$ can be thought of as the moment generating function for $\phi(t)$. Recall that to obtain the mean we can differentiate $h(s)$ with respect to s and then set $s = 0$ to obtain an expression for the mean acceleration time

$$\langle t \rangle = \int_0^\infty t\phi(t) dt = \frac{\partial}{\partial s} h(0)$$

$$= \frac{3}{u_1 - u_2} \int_{p_0}^{p_1} \left(\frac{\kappa_1}{u_1} + \frac{\kappa_2}{u_2} \right) \frac{dp}{p}. \tag{5.60}$$

Thus, the important conclusion is that the time scale for the acceleration of particles of momentum p at a shock not mediated by cosmic rays is simply

$$\tau_{acc}(p) = \frac{3}{u_1 - u_2} \left(\frac{\kappa_1}{u_1} + \frac{\kappa_2}{u_2} \right). \tag{5.61}$$

Exercises

1. Suppose that an upstream energetic particle distribution proportional to p^{-a} is convected into a shock with compression ratio r from upstream. In the absence of particle injection at the shock itself, calculate the reaccelerated downstream energetic particle spectrum, and explain what happens if $a < q = 3r/(r-1)$ or $a > q$.
2. Suppose that a shock of compression ratio r accelerates n cm^{-3} particles injected as a monoenergetic source $\delta(p - p_0)$ at the shock, so producing a downstream energetic particle spectrum $\propto p^{-q}$. Now suppose the shock propagates out of the system and the compressed gas relaxes back to the ambient state. Let another shock propagate into the system and suppose that this shock reaccelerates the decompressed accelerated power law spectrum that was accelerated earlier. Assume no additional injection of particles into the diffusive shock acceleration process. Compute the energetic particle distribution reaccelerated at the second shock. Again, suppose that the second shock disappears out of the system and the energetic particle decompresses again. Derive the energetic particle spectrum if a third shock reaccelerates the previously accelerated spectrum of particles. What can you infer about the effect of multiple accelerations and decompressions of a spectrum of energetic particles by multiple shock waves?

5.8 Application 2: The Modulation of Cosmic Rays by the Solar Wind

The fundamental concepts underlying the modulation of galactic cosmic rays by the solar wind can be developed on the basis of a simplified form of the cosmic ray transport equation. The solar wind flows supersonically and nearly radially outward from the sun and carries the heliospheric magnetic field. The large-scale magnetic field follows the Parker spiral. On smaller scales, as discussed, the solar wind convects magnetic irregularities – magnetic turbulence – that are responsible for scattering galactic cosmic rays. The charged particles gyrate about the mean magnetic field but experience pitch-angle scattering due to the magnetic turbulence, meaning that the cosmic ray transport equation is a suitable description of particle transport for galactic cosmic rays attempting to enter the heliosphere. That cosmic rays experience scattering in the outwardly flowing solar wind means that they experience considerable difficulty in reaching the inner heliosphere. Consequently, the intensity of cosmic rays in the inner heliosphere will be much lower than in the outer heliosphere.

To ensure a tractable description, consider the cosmic ray transport equation in the absence of a large-scale magnetic field and adopt a 1D spherically symmetric geometry. For a constant radial solar wind speed u, the steady-state spherically symmetric 1D cosmic ray transport equation becomes

$$u\frac{\partial f}{\partial r} - \frac{p}{3}\frac{1}{r^2}\frac{\partial(r^2 u)}{\partial r}\frac{\partial f}{\partial p} = \frac{1}{r^2}\frac{\partial}{\partial r}\left(r^2 \kappa_{rr}\frac{\partial f}{\partial r}\right),$$

where κ_{rr} denotes the radial diffusion coefficient and $f = f(t, r, p)$. This form of the transport equation does not have a simple analytic equation and is typically solved numerically for appropriate solar wind conditions. The transport equation can be rewritten as

$$r^2 u\frac{\partial f}{\partial r} + \frac{\partial(r^2 u)}{\partial r}Cf = \frac{\partial}{\partial r}\left(r^2 \kappa_{rr}\frac{\partial f}{\partial r}\right),$$

where the *Compton-Getting factor* C has been introduced,

$$C \equiv -\frac{p}{3}\frac{1}{f}\frac{\partial f}{\partial p}.$$

The Compton-Getting coefficient has been extensively studied, although not initially in the context of an expression in the cosmic ray transport equation. For galactic cosmic rays, C is a slowly varying function of r and p, and it may be approximated as $C = 1$ in the energy regime appropriate to cosmic ray modulation. Subject to this approximation, the transport equation reduces to

$$\frac{d}{dr}\left[r^2\left(uf - \kappa_{rr}\frac{df}{dr}\right)\right] = 0.$$

The equation in the inner brackets is simply

$$\frac{df}{dr} = \frac{u}{\kappa_{rr}}f,$$

which has the solution

$$f(r, p) = f(\infty, p)\exp\left[-\int_r^{R_0}\frac{u}{\kappa_{rr}}ds\right],$$

where $f(\infty, p)$ is the isotropic galactic cosmic ray distribution function beyond the heliosphere and R_0 is the radius of the modulation region – this is sometimes called the cosmic ray modulation boundary but it has little physical motivation within this formulation. The modulation of galactic cosmic rays within the heliosphere is therefore controlled by the modulation parameter

$$\Phi_{CR} = \int_r^{R_0}\frac{u}{\kappa_{rr}}ds,$$

which is a function of the solar wind properties through the solar wind radial speed u, and hence of the 11-year solar cycle. Consequently, the cosmic ray intensity exhibits a solar cycle variation.

References

M. Abramowitz, I.A. Stegun, *Handbook of Mathematical Functions* (Dover, New York, 1974)

W.I. Axford, in *Proceedings of the 10th Texas Symposium on Relativistic Astrophysics*, Baltimore, 1981. Annals of the New York Academy of Sciences, vol. 375, pp. 297–313

W.I. Axford, E. Leer, G. Skadron, in *Proceedings of the 15th International Cosmic Ray Conference*, Plovdiv, 1977, vol. 11, pp. 132–137

G.K. Batchelor, *The Theory of Homogeneous Turbulence* (Cambridge University Press, Cambridge, 1953)

J.W. Belcher, L. Davis, J. Geophys. Res. (Space) **76**, 3534 (1971)

A.R. Bell, Mon. Not. R. Astron. Soc. **182** 147–156 (1978)

R.D. Blandford, J.P. Ostriker, Astrophys. J. **221**, L29–L32 (1978)

P.J. Coleman, Astrophys. J. **153**, 371 (1968)

S. Corrsin, Progress report on some turbulent diffusion research, in *Atmospheric Diffusion and Air Pollution*, ed. by F. Frenkiel, P. Sheppard. Advances in Geophysics, vol. 6 (Academic, New York, 1959)

L.O'C. Drury, An introduction to the theory of diffusive shock acceleration of energetic particles in tenuous plasmas. Rep. Prog. Phys. **46**, 973–1027 (1983)

L.O'C. Drury, S.A.E.G. Falle, Mon. Not. R. Astron. Soc. **223**, 353 (1986)

M.A. Forman, G.M. Webb, Acceleration of energetic particles, in *Collisionless Shocks in the Heliosphere: A Tutorial Review*, ed. by R.G. Stone, B.T. Tsuratani. Monograph, vol. 34, (American Geophysical Union, Washington, DC, 1985), pp. 91–114

P.A. Isenberg, A hemispherical model of anisotropic interstellar pickup ions. J. Geophys. Res. **102**, 4719–4724 (1997)

J.D. Jackson, *Classical Electrodynamics* (Wiley, New York, 1975)

A.N. Kolmogorov, The local structure of turbulence in incompressible viscous fluid for very large Reynolds number. Dokl. Akad. Nauk. SSSR **30**, 310 (1941)

G.F. Krymsky, Sov. Phys.-Dokl. **23**, 327 (1977)

J.A. le Roux, G.M. Webb, Nonlinear cosmic ray diffusive transport in combined two-dimensional and slab magnetohydrodynamic turbulence: a BGK-Boltzmann approach. Astrophys. J. **667**, 930 (2007)

J.A. le Roux, G.M. Webb, A focused transport approach to the time-dependent shock acceleration of solar energetic particles at a fast traveling shock. Astrophys. J. **746**, 104 (2012)

J.A. le Roux, G.P. Zank, L.J. Milan, W.H. Matthaeus, Energetic charged particle transport and energization in dynamic two-dimensional turbulence. Astrophys. J. **602**, 396 (2004)

W.H. Matthaeus, C.W. Smith, Structure of correlation tensors in homogeneous anisotropic turbulence. Phys. Rev. A **24**, 2135 (1981)

W.H. Matthaeus, G. Qin, J.W. Bieber, G.P. Zank, Nonlinear collisionless perpendicular diffusion of charged particles. Astrophys. J. **590**, L53 (2003)

J. Skilling, Cosmic rays in the galaxy: convection or diffusion? Astrophys. J. **170**, 265 (1971)

A. Shalchi, *Nonlinear Cosmic Ray Diffusion Theories*. Astrophysics and Space Science Library (Springer, Berlin/Heidelberg, 2009). doi:10.1007/978-3-642-00309-7-2

A. Shalchi, J.W. Bieber, W.H. Matthaeus, Analytic forms of the perpendicular diffusion coefficient in magnetostatic turbulence. Astrophys. J. **604**, 675 (2004)

R.C. Tautz, I. Lerche, Magnetic field line random walk in non-axisymmetric turbulence. Phys. Lett. A **375**, 2587–2595 (2011)

G.M. Webb, Astron. Astrophys. **127**, 97 (1983)

G.P. Zank, Oscillatory cosmic ray shock structures. Astrophys. Space Sci. **140**, 301–324 (1988)

G.P. Zank, J.F. McKenzie, Short-wavelength compressive instabilities in cosmic ray shocks and heat conduction flows. J. Plasma Phys. **37**, 347–361 (1987)

G.P. Zank, W.I. Axford, J.F. McKenzie, Instabilities in energetic particle modified shocks. Astron. Astrophys. **233**, 275–284 (1990)

G.P. Zank, W.K.M. Rice, C.C. Wu, Particle acceleration at coronal mass ejection driven shocks: a theoretical model. J. Geophys. Res. (Space) **105**(A11), 25079–25095 (2000)

G.P. Zank, G. Li, V. Florinski, W.H. Matthaeus, G.M. Webb, J.A. le Roux, Perpendicular diffusion coefficient for charged particles of arbitrary energy. J. Geophys. Res. (Space) **109**, A04107 (2004)

G.P. Zank, G. Li, V. Florinski, Q. Hu, D. Lario, C.W. Smith, Particle acceleration at perpendicular shock waves: model and observations. J. Geophys. Res. (Space) **111**, A06108 (2006). doi:10.1029/2005JA011524

Chapter 6
The Transport of Low Frequency Turbulence

The transport of particles experiencing collisions with either other particles or with magnetic turbulence has been considered in the previous chapters. It is possible to utilize related techniques to investigate the transport of fields as well. Not surprisingly, the question of identifying appropriate closures becomes especially acute when considering the transport of fluctuating fields, and this is an enormous area of past and current research and we cannot begin to provide even the most cursory overview.[1] Instead, we shall focus on the transport of magnetized turbulence in an inhomogeneous flow such as a solar or stellar wind and utilize some quite simple closure approximations.

From almost the earliest times of spacecraft observation of the solar wind, there was strong evidence suggesting that fluctuations in the interplanetary medium could be interpreted in terms of a low frequency magnetohydrodynamic description. Furthermore, in a classic paper, Coleman (1968) found that spacecraft frame temporal fluctuations of the solar wind plasma velocity admitted power law spectral distributions, which in view of the super-Alfénic character of the mean interplanetary outflow from the Sun and the Taylor "frozen-in flow" condition, implies a power law distribution in wave number space very reminiscent of the Kolmogorov description of hydrodynamic turbulence. By contrast, related observations by Coleman (1968) and Belcher and Davis (1971) showed that velocity and magnetic field fluctuations are often highly correlated, suggesting the presence of low-frequency MHD waves propagating outward in the solar wind.

The above two sets of observations have more-or-less defined the two competing and conflicting interpretations of fluctuations in the solar wind. On the one hand, the existence of Kolmogorov-like spectra in solar wind magnetic field observations has been interpreted in terms of the in situ generation of a turbulent cascade by

---

[1] Very useful overviews can be found in e.g., McComb (1990), Frisch (1995), and Chassaing et al. (2002).

dynamical processes[2] suggesting that the solar wind is an evolving turbulent flow, driven by the injection of turbulent energy by processes such as stream shear, shock waves, pick-up of particles, etc. In marked contrast to this tangled, highly nonlinear picture of fluctuations in the interplanetary medium, the observations of Belcher and Davis suggested instead a picture corresponding to superimposed, non-interacting MHD waves, primarily Alfvénic, that are probably remnants of coronal processes.

The Alfvén mode description was for a long time the most widely accepted perspective until further sophisticated and detailed observations were presented which could be interpreted only on the basis of a turbulent solar wind. It was found for example that on average the frequency of occurrence of Alfvénic periods is greatest in the inner heliosphere, after which it decreases substantially with increasing heliocentric distance. By ~ 2 AU, as many "inwardly" as "outwardly" propagating fluctuations are observed. This is interpreted as a consequence of the in situ generation of turbulence by stream shear-driving and the dynamical evolution and coupling of fluctuations. Other dynamically changing quantities include the decrease of the "Alfvén ratio," the ratio of kinetic to magnetic energy in fluctuations, with increasing heliocentric distance, the tendency of the "cross helicity" to approach zero, indicating that the energy in inward and outward propagating modes approaches equality. Another critical observation is the non-adiabatic radial temperature of the solar wind, and this has long been interpreted as evidence for heating by turbulent dissipation or possibly shock waves. This led to growing consensus that the solar wind was an excellent example of a turbulent magnetofluid. However, the major advantage of the linear wave description was the development of an extremely tractable and simple theory describing the radial evolution of solar wind fluctuations. This was based on small amplitude or linear wave propagation in a slowly varying inhomogeneous background, for which a JWKB (Jeffreys-Wentzel-Kramers-Brillioun) expansion approach could be used, and is thus generally called the WKB theory. Despite the inability of WKB theory for linear Alfvén waves to explain observed turbulence properties of the interplanetary medium, it nonetheless proved remarkably accurate in describing the evolution of fluctuating magnetic power with increasing heliocentric distance, at least within about 8 AU. The apparent agreement between WKB theory and the observed heliocentric evolution of magnetic power provides perhaps the most striking argument in its favor. However, the lack of equipartition between magnetic and kinetic energy fluctuations and the observation that fluctuations become progressively less Alfvénic with increasing heliocentric distance casts doubt on the relevance of the WKB description. A resolution to these fundamental contradictions, at least for the outer heliosphere, was presented by Zank et al. (1996) and more modern models now extend this approach to all regions of the heliosphere.

---

[2]Matthaeus and Goldstein (1982).

6.1 Basic Description of Low-Frequency Turbulence

The development of turbulence models has been based largely on hydrodynamics in the incompressible regime. Here we very briefly review standard ideas about hydrodynamic turbulence based on the Kolmogorov theory, and apply this to magnetohydrodynamic turbulence.

Hydrodynamic turbulence proceeds primarily on the basis of the continuity and the Navier-Stokes form of the momentum equation in the limit of incompressibility. This reduces the governing equations to

$$\nabla \cdot \mathbf{u} = 0; \tag{6.1}$$

$$\frac{\partial \mathbf{u}}{\partial t} + \mathbf{u} \cdot \nabla \mathbf{u} = -\nabla \left(\frac{P}{\rho}\right) + \nu \nabla^2 \mathbf{u}. \tag{6.2}$$

In the incompressible formulation of hydrodynamics, the pressure P behaves as a passive scalar that is determined from the Poisson equation that results from taking the divergence of the momentum equation (6.2). Despite their apparent simplicity, the quadratic nonlinearity of the incompressible equations (6.1) and (6.2) admit no known general solution. The advection term introduces nonlinearity into the system, which can correspond to wave steepening and energy transfer e.g., suppose $u \propto \sin kx$. Then, $\mathbf{u} \cdot \nabla \mathbf{u} \sim \sin kx k \cos kx = k \sin 2kx$. Thus, the wave number doubles. An eddy with vorticity $\omega \equiv \nabla \times \mathbf{u}$ is enhanced by advection, and this is termed as vortex stretching.

We have seen that the Reynolds number Re is the magnitude of the advection to dissipation term,

$$Re = \frac{|\mathbf{u} \cdot \nabla \mathbf{u}|}{|\nu \nabla^2 \mathbf{u}|} \simeq \frac{U^2/L}{\nu U/L^2} = \frac{UL}{\nu},$$

where U and L denote the characteristic speed and length scale of the flow. The transition from laminar to turbulent flow corresponds to the transition from low Reynolds number flow to high Reynolds number flow.

Because of the quadratic nonlinearity, two modes can interact with one another to excite another. The mode-mode coupling leading to the excitation of a high frequency or larger wave number mode can proceed successively in the sense of a cascade. This concept of energy transfer from large to short scales is analogous to the physical picture introduced by Richardson in which large-scale eddies separate into smaller eddies, and so on into ever smaller eddies. Since energy migrates or cascades from larger scales to smaller scales, there does not exist a characteristic size for the eddies since all scales exist simultaneously. Such a cascade description can only be valid if there is a well separated region in which energy is deposited into the system. There also has to be well separated region in which nonlinearity and dissipation are comparable. In this regime, the nonlinear coupling and the dissipation proceed at the same rate, or dissipation begins to dominate, and leads to the conversion of kinetic energy to thermal energy.

In his classic analysis,[3] Kolmogorov separated the spectral range of turbulence into three spatial regimes, these being the injection or energy range, the inertial range, and the dissipation range, as illustrated in Fig. 5.2.

The smallest wave number regime (largest scales) is the injection or energy range, and corresponds to the regime in which turbulent energy is deposited. In the solar wind, this can be at frequencies as low as the 27 day solar rotation period, it can be due to the turbulence generated by shear and instabilities at the boundary regions of fast and slow streams, to the fluctuations excited by the creation of pickup ions, and sundry other instabilities. The energy range is the energetically dominant part of the fluctuation spectrum.

At the smallest scales, corresponding to wave numbers greater than some characteristic dissipation wave number k_d, the dissipative process due to the fluid viscosity dominates and fluctuation energy is converted to thermal energy. This regime is called the dissipation range. The spectral energy experiences a sharp decrease and steepening of the spectrum in the dissipation range.

Kolmogorov's key insight was to recognize that the inertial range that lies between the energy-containing and dissipation ranges results from the balancing of energy input into the inertial range and energy loss into the dissipation range. Let us examine this idea more closely in the context of hydrodynamics and MHD.

For hydrodynamics, we may consider the incompressible momentum equation (6.2), and for the energy-containing and inertial range, we can neglect the viscosity. We also neglect the thermal pressure. We therefore have

$$\frac{\partial \mathbf{u}}{\partial t} = -\mathbf{u} \cdot \nabla \mathbf{u} = NL,$$

where NL denotes the nonlinear term. Let the energy-containing eddies possess a characteristic velocity $\langle u^2 \rangle^{1/2}$ and length ℓ. The characteristic decay time due to nonlinear spectral transfer is then

$$\tau_{nl} \sim \frac{\langle u^2 \rangle^{1/2}}{\ell}.$$

For the kinetic energy in the fluctuations,

$$\frac{1}{2}\frac{\partial}{\partial t}\langle u^2 \rangle = -\mathbf{u} \cdot (\mathbf{u} \cdot \nabla \mathbf{u}),$$

we may approximate the triple correlation time by $\langle u^2 \rangle^{1/2} \langle u^2 \rangle / \ell$, which yields

$$\frac{d}{dt}\langle u^2 \rangle \simeq -\frac{\langle u^2 \rangle^{3/2}}{\ell},$$

---

[3] Kolmogorov (1941a,b).

6.1 Basic Description of Low-Frequency Turbulence

where the constant 2 has been absorbed into ℓ. This is the well-known Kolmogorov estimate for the turbulent decay of energy-containing eddies.

The analogous form of the "decay" or non-linear spectral time is rather more subtle and complicated for a magnetized gas in the presence of a mean magnetic field and inhomogeneous large-scale background flow. The identification and relationship of the various characteristic non-linear spectral and fluctuation time scales has been addressed for incompressible MHD by many authors.[4] Recall that for incompressible MHD, the introduction of the variables

$$\mathbf{Z}^\pm \equiv \mathbf{U} \pm \mathbf{B}/\sqrt{4\pi\rho},$$

(using the exact quantities now and not the fluctuating part) allows the fully 3D incompressible MHD equations to be expressed exactly as

$$\frac{\partial \mathbf{Z}^\pm}{\partial t} + \mathbf{Z}^\mp \cdot \nabla \mathbf{Z}^\pm = -\frac{1}{\rho}\nabla\left(P + \frac{B^2}{8\pi}\right), \tag{6.3}$$

after adding and subtracting the momentum and Faraday's equation respectively. Assume that the dissipation of magnetized fluctuations or turbulence is local and in approximate local statistical (quasi-)equilibrium on sufficiently small scales and that the system is locally homogeneous. By separating $\mathbf{Z}^\pm$ into a mean and fluctuating part, $\mathbf{Z}^\pm = \mathbf{Z}_0^\pm + \mathbf{z}^\pm$, we have locally

$$\frac{\partial \mathbf{z}^\pm}{\partial t} + \mathbf{Z}_0^\mp \cdot \nabla \mathbf{z}^\pm = -\mathbf{z}^\mp \nabla \mathbf{z}^\pm. \tag{6.4}$$

On neglecting the convection term in (6.4) above, we obtain

$$\frac{\partial}{\partial t}\left\langle z^{+2}\right\rangle = -2\left\langle \mathbf{z}^+ \cdot \left(\mathbf{z}^- \cdot \nabla \mathbf{z}^+\right)\right\rangle$$

$$\simeq -\left\langle z^{+2}\right\rangle \frac{\left\langle z^{-2}\right\rangle^{1/2}}{\lambda^+}, \tag{6.5}$$

by analogy with the von Karman-Howarth-Batchelor one-point closure for hydrodynamics (after absorbing the factor of 2). The non-linear spectral transfer term therefore couples the decay of the energy in the forward modes to those in the backward propagating modes via a characteristic scale length associated with the $\mathbf{z}^\pm$ modes. Similarly, in the absence of convection or zero-order propagation effects,

$$\frac{\partial}{\partial t}\left\langle z^{-2}\right\rangle \simeq -\left\langle z^{-2}\right\rangle \frac{\left\langle z^{+2}\right\rangle^{1/2}}{\lambda^-}.$$

---

[4] See the review by Zhou et al. (2004).

Consequently, the non-linear term due to spectral transfer in the absence of propagation effects is

$$NL_\pm = -\mathbf{z}^\pm \frac{\langle \mathbf{z}^{\mp 2} \rangle^{1/2}}{\lambda^\pm}, \qquad (6.6)$$

with a characteristic "eddy turn-over" time

$$\frac{1}{\tau_{nl}^\pm} \sim \frac{\langle \mathbf{z}^{\mp 2} \rangle^{1/2}}{\lambda^\pm}. \qquad (6.7)$$

The length scale $\lambda^\pm$ will correspond essentially to the correlation length of the energy in the forward and backward propagating Elsässer variables.

The modeling of the non-linear term is further complicated by the inclusion of propagation effects.[5] There are several subtleties with propagation effects that are not easily addressed. Alfvén wave propagation introduces an Alfvénic time scale

$$\frac{1}{\tau_A^\pm} \sim \frac{|\mathbf{V}_{A0}|}{\lambda_\parallel^\pm}. \qquad (6.8)$$

The time scale (6.8) yields the well-known Iroshnikov-Kraichnan spectrum for the energy density of the forward and backward propagating Elsässer fluctuations whereas the non-linear time scale implies a Kolmogorov spectrum for the energy in the $\mathbf{z}^+$ and $\mathbf{z}^-$ modes. This is seen as follows.

Use of the non-linear and Alfvén time scales $\tau_{nl}^\pm$ and $\tau_A^\pm$ yields the Kolmogorov and Iroshnikov-Kraichnan energy density spectra for $\langle z^{+2} \rangle$ and $\langle z^{-2} \rangle$ directly.[6] Recall that by making the Kolmogorov assumption that the energy flux = dissipation rate = $\varepsilon(k)$, one has for a hydrodynamical system that $\varepsilon \sim u^3/\ell$, where u is a characteristic velocity fluctuation and ℓ a characteristic length scale. If we introduce a characteristic triple correlation time τ_3, we may write $\varepsilon \simeq \tau_3 u^4/\ell^2$. Since $u_k \sim (kE_k)^{1/2}$, where E_k is the energy per unit wavelength, we may express

$$\varepsilon(k) \simeq \tau_3 E_k^2(k) k^4. \qquad (6.9)$$

By adopting the dimensional scaling (6.9), and identifying the relevant triple correlation time τ_3, we can recover either the Kolmogorov or Iroshnikov-Kraichnan forms of the energy spectrum provided $\varepsilon = $ constant. For example, if we assume that

---

[5] Iroshnikov (1963) and Kraichnan (1965).

[6] An excellent and much more extensive discussion can be found in Appendix A of Zhou et al. (2004).

the non-linear time scale describes τ_3, i.e., $\tau_3^{-1} = ku_k$, then we obtain immediately the Kolmogorov spectrum for the inertial range

$$E_k \simeq \varepsilon^{2/3} k^{-5/3}, \tag{6.10}$$

whereas taking $\tau_3^{-1} = kV_{A0}$ yields the Iroshnikov-Kraichnan spectrum

$$E_k \simeq (\varepsilon V_{A0})^{1/2} k^{-3/2}. \tag{6.11}$$

It is possible to use a more complicated form of the triple correlation time scale $\left(\tau_3^\pm\right)^{-1} = \left(\tau_{nl}^\pm\right)^{-1} + \left(\tau_A^\pm\right)^{-1}$ to obtain a spectrum $E_k^\pm$ that yields features of both the Kolmogorov and Iroshnikov-Kraichnan scalings in different regimes.

A generalization[7] of (6.9) is

$$\varepsilon^\pm(k) \simeq \tau_3^\pm \left(E_k^+(k)\right)\left(E_k^-(k)\right) k^4. \tag{6.12}$$

Use of the non-linear time scale above for $\tau_3^\pm$ yields the Kolmogorov spectrum

$$E_k^\pm = \left(\varepsilon^\pm/(\varepsilon^\mp)^{1/2}\right)^{4/3} k^{-5/3}, \tag{6.13}$$

together with the ratio,

$$E_k^+/E_k^- = \left(\varepsilon^+/\varepsilon^-\right)^2. \tag{6.14}$$

Unlike the scaling used above, use of the Alfvén time scale yields instead

$$E_k^+ E_k^- = V_{A0}\varepsilon^+ k^{-3}, \qquad E_k^+ E_k^- = V_{A0}\varepsilon^- k^{-3}. \tag{6.15}$$

The individual spectra then satisfy $E_k^+ \simeq k^{m_+}$ and $E_k^- \simeq k^{m_-}$ provided $m_+ + m_- = -3$. However, there is no expectation that the Iroshnikov-Kraichnan $-3/2$ spectrum will emerge for either the forward or backward energy spectra, and there is no estimate for the ratio of spectral energies E_k^+/E_k^-.

6.2 Mean Field Description of MHD Fluctuations

Let us now consider the transport of fluctuations in an expanding magnetized flow. To a leading order, low-frequency approximation, the solar wind may be described adequately on the basis of the MHD equations. A mean field decomposition of

---

[7]Dobrowonly et al. (1980a,b)

the MHD equations in the presence of an inhomogeneous large-scale flow will be utilized.[8] The compressible MHD equations are

$$\frac{\partial \rho}{\partial t} + \nabla \cdot \rho \mathbf{u} = 0;$$

$$\frac{\partial \mathbf{u}}{\partial t} + \mathbf{u} \cdot \nabla \mathbf{u} = -\frac{1}{\rho} \nabla \left(P + \frac{B^2}{4\pi} \right) + \frac{1}{4\pi\rho} (\mathbf{B} \cdot \nabla) \mathbf{B};$$

$$\frac{\partial \mathbf{B}}{\partial t} + \mathbf{u} \cdot \nabla \mathbf{B} = (\mathbf{B} \cdot \nabla) \mathbf{u} - (\nabla \cdot \mathbf{u}) \mathbf{B},$$

and of course $\nabla \cdot \mathbf{B} = 0$ and we have used $\mathbf{J} = (c/4\pi)\nabla \times \mathbf{B}$. Observe that we can identify $P^T \equiv P + \frac{B^2}{4\pi}$ as the total pressure, where P is the thermal plasma pressure. The solar wind, as indeed most flows, varies relatively smoothly on large scales associated with the local heliocentric radial coordinate R. Fluctuations typically possess correlation scales that are much smaller than R. Most of the turbulence "activity" occurs on scales that are smaller than a characteristic correlation length scale and hence significantly less than heliocentric length scales R. The MHD turbulence activity that we wish to explore is therefore well separated in length scale from the large-scale solar wind inhomogeneity, which is essentially reproducible and can be described in terms of mean field variables. By contrast, the small-scale fluctuating fields behave as random variables and so require a statistical description. Accordingly, we decompose the fields according to

$$\mathbf{u} = \mathbf{U} + \mathbf{u}; \quad \mathbf{B} = \mathbf{B}_0 + \mathbf{b}; \quad \rho = \rho_0 + \delta\rho; \quad P^T = P_0^T + \delta p^T,$$

where the existence of an appropriate averaging operator is assumed such that $\langle \mathbf{u} \rangle = \mathbf{U}$, $\langle \mathbf{B} \rangle = \mathbf{B}_0$, $\langle \rho \rangle = \rho_0$, and $\langle P^T \rangle = P_0^T$. The mean fields correspond to the large-scale inhomogeneity. Because of the inhomogeneous background, we need to regard the spatial coordinate $\mathbf{X}$ as comprising both a slowly varying part $\mathbf{R}$ and a local rapidly varying coordinate $\mathbf{x}$.[9] Formally, the averaging operator $\langle \cdots \rangle$ averages over $\mathbf{x}$ at a fixed $\mathbf{R}$, and hence no fast variations remain after averaging i.e.,

$$\frac{\partial}{\partial x_i} \langle F(\mathbf{R}, \mathbf{x}) \rangle = 0,$$

for each component x_i and any function F. Similarly, we also require that the average of any quantity vanish whenever that quantity may be written as a derivative with respect to the fast coordinate i.e.,

---

[8]This approach for solar wind fluctuations was developed in a series of papers by Ye Zhou, W.H. Matthaeus, C.-Y. Tu, and E. Marsch (Zhou and Matthaeus 1990, 1999; Matthaeus et al. 1994; Marsch and Tu 1989; Tu and Marsch 1990).

[9]For a very detailed discussion of a two-scale separation applied to the inhomogeneous MHD equations, see Hunana and Zank (2010).

6.2 Mean Field Description of MHD Fluctuations

$$\left\langle \frac{\partial F}{\partial x_i} \right\rangle = 0,$$

for any F and each x_i. This ensures that a quantity such as

$$\langle \nabla F \rangle = \frac{\partial F}{\partial \mathbf{R}}$$

can be regarded as a slowly varying spatial gradient of the locally averaged value of the function F.

A common assumption adopted in turbulence modeling is to assume that small scale fluctuations are incompressible, so that the mean density ρ_0 varies slowly and $\delta \rho = 0$. The fluctuating velocity field becomes solenoidal i.e.,

$$\nabla \cdot \mathbf{u} = 0.$$

The validity of the incompressible description in both hydrodynamics and MHD has been examined only recently[10] and is quite subtle, particularly for MHD. In essence, all high frequency fluctuations must vanish if the compressible equations are to converge properly to an incompressible description.

The fast scale averaged equations are easily derived from the momentum and induction equations, yielding

$$\frac{\partial \mathbf{U}}{\partial t} + \mathbf{U} \cdot \nabla \mathbf{U} + \langle \mathbf{u} \cdot \nabla \mathbf{u} \rangle - \frac{1}{4\pi \rho_0} [\mathbf{B}_0 \cdot \nabla \mathbf{B}_0 + \langle \mathbf{b} \cdot \nabla \mathbf{b} \rangle] = -\nabla P_0^T;$$

$$\frac{\partial \mathbf{B}_0}{\partial t} + \mathbf{U} \cdot \nabla \mathbf{B}_0 + \langle \mathbf{u} \cdot \nabla \mathbf{b} \rangle - [\mathbf{B}_0 \cdot \nabla \mathbf{U} + \langle \mathbf{b} \cdot \nabla \mathbf{u} \rangle]$$
$$= -[(\nabla \cdot \mathbf{U}) \mathbf{B}_0 + \langle (\nabla \cdot \mathbf{u}) \mathbf{b} \rangle].$$

Recall that the fluctuating velocity field is denoted by $\mathbf{u}$. The fast-scale equations describing the fluctuating fields are obtained by subtracting the above equations from the original momentum and induction equations. The turbulent velocity and magnetic fields are governed by

$$\frac{\partial \mathbf{u}}{\partial t} + \mathbf{U} \cdot \nabla \mathbf{u} + \mathbf{u} \cdot \nabla \mathbf{U} - \frac{1}{4\pi \rho_0} [\mathbf{B}_0 \cdot \nabla \mathbf{b} + \mathbf{b} \cdot \nabla \mathbf{B}_0] = -\frac{1}{\rho_0} \nabla \delta p^T + \mathbf{N}^u;$$
(6.16)

$$\frac{\partial \mathbf{b}}{\partial t} + \mathbf{U} \cdot \nabla \mathbf{b} + \mathbf{u} \cdot \nabla \mathbf{B}_0 - \mathbf{B}_0 \cdot \nabla \mathbf{u} - \mathbf{b} \cdot \nabla \mathbf{U} = -(\nabla \cdot \mathbf{U}) \mathbf{b}$$
$$- (\nabla \cdot \mathbf{u}) \mathbf{B}_0 + \mathbf{N}^b;$$
(6.17)

---

[10] Zank and Matthaeus (1991, 1993).

where

$$\mathbf{N}^u = -[\mathbf{u} \cdot \nabla \mathbf{u} - \langle \mathbf{u} \cdot \nabla \mathbf{u} \rangle] + \frac{1}{4\pi\rho_0}[\mathbf{b} \cdot \nabla \mathbf{b} - \langle \mathbf{b} \cdot \nabla \mathbf{b} \rangle]; \quad (6.18)$$

$$\mathbf{N}^b = -[\mathbf{u} \cdot \nabla \mathbf{b} - \langle \mathbf{u} \cdot \nabla \mathbf{b} \rangle] + [\mathbf{b} \cdot \nabla \mathbf{u} - \langle \mathbf{b} \cdot \nabla \mathbf{u} \rangle]. \quad (6.19)$$

The equations for the mean field and fluctuating density are given by

$$\frac{\partial \rho_0}{\partial t} + \nabla \cdot (\rho_0 \mathbf{U} + \langle \delta\rho \mathbf{u} \rangle) = 0;$$

$$\frac{\partial \delta\rho}{\partial t} + \nabla \cdot (\delta\rho \mathbf{U} + \rho_0 \mathbf{u} + \delta\rho \mathbf{u} - \langle \delta\rho \mathbf{u} \rangle) = 0.$$

The assumption of incompressibility then yields the constraint

$$\nabla \cdot (\rho_0 \mathbf{u}) = 0. \quad (6.20)$$

Since the small-scale field is assumed to be incompressible, we have

$$\nabla_x \cdot \mathbf{u} = 0,$$

where the divergence is with respect to the fast-scale variable $\mathbf{x}$. On slow scales, $\nabla \cdot \mathbf{u} \neq 0$ since this would violate the constraint (6.20). The small scale divergence can be neglected in the transport equation for $\mathbf{b}$, i.e., on the right-hand-side of (6.17).

The spatial derivatives in the equations for the dynamical small-scale variables $\mathbf{u}$ and $\mathbf{b}$, (6.16) and (6.17), contain both slow and fast scale variation. We assume that the variables are functions of both the $\mathbf{R}$ and $\mathbf{x}$, e.g.,

$$f = f^0(\mathbf{R}, \mathbf{x}, t) + \varepsilon f^1(\mathbf{R}, \mathbf{x}, t) + \varepsilon^2 f^2(\mathbf{R}, \mathbf{x}, t) + \cdots,$$

and introduce the multiple scales

$$\mathbf{R} = \mathbf{X}, \quad \mathbf{x} = \frac{\mathbf{X}}{\varepsilon} \implies \nabla_i = \frac{\partial}{\partial R_i} + \frac{1}{\varepsilon}\frac{\partial}{\partial x_i},$$

where ε is a small parameter. Let $\nabla_R \equiv \partial/\partial \mathbf{R}$ and $\nabla_x \equiv \partial/\partial \mathbf{x}$. The Equations (6.16) and (6.17) can be combined as a single equation with the introduction of the *Elsässer variables*

$$\mathbf{z}^{\pm} = \mathbf{u} \pm \frac{\mathbf{b}}{\sqrt{4\pi\rho_0}}.$$

The second term is the fluctuating magnetic field expressed in Alfvén speed units. The Elsässer variables can be interpreted as forward (the positive) and backward

6.2 Mean Field Description of MHD Fluctuations

(the negative) propagating modes. The large-scale magnetic field can be expressed similarly in terms of the Alfvén velocity

$$\mathbf{V}_A = \frac{\mathbf{B}_0}{\sqrt{4\pi\rho_0}}.$$

The Elsässer variables are very convenient in studying incompressible MHD turbulence since $\nabla \cdot \mathbf{z}^\pm = 0$ (although this is not in general true for compressible turbulence). In the present context of fast- and slow time variation, $\nabla_x \cdot \mathbf{z}^\pm = 0$ but

$$\nabla \cdot \mathbf{z}^\pm = \nabla_R \cdot \mathbf{u} \mp \frac{1}{2}\frac{\mathbf{b}}{\sqrt{4\pi\rho_0}} \cdot \nabla_R \ln \rho_0 \neq 0,$$

where we have used $\nabla \cdot \mathbf{b} = 0$. The slowly varying inhomogeneous background therefore ensures that, despite the assumed incompressibility of the small-scale fluctuations, the Elsässer variables are not solenoidal. Before combining the dynamical equations, we note the following relations,

$$\frac{\mathbf{U}}{\rho_0} \cdot \nabla \rho_0 = -\nabla \cdot \mathbf{U};$$

$$\mathbf{U} \cdot \nabla \frac{1}{\sqrt{4\pi\rho_0}} = \frac{1}{\sqrt{4\pi\rho_0}} \nabla \cdot \frac{\mathbf{U}}{2};$$

$$\mathbf{B}_0 \cdot \nabla \frac{1}{\sqrt{4\pi\rho_0}} = \nabla \cdot \mathbf{V}_A,$$

where we assumed that the background density ρ_0 was steady. On using the results above, Eqs. (6.16) and (6.17) can be combined in terms of the Elsässer variables as (Exercise)

$$\frac{\partial \mathbf{z}^\pm}{\partial t} + (\mathbf{U} \mp \mathbf{V}_A) \cdot \nabla \mathbf{z}^\pm + \frac{\mathbf{z}^\pm - \mathbf{z}^\mp}{2} \nabla \cdot (\mathbf{U}/2 \pm \mathbf{V}_A) + \mathbf{z}^\mp \cdot \left[\nabla \mathbf{U} \pm \frac{\nabla \mathbf{B}}{\sqrt{4\pi\rho_0}}\right]$$
$$= NL_\pm + S^\pm$$

which is equivalent to

$$\frac{\partial \mathbf{z}^\pm}{\partial t} + (\mathbf{U} \mp \mathbf{V}_A) \cdot \nabla \mathbf{z}^\pm + \frac{1}{2}\nabla \cdot (\mathbf{U}/2 \pm \mathbf{V}_A)\,\mathbf{z}^\pm + \mathbf{z}^\mp \cdot \left[\nabla \mathbf{U} \pm \frac{\nabla \mathbf{B}}{\sqrt{4\pi\rho_0}}\right.$$
$$\left. - \frac{1}{2}I\nabla \cdot (\mathbf{U}/2 \pm \mathbf{V}_A)\right] = NL_\pm + S^\pm, \tag{6.21}$$

where I is the identity matrix. $NL_\pm$ are nonlinear terms that we shall later model as dissipation terms, and

$$NL_\pm = \mathbf{N}^u \pm \frac{1}{\sqrt{4\pi\rho_0}}\mathbf{N}^b.$$

Note that in (6.21), the total pressure gradient ∇p^T has been neglected. This is due in part to the assumption of small scale incompressibility. In an incompressible hydrodynamic or MHD fluid, the pressure is a constraint, determined by the solution to the divergence of the incompressible momentum equation (a Poisson equation), and is determined by a combination of the fluid velocity and magnetic pressure. Thus, correlations that include the pressure will be of the third-order and similar therefore to the terms contained in $NL_\pm$. Since we will model correlations of the nonlinear terms on the right-hand-side of (6.21), we do not include the total pressure explicitly.

Exercises

1. Complete the derivation of the transport equation for the Elsässer variables $\mathbf{z}^\pm$, Eq. (6.21).

6.3 The Transport Equation for the Magnetic Energy Density

We will follow the original derivation given in Zank et al. (1996). although this has been extended recently by Breech et al. (2008) and Zank et al. (2012). Although we will not use all the following notation, these are quantities that appear typically in the analysis of turbulence in the solar wind, both observationally and theoretically. These quantities correspond to taking moments of the Elsässer variables in much the same way that we took moments of the particle distribution functions. There is a corresponding closure problem since the nonlinear terms introduce higher order moments in each derivation of a moment equation.

Introduce the following important moments of the Elsässer variables:

$$E_T \equiv \frac{\langle \mathbf{z}^+ \cdot \mathbf{z}^+ \rangle + \langle \mathbf{z}^- \cdot \mathbf{z}^- \rangle}{2} = \langle u^2 \rangle + \langle b^2/4\pi\rho_0 \rangle; \qquad (6.22)$$

$$E_C \equiv \frac{\langle \mathbf{z}^+ \cdot \mathbf{z}^+ \rangle - \langle \mathbf{z}^- \cdot \mathbf{z}^- \rangle}{2} = \langle \mathbf{u} \cdot \mathbf{b}/\sqrt{4\pi\rho_0} \rangle; \qquad (6.23)$$

$$E_D \equiv \langle \mathbf{z}^+ \cdot \mathbf{z}^- \rangle = \langle u^2 \rangle - \langle b^2/4\pi\rho_0 \rangle, \qquad (6.24)$$

where the first is twice the *total energy* in the fluctuations (the sum of kinetic and magnetic energy), the second is the *cross helicity*, the difference in energy between the forward and backward propagating modes, and the energy difference i.e., the difference between twice the fluctuation kinetic energy and magnetic energy (measured in Alfvén speed units) densities, sometimes called the residual energy. These are all useful and measurable quantities describing turbulence in the solar wind. By combining the moments, (6.22)–(6.24), we also have the following useful relations,

6.3 The Transport Equation for the Magnetic Energy Density

$$r_A \equiv \frac{\langle u^2 \rangle}{\langle b^2/4\pi\rho_0 \rangle} = \frac{E_T + E_D}{E_T - E_D};$$

$$\langle z^{+2} \rangle = E_T + E_C; \qquad \langle z^{-2} \rangle = E_T - E_C;$$

$$H_C = \frac{E_C}{E_T}; \qquad H_D = \frac{E_D}{E_T};$$

$$\langle u^2 \rangle = \frac{E_T + E_D}{2}; \qquad \langle b^2/4\pi\rho \rangle = \frac{E_T - E_D}{2}.$$

Here r_A denotes the Alfvén ratio, and H_C and H_D are the normalized cross-helicity and energy difference or residual energy respectively.

A very general set of transport equations can be derived from (6.21) in terms of the above moments together with correlation length equations. The physical content is sometimes difficult to extract, so we make the following assumptions that are quite reasonable beyond some 1–2 AU in the solar wind.

1. The Alfvén ratio is assumed to be constant i.e., the ratio of kinetic to magnetic energy in fluctuations is constant. This is quite well supported observationally in the solar wind for suitably large heliocentric distances from the Sun.
2. The cross-helicity is assumed to be zero i.e., the energy in inward and outward propagating modes is equal.
3. We introduce a *structural similarity hypothesis* that essentially imposes specific symmetries on the turbulence in the sense that non-diagonal correlations can be expressed as a linear function of the trace of the corresponding correlation tensor i.e., we approximate the product $z_i^\eta z_j^\xi = a \mathbf{z}^\eta \cdot \mathbf{z}^\xi$ where a is a scalar constant and η and ξ can be + or −. This assumption amounts to a closure assumption since it relates certain unknown moments to the smaller subset (6.22)–(6.24). Unlike the other assumptions above, it does not appear possible to weaken this assumption in the more general theory of Zank et al. (2012).
4. One further assumption is needed, this related to the ratio of the velocity and magnetic field fluctuation correlation lengths, but we defer further discussion until the appropriate section.

In the transport equation for the Elsässer variables $\mathbf{z}^\pm$, we express

$$\frac{1}{\sqrt{4\pi\rho}} (\mathbf{z}^\mp \cdot \nabla \mathbf{B}) = \mathbf{z}^\mp \cdot \mathbf{V}_A + \frac{1}{2} \mathbf{V}_A \frac{1}{\rho} \mathbf{z}^\mp \cdot \nabla \rho.$$

An evolution equation for E_T can be constructed (by taking the dot product of the evolution equation for $\mathbf{z}^\pm$ with $\mathbf{z}^\pm$ i.e., $\mathbf{z}^\pm \cdot \partial \mathbf{z}^\pm/\partial t$ etc. and then adding the two equations – Exercise) which yields

$$\frac{\partial E_T}{\partial t} + \mathbf{U} \cdot \nabla E_T + \frac{1}{2} \nabla \cdot \mathbf{U} E_T - \mathbf{V}_A \cdot \nabla E_C + \nabla \cdot \mathbf{V}_A E_C + \langle (\mathbf{z}^- \cdot \nabla \mathbf{U}) \cdot \mathbf{z}^+ \rangle$$
$$+ \langle (\mathbf{z}^+ \cdot \nabla \mathbf{U}) \cdot \mathbf{z}^- \rangle + \langle (\mathbf{z}^- \cdot \nabla \mathbf{V}_A) \cdot \mathbf{z}^+ \rangle - \langle (\mathbf{z}^+ \cdot \nabla \mathbf{V}_A) \cdot \mathbf{z}^- \rangle$$

$$+ \frac{1}{2}\frac{1}{\rho}\left[\langle(\mathbf{V}_A \cdot \mathbf{z}^+)(\mathbf{z}^- \cdot \nabla\rho)\rangle - \langle(\mathbf{V}_A \cdot \mathbf{z}^-)(\mathbf{z}^+ \cdot \nabla\rho)\rangle\right] - \frac{1}{2}\nabla \cdot \mathbf{U}E_D$$

$$= \langle \mathbf{z}^+ \cdot NL_+ \rangle + \langle \mathbf{z}^- \cdot NL_- \rangle + \langle \mathbf{z}^+ \cdot S^+ \rangle + \langle \mathbf{z}^- \cdot S^- \rangle, \qquad (6.25)$$

where turbulence source terms $S^\pm$ have been introduced.

To deal with the mixed terms, we invoke the structural similarity hypothesis and approximate $z_i^+ z_j^- = a\mathbf{z}^+ \cdot \mathbf{z}^-$ for some constant a. Notice that

$$z_i^+ z_j^- - z_i^- z_j^+ = 2\left(\frac{b_i}{\sqrt{4\pi\rho_0}}u_j - \frac{b_j}{\sqrt{4\pi\rho_0}}u_i\right)$$

so if the fluctuations are purely Alfvénic, the coefficients introduced by the mixed terms vanish identically.

By considering specific terms, the structural similarity hypothesis implies

$$(\mathbf{z}^- \cdot \nabla\mathbf{U}) \cdot \mathbf{z}^+ + (\mathbf{z}^+ \cdot \nabla\mathbf{U}) \cdot \mathbf{z}^- = 2a\mathbf{z}^- \cdot \mathbf{z}^+ \nabla \cdot \mathbf{U} + 2a\mathbf{z}^- \cdot \mathbf{z}^+ S_x^u;$$

$$(\mathbf{z}^- \cdot \nabla\mathbf{V}_A) \cdot \mathbf{z}^+ - (\mathbf{z}^+ \cdot \nabla\mathbf{V}_A) \cdot \mathbf{z}^- = 0;$$

$$\left[\langle(\mathbf{V}_A \cdot \mathbf{z}^+)(\mathbf{z}^- \cdot \nabla\rho)\rangle - \langle(\mathbf{V}_A \cdot \mathbf{z}^-)(\mathbf{z}^+ \cdot \nabla\rho)\rangle\right] = 0,$$

where $S_x^u = \sum_{i,j;i\neq j} \partial U_i/\partial x_j$ is the sum of shear velocity gradient terms. This then yields the total energy density transport equation in the form

$$\frac{\partial E_T}{\partial t} + \mathbf{U} \cdot \nabla E_T + \frac{1}{2}\nabla \cdot \mathbf{U}E_T - \mathbf{V}_A \cdot \nabla E_C + \nabla \cdot \mathbf{V}_A E_C$$

$$+ \left(2a - \frac{1}{2}\right)\nabla \cdot \mathbf{U}E_D + 2aS_x^u E_D$$

$$= \langle \mathbf{z}^+ \cdot NL_+ \rangle + \langle \mathbf{z}^- \cdot NL_- \rangle + \langle \mathbf{z}^+ \cdot S^+ \rangle + \langle \mathbf{z}^- \cdot S^- \rangle. \qquad (6.26)$$

The nonlinear dissipation terms are evaluated separately below. The first three terms describe the WKB-like transport terms that arise in linear wave theory, and exhibit a form that resembles a thermal pressure equation with adiabatic index 1/2. The remaining terms describe the mixing of the forward and backward Elsässer modes due to large scale inhomogeneity of the background plasma flow associated with expansion/compression, and shear terms. As expressed in the transport equation, this coupling is through the cross-helicity energy density E_C and the energy difference E_D terms.

To simplify this equation further, as discussed above, we assume that the cross-helicity $E_C = 0$, and that the Alfvén ratio is constant. Hence, using the relations

$$E_T = \langle u^2 \rangle + \langle b^2/4\pi\rho_0 \rangle = \langle b^2/4\pi\rho_0 \rangle \left(\frac{\langle u^2 \rangle}{\langle b^2/4\pi\rho_0 \rangle} + 1\right) = E_b(r_A + 1);$$

6.4 Modeling the Dissipation Terms

$$H_D = \frac{E_D}{E_T} = \frac{\langle u^2 \rangle - \langle b^2/4\pi\rho_0 \rangle}{\langle u^2 \rangle + \langle b^2/4\pi\rho_0 \rangle} = \frac{r_A - 1}{r_A + 1} \Rightarrow E_D = H_D E_T = \frac{r_A - 1}{r_A + 1} E_T,$$

reduces the transport equation (6.26) to

$$\frac{\partial E_b}{\partial t} + \mathbf{U} \cdot \nabla E_b + \frac{1}{2}\nabla \cdot \mathbf{U}E_b + \left(2a - \frac{1}{2}\right)\nabla \cdot \mathbf{U}H_D E_b + 2aS_x^u H_D E_b$$
$$= \langle \mathbf{z}^+ \cdot NL_+ \rangle + \langle \mathbf{z}^- \cdot NL_- \rangle + \langle \mathbf{z}^+ \cdot S^+ \rangle + \langle \mathbf{z}^- \cdot S^- \rangle. \tag{6.27}$$

Neglect of the right-hand-side of (6.26) and the *mixing term* proportional to H_D yields the WKB equation

$$\frac{\partial E_b}{\partial t} + \mathbf{U} \cdot \nabla E_b + \frac{1}{2}\nabla \cdot \mathbf{U}E_b = 0,$$

for the magnetic energy density in linear magnetic fluctuations $\langle b^2/4\pi\rho_0 \rangle$ in the solar wind.

Exercises

1. Complete the derivation of the transport equation for E_T, Eq. (6.25) and hence derive the final form of the transport equation (6.26).
2. Solve the steady-state WKB equation for the energy density of magnetic field fluctuations

$$\frac{\partial E_b}{\partial t} + \mathbf{U} \cdot \nabla E_b + \frac{1}{2}\nabla \cdot \mathbf{U}E_b = 0,$$

in a steady spherically symmetric steady flow for which $\mathbf{U} = U_0 \hat{\mathbf{r}}$, $U_0 = \text{const.}$, $\rho_0 = \rho_{00}(R_0/r)^2$ where ρ_{00} is the density at a heliocentric distance R_0, and hence show that $b^2/b_0^2 = (R_0/r)^3$.

6.4 Modeling the Dissipation Terms

Consider now the nonlinear terms that describe dissipation on the right-hand-side of the transport equation (6.26) or (6.27). We adopt a simple one-point closure model for energy decay similar to those used in hydrodynamics. The analysis here is a little more general than that given in Zank et al. 1996. As discussed above, assume that the non-linear decay terms are exponential in form with an appropriate non-linear spectral cascade time $\tau_{nl}^\pm$ i.e.,

$$NL_\pm \propto \frac{\mathbf{z}^\pm}{\tau_{nl}^\pm}.$$

The constant of proportionality is of order unity but we do not worry about this for the present. Since the spectral cascade is mediated by energy in oppositely directed modes, we assume that

$$\frac{1}{\tau_{nl}^{\pm}} \sim \frac{|\mathbf{z}^{\mp}|}{\lambda^{\pm}},$$

and $\lambda^{\pm}$ is a characteristic length scale consistent with the one-point closure, and will correspond essentially to a correlation length for the energy in the forward and backward propagating Elsässer variables.

From the evolution equation for the total energy E_T,

$$\text{Dissipation term} = -\langle \mathbf{z}^{+} \cdot NL_{+} \rangle - \langle \mathbf{z}^{-} \cdot NL_{-} \rangle$$

$$= -\frac{z^{+2}}{\lambda^{+}}|\mathbf{z}^{-}| - \frac{z^{-2}}{\lambda^{-}}|\mathbf{z}^{+}|$$

$$= -\left(E_T^2 - E_C^2\right)^{1/2}\left[\frac{(E_T + E_C)^{1/2}}{\lambda^{+}} + \frac{(E_T - E_C)^{1/2}}{\lambda^{-}}\right].$$

If we assume that $E_C = 0$ again, the dissipation term reduces to the simpler form

$$\text{Dissipation term} = -E_b^{3/2}\left(\frac{1}{\lambda^{+}} + \frac{1}{\lambda^{-}}\right) = -\frac{E_b^2}{\lambda}. \tag{6.28}$$

To close the turbulence model, the dynamical behavior of λ needs to be determined. We proceed by analogy with Batchelor (1953) and introduce three correlation lengths through the covariances

$$L^T \equiv \int \left(\langle \mathbf{z}^{+} \cdot \mathbf{z}^{+\prime}\rangle + \langle \mathbf{z}^{-} \cdot \mathbf{z}^{-\prime}\rangle\right) dr = \left(\langle \mathbf{z}^{+} \cdot \mathbf{z}^{+}\rangle + \langle \mathbf{z}^{-} \cdot \mathbf{z}^{-}\rangle\right)\lambda^T$$

$$= \left(|z^{+}|^2 + |z^{-}|^2\right)\lambda^T = 2E_T\lambda^T;$$

$$L^C \equiv \int \left(\langle \mathbf{z}^{+} \cdot \mathbf{z}^{+\prime}\rangle - \langle \mathbf{z}^{-} \cdot \mathbf{z}^{-\prime}\rangle\right) dr = \left(\langle \mathbf{z}^{+} \cdot \mathbf{z}^{+}\rangle - \langle \mathbf{z}^{-} \cdot \mathbf{z}^{-}\rangle\right)\lambda^C$$

$$= \left(|z^{+}|^2 - |z^{-}|^2\right)\lambda^C = 2E_C\lambda^C;$$

$$L^D \equiv \int \langle \mathbf{z}^{+} \cdot \mathbf{z}^{-\prime} + \mathbf{z}^{+\prime} \cdot \mathbf{z}^{-}\rangle dr = E_D\lambda^D,$$

Note that the prime denotes the spatially lagged Elsässer variable in the coordinate r, which unfortunately introduces a particular direction since the translation is along one particular Cartesian direction. By using, for example, $\mathbf{z}^+ \cdot \partial \mathbf{z}^{+\prime}/\partial t + \mathbf{z}^{+\prime} \cdot \partial \mathbf{z}^+/\partial t = \partial\left(\mathbf{z}^+ \cdot \mathbf{z}^{+\prime}\right)/\partial t$ etc., we can construct transport equations for L^T and L^C. Not surprisingly, we have to introduce yet another form of the structural similarity hypothesis, this time assuming that we can approximate

$$z_i^+ z_j^{-\prime} = d\mathbf{z}^+ \cdot \mathbf{z}^{-\prime}; \qquad z_i^{+\prime} z_j^- = e\mathbf{z}^{+\prime} \cdot \mathbf{z}^-,$$

and, again for simplicity, we assume that $d = e = a$. This yields the covariance transport equation for L^T as

$$\frac{\partial L^T}{\partial t} + \mathbf{U} \cdot \nabla L^T - \mathbf{V}_A \cdot \nabla L^C + \nabla \cdot (\mathbf{U}/2) L^T + \nabla \cdot \mathbf{V}_A L^C$$
$$+ 2\left(a - \frac{1}{4}\right) \nabla \cdot \mathbf{U} L^D + 2a S_x^u L^D = 0.$$

Since $L^C = E_C \lambda^C$ and $E_C = 0$, the covariance equation for L^T reduces to

$$\frac{\partial L^T}{\partial t} + \mathbf{U} \cdot \nabla L^T + \frac{1}{2} \nabla \cdot \mathbf{U} L^T + 2\left(a - \frac{1}{4}\right) \nabla \cdot \mathbf{U} L^D + 2a L^D S_x^u = 0. \quad (6.29)$$

Zank et al. (1996) argued that the velocity and magnetic field fluctuations possess equal areas under their respective correlation functions, from which we can infer that $\lambda^D = 0$. This is a somewhat severe restriction but it maintains some tractability in the turbulence model. By using $\lambda^D = 0$ and recalling that $L^T = E_T \lambda^T = (r_A + 1) E_b \lambda^T$ and identifying λ^T with 2λ yields (Exercise)

$$\frac{\partial \lambda}{\partial t} + \mathbf{U} \cdot \nabla \lambda + \left[2a S_x^u - \left(2a - \frac{1}{2}\right)\right] H_d \lambda = \frac{E_b^{1/2}}{2} - \frac{S\lambda}{2E_b}. \quad (6.30)$$

With the exception of identifying the source terms, this completes the derivation of the turbulence transport model. Under the assumptions listed above, the transport of the energy density in magnetic field fluctuations E_b is governed by the Eqs. (6.27) and (6.30). These equations describe the convection of magnetic energy and its evolution in an inhomogeneous flow, while experiencing the dissipation of magnetic energy into the plasma, as well as driving by sources.

Zank et al. (1996) considered three possible sources of turbulence in the solar wind; driving by stream-stream interactions, interplanetary shock waves, or in the outer heliosphere by ionization of interstellar neutrals. They solved the steady-state spherically symmetric form of the transport equations (6.27) and (6.30),

$$U \frac{\partial E_b}{\partial r} + \frac{U}{r} E_b - \Gamma \frac{U}{r} E_b = -\frac{E_b}{\lambda} + S;$$

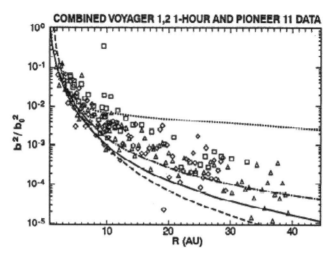

Fig. 6.1 Semilog plot of b^2/b_0^2 for the combined Voyager 1 and 2 and Pioneer 11 data set (normalized to 1 AU) and four theoretical models as a function of heliocentric distance. The *solid curve* corresponds to a WKB solution, the *dotted curve* to a WKB solution with pickup ion driving, the *dashed curve* to turbulence dissipative solution with driving by stream interactions, and the *dashed-dotted curve* to a turbulence dissipative solution with driving by stream interactions and pickup ions. The *triangles* and *diamonds* denote Voyager 1 and 2 1-h data respectively and the squares identify the Pioneer 15 min data. A moderate driving parameter of $\Gamma = 0.2$ was used (Zank et al. 1996)

$$U\frac{\partial \lambda}{\partial r} + \Gamma \frac{U}{r}\lambda = \frac{E_b^{1/2}}{2} - \frac{S}{2E_b}\lambda,$$

where $0 \leq \Gamma \leq 1$ expresses the mixing and is treated parametrically. Solutions of these equations were obtained numerically and compared to observations. Illustrated in Fig. 6.1 are four theoretical models. The solid line depicts the well-known WKB solution, the dotted line corresponds to a WKB solution with pickup ion driving, the dashed line illustrates the dissipative turbulence solution with stream-driving only, and finally, the dashed-dotted line depicts the dissipative turbulence model with driving by both streams and pickup ions. It is apparent that there is little to choose between the four solutions at heliocentric distances within some 6–10 AU. All appear to describe the overplotted data adequately. From ∼7 AU outward, the undriven WKB model and the stream-driving-only model underestimate the observed power in magnetic field fluctuations, and the pickup ion driven WKB model is clearly inappropriate. The choice of reasonable parameters yields a dissipative, stream- and pickup ion driven model that is in good agreement with observations from 1 to 40 AU. Other authors have applied this model successfully to distances as large as 80 AU and in different latitudinal regions of the heliosphere.[11]

---

[11] See e.g., Smith et al. (2001).

Exercises

1. Complete the derivation of the correlation length equation (6.30).
2. Integrate the steady-state spherically symmetric form of the transport equations (6.27) and (6.30),

$$U\frac{\partial E_b}{\partial r} + \frac{U}{r}E_b - \Gamma\frac{U}{r}E_b = -\frac{E_b}{\lambda};$$

$$U\frac{\partial \lambda}{\partial r} + \Gamma\frac{U}{r}\lambda = \frac{E_b^{1/2}}{2},$$

analytically if $E_b(r = R_0) = E_{b0}$ and $\lambda(r = R_0) = \lambda_0$. Hence show that asymptotically, in the limit of no mixing $\Gamma = 0$ (which is appropriate for either 2D or slab turbulence), one obtains the estimates

$$b^2/b_0^2 \sim (R_0/r)^{3.5}, \qquad \lambda/\lambda_0 \sim (r/R_0)^{1/4}.$$

This model corresponds to Kolmogorov/von Karman turbulence in an expanding medium. Show that in the opposite limit of strong turbulence ($\Gamma = 1$), the solutions reduce asymptotically to

$$b^2/b_0^2 \sim (R_0/r)^4, \qquad \lambda/\lambda_0 \sim \text{constant}.$$

This solution describes Taylor turbulence in a non-expansive medium.

3. Determine the general solution to the stream-driven steady-state spherically symmetric form of the transport equations (6.27) and (6.30),

$$U\frac{\partial E_b}{\partial r} + \frac{U}{r}E_b - \Gamma\frac{U}{r}E_b = -\frac{E_b}{\lambda} + C_{sh}\frac{U}{r}E_b;$$

$$U\frac{\partial \lambda}{\partial r} + \Gamma\frac{U}{r}\lambda = \frac{E_b^{1/2}}{2} - C_{sh}\frac{U}{2r}\lambda,$$

analytically if $E_b(r = R_0) = E_{b0}$ and $\lambda(r = R_0) = \lambda_0$. Hence find asymptotic solutions for weak and strong mixing.

References

G.K. Batchelor, *Theory of Homogeneous Turbulence* (Cambridge University Press, New York, 1953)

J.W. Belcher, L. Davis, Large-amplitude waves in the solar wind, 2. J. Geophys. Res. (Space) **76**, 3534 (1971)

B. Breech et al. Turbulence transport throughout the heliosphere. J. Geophys. Res. (Space) **113**, A08105 (2008)

P. Chassaing, R.A. Antonia, F. Anselmet, L. Joly, S. Sarkar, *Variable Density Fluid Turbulence*, (Kluwer Academic, Dordrecht/Boston, 2002)
P.J. Coleman, Turbulence, viscosity, and dissipation in the solar wind plasma. Astrophys. J. **153**, 371 (1968)
M. Dobrowolny, A. Mangeney, P. Veltri, Phys. Rev. Lett. **45**, 144 (1980a)
M. Dobrowolny, A. Mangeney, P. Veltri, Astron. Astrophys. **83**, 26 (1980b)
U. Frisch, *Turbulence*. (Cambridge University Press, New York, 1995)
P. Hunana, G.P. Zank, Inhomogeneous nearly incompressible description of magnetohydrodynamic turbulence. ApJ **718**, 148 (2010)
P.S. Iroshnikov, AZh **40**, 742 (1963)
A.N. Kolmogorov, The local structure of turbulence in incompressible viscous fluid for very large Reynolds number. Dokl. Akad. Nauk. SSSR. **30**, 299–303 (1941a)
A.N. Kolmogorov, Dissipation of energy in locally isotropic turbulence. Dokl. Akad. Nauk. SSSR. **32**, 19–21 (1941b)
R.H. Kraichnan, Phys. Fluids **8**, 1385 (1965)
E. Marsch, C.-Y. Tu, Dynamics of correlation functions with Elsässer variables for inhomogeneous MHD turbulence. J. Plasma Phys. **41**, 479 (1989)
W.H. Matthaeus, M.L Goldstein, Measurements of the rugged invariants of magnetohydrodynamic turbulence in the solar wind. J. Geophys. Res. (Space) **87**, 6011 (1982)
W.H. Matthaeus, Y. Zhou, G.P. Zank, S. Oughton, Transport theory and the WKB approximation for interplanetary MHD fluctuations. J. Geophys. Res. (Space) **99**, 23421 (1994)
W.D. McComb, *The Physics of Fluid Turbulence*. Oxford Engineering Science Series, vol. 25 (Clarendon Press, Oxford, 1990)
C.W. Smith et al., Heating of the low-latitude solar wind by dissipation of turbulent magnetic fluctuations. J. Geophys. Res. (Space) **106**, 8253–8272 (2001)
C.-Y. Tu, E. Marsch, Transfer equations for spectral densities of inhomogeneous MHD turbulence. J. Plasma Phys. **44**, 103 (1990)
G.P. Zank, W.H. Matthaeus, The equations of nearly incompressible fluids: I. Hydrodynamics, turbulence and waves. Phys. Fluids A. **3**, 69–82 (1991)
G.P. Zank, W.H. Matthaeus, Nearly incompressible fluids II: magnetohydrodynamics, turbulence and waves. Phys. Fluids A **5**, 257–273 (1993)
G.P. Zank, W.H. Matthaeus, C.W. Smith, Evolution of turbulent magnetic fluctuation power with heliospheric distance. J. Geophys. Res. (Space) **101**, 17093 (1996)
G.P. Zank et al., The transport of low frequency turbulence in the corona, supersonic solar wind, and outer heliosheath. Astrophys. J. **745**, 35 (2012). doi:10.1088/0004-637X/745/1/35
Y. Zhou, W.H. Matthaeus, Transport and turbulence modeling of solar wind fluctuations. J. Geophys. Res. (Space) **95**, 10291 (1990)
Y. Zhou, W.H. Matthaeus, Non-WKB evolution of solar wind: a turbulence modeling approach. Geophys. Res. Lett. **16**, 755 (1999)
Y. Zhou, W.H. Matthaeus, P. Dmitruk, Rev. Mod. Phys. **76**, 1015 (2004)

Index

anomalous transport theory for plasmas, 3
classical collisional theory for plasmas, 3
neoclassical transport theory for plasmas, 3

A
autocovariance, 66

B
Burgers' equation, 109, 110, 247
Burgers' equation
    Cole-Hopf transformation, 110
    inviscid, 100, 102, 104, 106

C
central limit theorem, 57, 60
Chapman-Enskog expansion, 92, 93, 155, 156
Chapman-Kolmogorov equation, 129, 131–133, 227
characteristic function, 30, 239
collisions
    BGK operator, 162
    BGK scattering operator, 91, 92
    binary charged particle collisions, 136, 137
    binary collision, 74
    Boltzmann collision integral, 76
    Boltzmann collision operator, 60, 74
    central force, 79
    Chandrasekhar function, 148
    charge-exchange, 1
    collision frequency, 147, 155, 156, 159, 164
    collision operator, 73, 94, 122
    collision operator moments, 123
    collisional invariants, 84
    collisional plasma, 121
    collisional time, 128, 133
    collisions, 155
    Coulomb, 132
    differential cross section, 76
    diffusion coefficients, 134, 147
    diffusion scattering frequencies, 149
    Einstein relation, 141
    electron-electron, 155
    electron-proton, 142, 143, 155
    electron-proton collision frequency, 145, 154
    energetic ion collision operator, 152
    energetic ion-electron, 150
    energetic ions-protons, 151
    energetic particles, 150
    fast ions, 150
    Fokker-Planck equation, 134
    Fokker-Planck operator, 122, 139–141
    frictional force, 128, 134
    fundamental collision frequency, 149
    Galilean invariance, 123
    H-theorem, 85
    hard sphere scattering, 76
    Landau operator, 141
    local thermodynamical equilibrium, 123, 155
    logarithmic integral cut-off, 139
    Lorentz operator, 144
    Markov process, 129
    Markovian, 133
    Maxwell-Boltzmann distribution, 86, 134
    Maxwellian background, 145
    molecular chaos, 75
    momentum and energy transfer, 125
    multiple species, 135
    particles, 60, 127, 135
    proton-electron, 152, 168

collisions (*cont.*)
    proton-proton, 155
    relaxation time operator, 91, 92, 162
    Rosenbluth potentials, 140, 141, 147
    shell distribution, 143, 147
    streaming scattering frequency, 148
    transition matrix, 75
    velocity diffusion frequency, 147
conservation
    angular momentum, 80
    charge, 88, 167
    collision operator, 126, 142
    energy, 76, 80, 82, 84, 89, 90, 124, 170
    entropy, 127, 171
    frozen-in flux theorem, 171
    hydrodynamics, 90
    magnetic flux, 171
    magnetohydrodynamics, 124, 169, 171
    mass, 84, 88, 124, 167
    MHD shocks, 175
    momentum, 76, 82, 84, 88, 124, 169
    Rankine-Hugoniot, 104, 175
    weak solutions, 102, 104, 171
contact discontinuity, 88, 105
correlation tensor, 232
correlation tensor
    axisymmetric, 233
    dynamical, 216, 239
    Fourier transform, 215
    isotropic, 218
    magnetic, 215
    magnetostatic, 216
    slab, 221
    stationary, 239
    two-component, 219
    two-D, 224
    wave number space, 216
Corrsin's independence hypothesis, 239
Coulomb logarithm, 139
covariance, 27, 36, 41, 64–66, 69, 276, 277
covariance
    decay, 66
    joint, 65, 66, 69
cross-spectral density, 66

D
Debye radius, 139
differential form, 102
diffusion
    ballistic or free streaming, 235
    Kubo form, 237
    Markovian, 235
    perpendicular, 240
    regular, 235
    running coefficient, 237
    subdiffusion, 235
    superdiffusion, 235
distribution function, 15, 19, 21, 47, 62, 71–73, 127, 142, 228
distribution function
    anisotropic, 118
    binomial, 48
    bispherical, 2
    continuous, 16, 19
    cosmic rays, 250, 253, 256–258
    delta function, 142
    discrete, 15, 19
    electron, 149, 152, 153, 156, 159
    energetic particles, 150, 152, 186, 204, 250, 252, 253, 256, 257
    Gaussian, 53, 61, 239
    Gaussian, normal, 57
    gyrophase averaged, 111, 186, 187, 192, 193, 211, 214, 230
    gyrotropic, 144
    hemispherical, 192
    invariance of phase space volume, 72
    Iroshnikov-Kraichnan, 267
    isotropic, 111, 146, 152, 191, 201, 214, 238, 240, 251, 252
    Kolmogorov, 261, 267
    Lorentz, 68, 69
    Lorentz invariance, 72, 205
    Maxwell-Boltzmann, 57, 60, 82, 86, 87, 93, 134, 143, 145, 147, 162, 163, 165, 166
    mixed coordinates, 187
    moments, 23, 29–31, 64, 65, 84, 91, 92, 94, 121–124, 127, 130, 241, 272, 273
    multiple species, 135
    multiple variables, 22
    normal, 53
    Poisson, 51
    polynomial expansion, 158
    power law, 202, 253, 267
    shell, 2
    turbulent fluctuations, 261
    velocity, 111, 121
    velocity shifted, 123
    wave number, 261
Dreicer field, 149

E
Elsässer variables, 270–272, 274
ensemble average, 63–65, 215, 228–231, 239

Index 283

ergodic, 63
Euler equations, 90, 93, 94, 98, 104
expectation, 23, 24, 27–30, 37, 38, 40, 43, 62, 65, 87, 97, 133, 139, 140, 146
expectation
    conditional, 38, 40

F
field lines, 173
flux surface, 173

G
galactic cosmic rays
    Compton-Getting, 258
    knee, 254
    modulation, 257
    spectrum, 253, 254
Gibbs ensemble, 62
gyrofrequency, 155
gyroradius, 155
gyroviscosity, 164

H
H-theorem, 82, 85, 86
heat flux
    vector, 89, 94, 97
heat flux tensor, 125, 165
heat flux vector, 96, 123, 125, 161–163
heat flux vector
    diamagnetic, 160, 165
    electron, 159, 160
    perpendicular, 161
    proton, 159
heliosphere, 1
homogeneous
    random function, 66
    turbulence, 216–218, 221, 239, 265, 276

I
integral form, 102
internal energy, 89, 90, 104, 106
internal energy
    cosmic rays, 240
    energetic particles, 240, 242

K
kurtosis, 33

L
Larmor radius, 155
Liouville's theorem, 63, 129
Lorentz force, 126, 189, 207
Lorentz transformations, 204

M
magnetic helicity, 174
magnetic surface, 173
Maxwell's equations, 168
mean value, 25, 30, 49, 51, 54, 123
mode, 31
moment generating function, 28–30, 36, 37, 43, 45, 49, 51, 256
most probable value, 31
multiple scales method, 92, 108, 244, 270

N
Navier-Stokes equations, 90, 92, 97, 106–108, 263
noise, 68

P
pickup ions, 1
plasma beta, 181
power spectral density, 66
Prandtl number, 107, 109
pressure
    Chapman-Enskog expansion, 96
    cosmic rays, 242
    energetic particles, 242
    evolution, 170
    isotropic, 94, 124, 163
    MHD, 168
    scalar, 94, 124, 163
    tensor, 88, 90, 123, 162
    trace, 90
probability density function, 15, 16, 62, 127
probability density function
    conditional, 37, 38, 40, 43, 131
    joint, 23, 27, 34, 36, 40, 42, 43, 60, 65, 66, 75, 239
    joint conditional, 40
    marginal, 35–37, 40, 42, 43, 45, 46, 75
probability set function, 7, 8, 10, 12, 13, 15, 16, 19, 21, 22, 33
probability set function
    conditional, 33

R
random variable, 11–13, 15
random variable
 continuous, 11, 19
 discrete, 11, 19
 multiple variables, 16, 22
rate-of-strain tensor, 90, 96, 157, 160
reductive perturbation method, 244
Reynolds number, 107, 263
rotation tensor, 190
runaway electrons, 149

S
sample space, 7, 8, 10, 11, 13, 33
scalar potential, 174
scattering
 Alfvén waves, 186, 204, 214
 BGK scattering operator, 91
 collisional, 144
 collisional , 134
 collisionless scattering operator, 186
 cosmic rays, 256, 257
 Coulomb scattering cross section, 81
 differential cross section, 76, 81
 diffusion, 227
 diffusive shock acceleration, 253
 electron-proton, 144
 hard sphere, 76, 78
 impact parameter, 78
 low frequency turbulence, 185
 Markov process, 127
 non-resonant, 185
 parallel scattering, 147, 214
 pitch angle, 111, 127, 152, 214, 227, 232, 238
 pitch angle diffusion operator, 192
 plasma fluctuations, 185
 quasi-linear, 226
 resonant, 185, 204, 235
 Rutherford scattering cross section, 81
 scattering cross section, 76
 scattering frame, 186, 204
 scattering tensor, 226
 slab turbulence, 234
 time scales, 129
 turbulence, 202, 215, 218
 wave frame, 204
shear tensor, 190
shock waves, 88, 105
shock waves
 Alfvénic shock, 180
 breaking time, 101
 Burgers' equation, 109
 characteristics, 105
 co-planarity, 175
 cosmic ray mediated shocks, 244
 diffusive shock acceleration, 250, 252–254
 dissipation, 109, 110, 247
 energetic particle mediated shocks, 244
 entropy, 177
 fast mode, 180, 182
 gas dynamic Hugoniot, 105
 gas dynamics, 104
 Hugoniot equation, 105
 intermediate shock, 182
 inviscid Burgers' equation, 104
 jump conditions, 103, 104
 MHD jump conditions, 175
 MHD Rankine-Hugoniot conditions, 175
 MHD shock waves, 175
 multiple shocks, 257
 Navier-Stokes equations, 106
 parallel, 180
 perpendicular, 180
 quasi-parallel, 180
 quasi-perpendicular, 180
 Rankine-Hugoniot conditions, 103, 104
 shock adiabatic, 178, 179
 shock normal, 178
 shock polar relation, 179, 180, 184
 slow mode, 180, 182
 switch-off shock, 182
 switch-on shock, 182
 turbulence, 262, 277
 weak shocks, 98, 108, 179, 180, 244
Sirovich method, 115
skewness, 32
slip line, 105
Smoluchowsky equation, 131
solar energetic particles, 236, 254
sound speed
 acoustic, 108, 109
 acoustic Mach number, 182
 Alfvén, 180, 182
 fast, 180, 182, 244, 247
 intermediate, 180, 182
 magnetosonic, 169, 179
 slow, 180, 182, 244, 247
 acoustic, 91
spectrum
 bendover scale, 220, 226
 co-spectrum, 66, 69
 correlation length, 220, 226
 cosmic rays, 250, 253, 254
 decorrelation length, 226
 discrete, 117
 dissipation range, 220, 264

Index
285

energetic particles, 250, 253–255, 257
energy range, 220, 264
frequency, 67
inertial range, 220, 264, 267
Iroshnikov-Kraichnan, 266, 267
Kolmogorov, 220, 264, 266, 267
magnetic fluctuations, 220
minimum wave number, 226
power density, 67, 68
quadrature, 66, 69
slab turbulence, 220, 222
turbulence, 218
two-D turbulence, 220, 224
wave number, 67, 219, 220
spherical coordinates, 77, 144, 187, 206, 211, 229, 231
stochastic in dependence, 42
stochastic independence, 43–46, 48, 75
stress tensor
   hydrodynamics, 88
   Maxwell, 241

T

transport
   collisional, 135
   collisional coefficients, 135
   collisional plasma, 155
   collisionless plasma, 185
   collisions, 155
   cosmic ray diffusion tensor, 202
   diamagnetic flows, 161
   diffusion coefficients, 133, 134, 147, 201, 227, 234
   Elsässer variables, 271
   energy, 125, 161
   focusing length, 203
   heat, 164
   MHD turbulence, 261, 267
   momentum, 160, 161
   parallel diffusion, 230–232
   perpendicular **B**, 161
   perpendicular diffusion, 235, 240
   pitch angle diffusion, 192, 198
   spatial diffusion, 227, 235
   spatial diffusion tensor, 202, 203
   Taylor-Green-Kubo diffusion coefficient, 236, 237
transport equation
   advective-diffusive equation, 192, 202, 203, 214
   Boltzmann, 71, 76, 111
   Chapman-Kolmogorov, 227
   collisional plasma, 155

collisionless plasma, 185, 234
conservation form for MHD, 171
convective-diffusive equation, 202, 214
convective-diffusive transport equation, 214
correlation lengths, 277
cosmic ray, 191, 240, 252, 257
cosmic ray modulation, 257
cosmic ray transport equation, 214
current, 167
diffusion, 110
diffusive shock acceleration, 252, 254
dissipation, 275
energetic particle, 191
energetic particles, 240, 252
energy conservation, 170
entropy, 171
focussed transport, 186, 190, 191, 193, 200, 204, 214
Fokker-Planck, 111, 227, 234, 235
gyrophase averaged, 186, 190, 193, 200, 204, 211, 214
Hall term, 168
heat, 165
Legendre polynomial expansion, 193, 200
magnetic energy density, 272, 277
magnetic field lines, 173
magnetic flux, 171
magnetic helicity, 174
magnetohydrodynamics, 166
mass conservation, 169
mean field, 228
MHD, 166, 168, 171
MHD turbulence, 261, 267, 271, 273
MHD, ideal, 168, 171
momentum, 166
momentum conservation, 169
non-relativisitic, 190
non-relativistic, 190
non-relativistic particles, 186, 190, 191
Nonlinear Guiding Center - NLGC, 240
number density, 166
Ohm's law, 167
quasi-linear, 226, 227
relativistic particles, 204, 214
telegrapher, 110, 112, 113
total energy density, 274
turbulence, 261
two-fluid, 158, 166, 242, 244
Vlasov equation, 204
WKB, 274, 275
turbulence
   Alfvén ratio, 273
   Alfvénic time scale, 266

turbulence (*cont.*)
  axisymmetric, 218, 234
  correlation length, 220, 276
  cross helicity, 273
  decorrelation, 220
  eddy turn-over time , 266
  energy difference, 273
  fluid, 64
  Gaussian, 61
  homogeneous, 216, 217
  incompressible MHD, 269
  inhomogeneous flow, 261
  isotropic, 217
  Kolmogorov, 220, 263
  magnetic, 185, 217
  mean field decomposition, 267
  MHD, 185
  moments, 272
  one-point closure, 265, 275
  perpendicular, 185, 219
  residual energy, 273
  slab, 185, 219, 221, 234
  sources, 262, 277
  spectrum, 220
  stationary, 216
  structural similarity hypothesis, 273
  total energy density, 273, 274
  transport, 261, 274
  triple correlation time, 266
  two-component model, 219, 229
  two-D, 185, 219, 224
  WKB models, 262

V

variance, 25, 26, 49, 51, 54, 64, 235, 236
variance
  conditional, 38
vector potential, 173
viscosity tensor, 124, 160, 162, 164

W

weak solution, 101, 102, 104, 171
white noise, 68